Applied Demography Series

Volume 11

Series Editor
David A. Swanson, Edward A. Dickson Emeritus Professor, Department of Sociology, University of California Riverside, Riverside, CA, USA

The field of applied demography is largely driven by the quest for the knowledge required by clients, both in public and private sectors, to make good decisions within time and costs constraints. The book series, Applied Demography, provides a forum for illustrating and discussing the use of demographic methods, concepts, and perspectives in a wide range of settings – business, government, education, law, and public policy - as well as the influence of these settings on demographic methods, concepts, and perspectives. The books within the series can be used as resources for practitioners and as materials serving as case studies for pedagogical uses.

More information about this series at http://www.springer.com/series/8838

Billystrom Jivetti • Md. Nazrul Hoque
Editors

Population Change and Public Policy

Springer

Editors
Billystrom Jivetti
Geospatial & Population Studies
University of New Mexico
Albuquerque, NM, USA

Md. Nazrul Hoque
Hobby School of Public Affairs
University of Houston
Houston, TX, USA

ISSN 2352-376X ISSN 2352-3778 (electronic)
Applied Demography Series
ISBN 978-3-030-57068-2 ISBN 978-3-030-57069-9 (eBook)
https://doi.org/10.1007/978-3-030-57069-9

This Springer imprint is published by the registered company Springer Nature Switzerland AG
The registered company address is: Gewerbestrasse 11, 6330 Cham, Switzerland

Foreword

The 20 chapters found in this book not only provide a view of the scope of applied demography interests but also the range of the field's geographic reach. It starts with topical applications in the United States such as the 2020 census and political redistricting and concludes by addressing long-standing issues of public policy and population change in Australia, India, and the United States. In between, the book deals with both emerging and existing topics in health care and economic disparities, fertility and contraception, and immigration and resettlement.

If the first volume on applied demography, edited by Hallie Kintner, Tom Merrick, Paul Voss, and Peter Morrison (1994), serves as the initial twentieth-century "book-end" of the many applied demography books that have been released in the past quarter century, this volume serves as a fitting book-end to mark the end of the first 20 years of the twenty-first century.

As a participant in the conference, it was my pleasure to see the interest of so many graduate students in the field alongside the interest of many faces familiar to me. Because of their efforts, Dr. Billystrom Jivetti and his colleagues have made the conference come alive in the papers selected for publication in this volume.

We dedicate this book to our colleague, Dr. Adélamar "Dely" Alcántara, whose tireless efforts brought not only financial and administrative support from the University of New Mexico but also engaged a host of UNM faculty and graduate students who, otherwise, may not have thought about participating. She did not live to see the publication of this book, but, along with Dr. Jivetti, Robert Rhatigan did a terrific job in stepping in as director of UNM's Center for Geospatial and Population Studies on her behalf to make sure this volume was published. We all hope it brings her a smile in knowing that all of her efforts paid off. This book will be an important addition to the shelves of applied demographers around the world as well as the shelves of those who teach applied demography.

Edward A. Dickson Emeritus Professor
Department of Sociology
University of California Riverside
Riverside, CA, USA

David A. Swanson, Ph.D.

Reference

Kintner, H., Merrick, T., Morrison, P., & Voss, P. (Eds.). (1994). *Demographics: A casebook for business and government*. Boulder: Westview Press.

Contents

Part I
Reproductive Health and Fertility

Chapter 1
The Lasting Effects of Teen Pregnancy Programs: Evidence from a Regional Collaborative

Jonathan Bennett, Christopher N. King, Chelsea Joyner, Wesley James, and Karen C. Matthews

According to recent reports from the National Center for Health Statistics (Hamilton and Matthews 2016) the national teen pregnancy rate has been continually dropping every year and is at its lowest rate since 2009. The Centers for Disease Control (CDC 2019) reported that "in 2017, a total of 194,377 babies were born to women aged 15–19 years" and that this was a 7% drop in U.S. teen pregnancies compared to 2016. Even though pregnancy rates in the 15–19 age group have decreased dramatically from 57 per 1000 in 2010 to 18.8 per 1000 in 2017 (Sedgh et al. 2015; CDC 2019), the rate is still high relative to countries such as Korea and Switzerland, which reported teen pregnancy rates of 3 or less per 1000 in 2017 (The World Bank 2017). In particular, disparities persist along the lines of race, poverty, and rurality. According to the CDC (2019), teen pregnancies remain high among Hispanic teens (28.9) and non-Hispanic Black teens (27.5), which have birth rates more

---

The authors thank the following individuals for their useful feedback and assistance during the writing of this chapter: Rachel Arthur, Nikki Payne, Madeline Plaster, Karin Scott, Twanda Wadlington, and Nakisha Watts. We also thank Billystrom Jivetti, Nazrul Hogue, and the International Association of Applied Demography for the invitation to contribute to this edited volume, as well as the anonymous reviewer for helpful feedback.

This is an original work by the named authors and is being submitted exclusively to the International Association of Applied Demography for publication consideration into an edited book on Applied Demography by Springer Publishing Company.

J. Bennett (✉) · W. James
Center for Community Research and Evaluation, University of Memphis, Memphis, TN, USA
e-mail: jrbnnett@memphis.edu

C. N. King
University of Memphis, Memphis, TN, USA

C. Joyner
Department of Psychological Sciences, Kansas State University, Manhattan, KS, USA

K. C. Matthews
Delta Health Alliance, Stoneville, MS, USA

B. Jivetti, Md. N. Hoque (eds.), *Population Change and Public Policy*, Applied Demography Series 11, https://doi.org/10.1007/978-3-030-57069-9_1

than two times higher than that of non-Hispanic White teens (13.2). In the rural, impoverished, and majority-minority Delta region of northwest Mississippi, rates commonly exceed 50 per 1000 (MSTAHRS 2019).

Researchers have devoted extensive study to explaining why differences in teen pregnancy rates persist along racial, socioeconomic, and geographical lines (Maness et al. 2016; Penman-Aguilar et al. 2013; Skatrud et al. 1998). Yet, as teen pregnancy rates have plummeted, more and more studies have identified the increased use of contraceptives by individuals as the primary driver of the change (Boonstra 2019; Lindberg et al. 2016, 2018; Santelli et al. 2007). As a result, most teen pregnancy prevention curricula are developed from theory that focuses on changing the behavior of the individual through increasing the use of contraception and/or decreasing sexual activity and evaluate their effectiveness through their success in moving these outcomes (Garney et al. 2019). However, states like Mississippi find themselves in the paradoxical position of having an elevated teen pregnancy rate and exhibiting all the risk factors associated with teen pregnancy combined with very limited access to contraception in comparison to other states (Guttmacher Institute 2019a, b, c), yet following their trajectory of rapidly declining rates (CDC 2019). Studying the effectiveness of a teen pregnancy prevention program situated within this context can offer unique insight into the interplay between individual thoughts and behaviors and community factors with regard to teen pregnancy.

This paper explores the effectiveness of a comprehensive regional effort to improve the quality and scope of teen pregnancy programming in 13 counties of the Mississippi Delta. The *Delta Futures* project, funded by a grant by the Office of Population Affairs, Department of Health and Human Services, is an effort to implement evidence-based teen pregnancy curricula across the Mississippi Delta. The initiative strives to maximize the effectiveness of teen pregnancy programming by implementing evidence-based programs in areas of greatest need, adopting a systematic, regional approach that incorporates multiple settings (middle schools, high schools, and community spaces including rural health clinics) to extend the reach and effectiveness of the program. The initiative was conceived due to longstanding difficulty among school districts to select and implement appropriate, evidence-based TPP curricula, as well as a previous lack of outreach beyond public schools. The effort is coordinated by Delta Health Alliance (DHA), a non-profit that leads over 60 projects which strive to improve access to health services, promote healthier lifestyles, and expand educational opportunities for residents of the Mississippi Delta region.

By comparing participant assessment responses collected at the beginning and end of the curriculum, we explore whether the intervention has a meaningful impact on students' self-reported commitment to pursue abstinence and use birth control. We also capture changes in participant motivations by using a battery of questions on abstinence motivations, birth control motivations, and perceptions of the riskiness of sexual activity. We find evidence that the program improves participants' willingness to commit to using birth control, but not to pursue abstinence. However, for participants who have engaged in sexual activity within the 3 months prior to pre-test, we find the opposite result: commitment to abstinence improves, but not birth control. Importantly, we find evidence that improvements in birth control intentions are sustained for at least 6 months, and motivations relating to birth control for at least 12 months. In addition to these results, we present a comprehensive series of results for subgroup-level and curriculum-level effects.

Participants

We examine data from 3390 adolescents from dozens of programs in 13 Mississippi Delta counties who completed at least one pre-test or post-test between July 2016 and December 2017 at a program site sponsored by Delta Futures. The sample consists predominately of non-Hispanic African Americans (91.1%), with a small number of non-Hispanic Caucasians (6.5%) and other individuals (2.4%). Participant ages range from 9 to 19. Descriptive statistics are reported in Table 1.1.

Table 1.1 Descriptive statistics

Category	Classification	Percentage (%)
Type of program	Community/Clinic Program	34.6
	High School Program	32.5
	Middle School Program	32.9
Sex	Female	52.9
	Male	47.1
Race/ethnicity	African-American, Non-Hispanic	91.1
	Caucasian, Non-Hispanic	6.5
	Hispanic	1.3
	Other	1.1
Time	Second half of 2016	15.6
	First half of 2017	49.9
	Second half of 2017	34.6
County	Coahoma County	16.3
	Quitman County	11.7
	Sunflower County	20.6
	Yazoo County	17.8
	Other Mississippi Delta counties	33.6
Age	Age 11 or less	5.5
	Age 12	14.5
	Age 13	17.4
	Age 14	23.2
	Age 15	15.2
	Age 16	10.1
	Age 17	8.7
	Age 18 or more	5.4

Curricula

The TPP programs used varying curricula depending on the setting, preferences of the site staff, and state regulations. However, all curricula are evidence-based and approved by the Office of Population Affairs. Five curricula were deployed: Draw the Line, Respect the Line (DTLRTL), Making a Difference (MAD), Promoting Health Among Teens (PHAT), Making Proud Choices (MPC), and Reducing the Risk (RTR). DTLRTL, MPC, and RTR are evidence-based abstinence-plus programs (Coyle et al. 2004; Jemmott et al. 1998; Kirby et al. 1991), while MAD and PHAT are evidence-based abstinence-only programs (Jemmott et al. 1998, 2010). Characteristics of the ages served, settings served, and reach of each curriculum are detailed in Table 1.2.

Measures

Participants took a pre-intervention and post-intervention survey consisting of questions sourced from the Youth Risk Behavior Survey (CDC 2017), the National Longitudinal Study of Adolescent Health (Add Health 2019), as well as demographic information. Surveys given to all participants were identical, except that middle school students were asked about neither birth control motivations nor

Table 1.2 Descriptive statistics by curriculum

		Draw the Line/Respect the Line (DTLRTL)	Making a Difference (MAD)	Making Proud Choices (MPC)	Promoting Health Among Teens (PHAT)	Reducing the Risk (RTR)
Type of curriculum		Plus	Only	Plus	Only	Plus
N (# assessments)		2100	268	1536	2000	1390
Age (%)	11 or younger	12.3%	5.2%	5.3%	2.4%	0.0%
	12	31.2%	12.7%	11.3%	9.7%	0.0%
	13	37.0%	9.0%	14.2%	12.6%	0.1%
	14	16.5%	14.2%	15.4%	22.5%	44.7%
	15	2.9%	32.1%	15.0%	16.3%	29.2%
	16	0.2%	20.9%	14.8%	13.8%	12.2%
	17	0.0%	4.5%	13.5%	14.8%	8.5%
	18 or older	0.0%	1.5%	10.4%	7.9%	5.2%
Setting (%)	Community	0.0%	0.0%	84.6%	66.4%	0.0%
	High School	0.0%	69.4%	9.6%	23.2%	100.0%
	Middle School	100.0%	30.6%	5.7%	10.4%	0.0%

previous sexual history. Measures of attitudes towards birth control and abstinence, as well as intentions of using birth control or practicing abstinence are obtained from these surveys. Consent forms were obtained with parental signatures (or participant's signature if 18 or older) prior to pre- or post-intervention surveys. Participants were informed that all answers would be kept confidential and they could stop partaking in the survey or intervention at any time. Assessments were administered during the first and last session. Participants who enrolled in two or more programs during the 18 months of intervention, which may have occurred either by chance or as a result of one school curriculum comprising a multi-year sequence, are considered distinct participants; we feel this assumption is conservative as repeated intervention is likely less effective than initial intervention, and would therefore reduce the estimated impact size.

Survey responses are aggregated into five measures: (1) abstinence intentions; (2) birth control intentions; (3) abstinence motivations; (4) birth control motivations; and (5) perceptions of risk. All measures are coded on a 0–1 scale, where higher values indicate more favorable attitudes, intentions, or motivations. Abstinence intentions, or the extent to which participants are willing to commit to abstinence, is captured using the following question: "Do you intend to have sexual intercourse in the next year, if you have the chance?" Participants answered using one of the following ordinal responses: "*Yes, definitely*" (0), "*Yes, probably*" (1/3), "*No, probably not*" (2/3), "*No, definitely not*" (1), or "*Don't know*" (NA/Missing). Birth control intentions use the same answer choices for the following question: "If you were to have sexual intercourse in the next year, do you intend to use (or have your partner use) any of these methods of birth control: Condoms, Birth Control Pills, The Shot (DepoProvera), The Patch, The Ring (NuvaRing), IUD (Mirena or Paragard), Implant (Implanon)". Responses for birth control intentions are coded as follows: "*Yes, definitely*" (1), "*Yes, probably*" (2/3), "*No, probably not*" (1/3), "*No, definitely not*" (0), or "*Don't know*" (NA/Missing).

We also calculate three other indices that aggregate various attitudinal and motivation questions. The abstinence motivations and birth control motivations indices capture the extent to which students were motivated to abstain from sexual activity and use birth control, respectively. The perceptions of risk index captures the extent to which students perceived risk in regard to sexual activity. A full explanation of the calculation of these indices is included in Appendix 1. Copies of the survey instruments are provided in Appendix 2.

Research Design

Our research design is aimed at exploring change attributable to the intervention along our five dependent variables: abstinence intentions, birth control intentions, the abstinence motivations index, the birth control motivations index, and the perceptions of risk index. Specifically, for each of these five dependent variables, we ask the following research questions:

1. Is there change between pre-test and post-test in terms of the effectiveness of Delta Futures?
2. Are there differences between curricula in terms of the effectiveness of Delta Futures?
3. Are there differences among subgroups in terms of the effectiveness of Delta Futures?
4. Are changes sustained over time (6 months and/or 12 months)?

For the first research question, we explore whether there exist statistically significant levels of change on the dependent variables from pre-test to post-test using paired t-tests. For any particular variable, participants are only included in the analysis if they: (1) complete both a pre-assessment and post-assessment; (2) answer the question; and (3) answer something other than "Don't Know" both times.

For the second research question, we explore curriculum-level effects by comparing post-test individual-level outcomes between community settings using the abstinence-plus MPC curriculum and community settings using the abstinence-only PHAT curriculum. We use coarsened exact matching (Iacus et al. 2012) to ensure that both the MPC and PHAT groups are comparable on the basis of pre-test response, gender, sexual history, previous sexual activity, and age (integer), as well as two self-esteem variables (expects to live to age 35 [3/5 or more], expects to graduate college [3/5 or more]). We focus on community settings only, and thus exclude school settings and their associated curricula, to ensure that geographic disparities or differences among school districts or state regulations are not driving the results.

For the third research question, we explore differences among subgroups by estimating a set of difference-in-differences models – one for each dependent variable – that includes a random intercept for each student. For independent variables, the following variables are included, both individually and in interaction with post-test: female, age, age squared, age cubed, previous sexual history reported at pre-test, sexually active within previous 3 months reported at pre-test, expects to graduate college (3/5 or higher), expects to live to age 35 (3/5 or higher), and expects to be married by 25 (3/5). A dummy variable for posttest is also included. This framework is a quasi-experimental approach designed to capture the effect of each variable while controlling for differences in baseline values.

Finally, for the fourth research question, to explore whether changes are sustained over time, we utilize the fact that several program sites, especially school-based sites, were designed to provide repeated intervention over time and re-enroll students in subsequent programming 6 or 12 months after the initial enrollment. To do this, we compare a participant's initial pre-test from his or her first enrollment to an initial pre-test during a subsequent enrollment in the one (or two) half-year(s) subsequent to the initial enrollment. That is to say, the sample of cases included in the 12-month evaluation includes those who have a pre-test in both the second half of 2016 and the second half of 2017. The sample of cases included in the 6-month evaluation either have a pre-test in the second half of 2016 and the first half of 2017, or the first half of 2017 and the second half of 2017. Post-test data is ignored to avoid picking up the effect of additional intervention (although it is possible additional intervention occurred between the two assessments). Follow-up data must occur 330 days (12-month evaluation) or 150 days (6-month evaluation) after initial assessment. If multiple follow-up assessments are available, school assessments are preferred, and in the event of further ambiguity, the earliest is used. Baseline and follow-up data are compared using paired t-tests; however, because values of the dependent variable are highly correlated with age which is changing considerably over the course of 6 or 12 months, differences between outcomes are compared not to zero but to the amount of change expected as a result of the increase in age by 6 or 12 months. These expectations are derived from the random-intercept model used to answer the third research question.

Results

Before/After Attitudes

Table 1.3 shows the results of the paired t-tests between pre-test and post-test. Positive and statistically significant improvements are observed for all dependent variables except for abstinence intentions. That is to say, the program can be shown to effectively improve students' commitment to use birth control, but not to pursue abstinence. We observe improvement in attitudinal measures for all categories, but interestingly, improvement for motivations to pursue abstinence is very subdued relative to improvement in motivations to pursue birth control. All variables are coded on a scale from 0 to 1, with higher responses indicating favorable attitudes, intentions, or motivations.

The Role of Curricula

While statistically significant differences between the two curricula are not observed, the point estimates are paradoxical. Interestingly, students enrolled in MPC, the

Table 1.3 Results of paired t-tests for all dependent variables, pre-test vs. post-test, all programs pooled

Variable	Before	After	Change	N
Abstinence intentions	0.635	0.622	−0.013^	1846
Birth control intentions	0.820	0.877	0.058**	1942
Abstinence motivations	0.584	0.597	0.013**	2524
Birth control motivations	0.629	0.673	0.043**	1264
Perceptions of risk	0.583	0.622	0.041**	2646

^ $p < 0.10$, * $p < 0.05$, ** $p < 0.01$

Table 1.4 Pre-test and post-test means of dependent variables for participants receiving MPC (PHAT) curriculum, as well as corresponding comparison groups

	MPC (Abstinence plus)				PHAT (Abstinence only)			
Variable	N (matched)	N (total)	Pre-test	Post-test	N (matched)	N (total)	Pre-test	Post-test
Abstinence intentions	106	170	0.723	0.726	162	215	0.723	0.694
Birth control intentions	128	178	0.786	0.844	176	212	0.786	0.870
Abstinence motivations	113	224	0.574	0.590	133	259	0.561	0.552
Birth control motivations	87	195	0.593	0.622	109	222	0.579	0.613
Perceptions of risk	120	247	0.587	0.614	157	271	0.583	0.617

abstinence-plus curriculum, score higher on abstinence intentions (0.726 vs. 0.694) and lower on birth control intentions (0.844 vs. 0.870) than do students enrolled in PHAT, the abstinence-only curriculum. Birth control motivations is also higher in MPC than PHAT. Results are illustrated in Table 1.4.

Subgroups

As articulated in the Research Design section, differences among subgroups were explored by estimating a set of difference-in-differences models – one for each dependent variable – that includes a random intercept for each student. This framework is a quasi-experimental approach designed to capture the effect of each variable while controlling for differences in baseline values.

In Table 1.5, we highlight a summary of the key marginal effects of the five models. For example, females are likely to grow 0.035 points more on the abstinence intentions index than do males, a statistically significant value. In general, the marginal effects reveal greater impacts on the abstinence intentions variable for females, greater impacts on abstinence and risk perceptions measures for the most sexually active, and greater improvements in all categories except abstinence intentions for individuals scoring low on the self-esteem index, which is defined as either not expecting to graduate college or live to age 35.

Figure 1.1 highlights the expected effect on each variable by age. The relative placement of each effect on the Y-axis is contingent on the other variables – for this plot in particular, a male student with high self-esteem and no sexual history is assumed. The key observations from this figure are that the effectiveness of the program at improving birth control motivations, birth control intentions, and abstinence intentions is strongest for the youngest students, while the effects of the program for the remaining variables are relatively invariant to age.

Table 1.5 Marginal effects for key independent variables for five random-intercept models

Variable	Category	Abstinence intentions	Birth control intentions	Abstinence motivations	Birth control motivations	Perceptions of risk
Sex	Females (relative to males)	0.035*	0.030^	0.004	0.003	0.014
Sexual history	Ever had sex (relative to never)	0.046	−0.029	0.020	0.001	0.010
	Had sex in last 3 months (relative to never)	0.101**	−0.015	0.059**	0.000	0.034*
Self-esteem	Low self-esteem (relative to high)	0.000	0.117**	0.022*	0.070**	0.029*

^ $p < 0.10$, * $p < 0.05$, ** $p < 0.01$

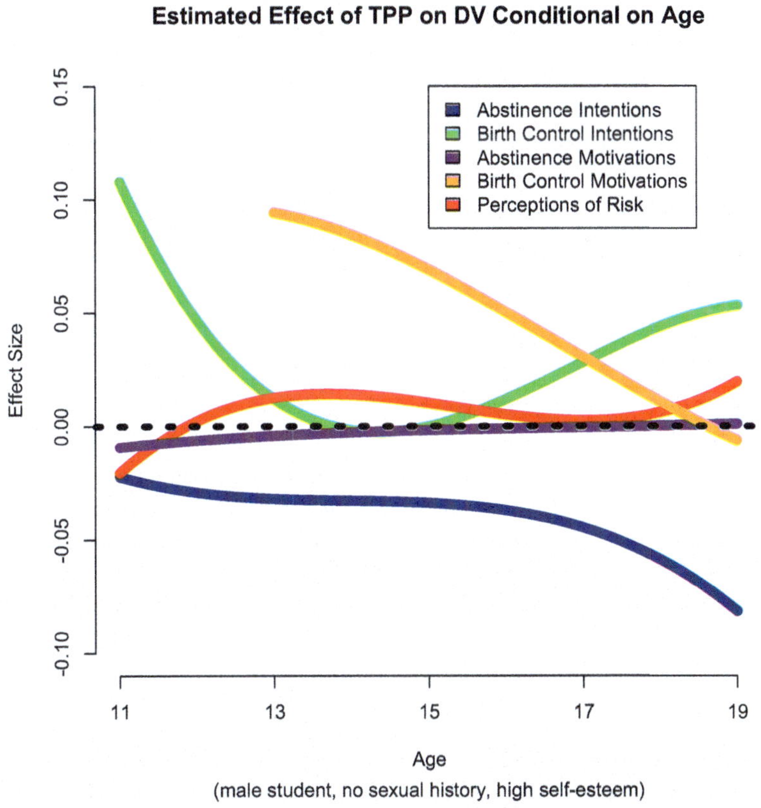

Fig. 1.1 Marginal effect of TPP intervention on each dependent variable, by age

Table 1.6 Results of longitudinal analysis

Variable	N (6 months)	Change after 6 months	Expected change after 6 months	N (12 months)	Change after 12 months	Expected change after 12 months
Abstinence intentions	161	+0.000	−0.031	89	−0.026	−0.048
Birth control intentions	167	+0.102***	+0.012	90	−0.022	+0.031
Abstinence motivations	230	−0.019	−0.014	123	−0.053	−0.023
Birth control motivations	76	+0.089**	+0.010	54	+0.174***	+0.020
Perceptions of risk	245	−0.009	+0.004	129	+0.011	+0.012

^ $p < 0.10$, * $p < 0.05$, ** $p < 0.01$, *** $p < 0.001$

Sustainability of Impacts

The results of the longitudinal analyses, which compare pre-test scores for students enrolled in multiple cohorts over time to expected change estimated from the random-intercept models derived in the analysis of subgroups, are detailed in Table 1.6. The analyses reveal sustained effects for birth control intentions after 6 months of initial intervention, and for birth control motivations for both 6 months and 12 months subsequent to initial intervention. These results suggest that that key attitudes and intentional commitments relating to birth control education can have a sustained impact well beyond initial intervention.

Discussion

Our analysis of Delta Futures reveals several findings that are relevant for collaboratives that strive to improve the quality of teen pregnancy prevention education and sexual risk avoidance programming. First, we find improvements in commitment to use birth control, but not commitment to abstinence, following other studies of similar programming in similar areas (Manaseri et al. 2019; Trenholm et al. 2007). Associated motivations have a similar result – while abstinence measures improve, they improve to a smaller degree than do birth control motivations and perceptions of the riskiness of sexual activity. Fortunately, we do see improvement in abstinence commitments (but unfortunately not birth control commitments) among the most at-risk participants, specifically, those who have engaged in sexual activity within the 3 months prior to pre-test. This nuanced result is informative, because while abstinence improvements may not be meaningful overall, the programming appears to be effective among the cohort that is in greatest need.

In addition, and contrary to expectations, we find limited evidence that suggests smaller declines in abstinence intentions, and greater increases in birth control intentions, among the cohort receiving an abstinence-plus curriculum, compared to a comparable cohort receiving an abstinence-only curriculum. While these differences are not statistically significant, the counter-intuitive point estimates suggest that abstinence-only curricula can nevertheless lead to meaningful improvements in attitudes relating to contraception, while curricula that de-emphasize abstinence can nevertheless be meaningful.

In general, we find more powerful impacts among females, sexually active participants, and individuals who have low self-esteem. Importantly, we find evidence that improvements in birth control intentions are sustained for at least 6 months, and birth control motivations for at least 12 months, even after controlling for the amount of change that would be expected as a result of participants aging. Because the objective of the intervention is to promote sustained, systematic change in behaviors and their related attitudes and motivations, this is a critical finding for the effectiveness of the intervention.

Implications

This study contributes to existing literature on teen pregnancy prevention in several ways. First, its position at the meso-level, in-between much larger analyses of secondary national or international data and more fine-grained analyses of individual curricula, small groups, or single-site programs allows for unique insights. A tighter view of the programming's effectiveness in a somewhat historically and geographically homogenous location, among an especially at-risk group, allows us to make conclusions that would typically be lost in larger studies. And the inclusion of multiple interconnected sites and community partnerships contributes to literature that focuses on the need for those partnerships to reinforce ecological engagement and "better address the social determinants of teen pregnancy" through "community-wide teen pregnancy prevention efforts," one of the main goals of the CDC and OAH partnership that initiated this wave of TPP programming (Fuller et al. 2018: 25).

Second, improved attitudes toward and commitments to use contraceptives among this group, specifically, suggest that knowledge or access may not tell the entire story of why and how teen pregnancy rates are declining. While declining teen pregnancy rates have been tied to increasing contraception use over and above abstinence, repeatedly (Lindberg et al. 2018), the explanations for why this is happening nationally are difficult to reconcile with the sociopolitical realities in the Mississippi Delta. The decline cannot be explained by the traditional social determinants. This uncoupling of teen pregnancy from social determinants follows the findings of Campa and Eckenrode (2006) in which, among a sample of daughters born to teen mothers, the mothers' SES and education level had no effect on whether or not their daughters gave birth in adolescence. Driscoll et al. (2005) similarly found that high educational expectations were protective of births for teens from low-SES families living in low opportunity communities, but not for other teens. In a study using the National Longitudinal Study of Adolescent Health, Pearson (2006) found that a sense of personal control and self-efficacy in situations of sexual negotiation were protective of sexual activity and sex without contraception, especially for females and non-Hispanic black respondents. The fact that the present study's result was achieved in schools restrictive of contraceptive education, in a rural region with limited access to the healthcare system including contraception-related care, through abstinence-plus *and* abstinence-only programming, and showed the strongest improvement among female (almost entirely non-Hispanic black) participants with low self-esteem at pre-test lends support to Pearson's hypothesis that "adolescents with an increased sense of general personal control and self-efficacy in sexual negotiation [are] less likely to have sexual intercourse and, if they [do] have sex, more likely to use a condom" (2006: 618).

Limitations

While the study is able to leverage a substantial amount of data across a wide variety of settings, there are several limitations. First, we only have data on participants in the program and do not have a control group. Although we model expected change over time for the longitudinal analysis, we are unable to directly observe expected change in the absence of intervention. Second, a randomization process to screen participants into intervention or to assign curricula to program sites is not used, introducing the possibility of selection bias if the control variable methods utilized are not used. Assignment of curricula in particular is notable because that is potentially driven by site-level differences, such as religious, cultural, or professional orientation. Third, attrition in the sample limits our analysis to individuals who successfully complete the intervention. Fourth, we do not currently have individual-level data for our participants on realized health outcomes which are expected to improve as a result of improvement in attitudes and motivations, such as pregnancy, contraception usage, or STIs. Fifth, the focus on African-Americans in the Mississippi Delta raise potential external validity concerns. Sixth, the deployment of abstinence-plus curricula required modifications in school settings to comply with local and state regulations.

Future research would do well to continue to address each of these limitations, namely the use of experimental design with a protocol for follow-up to determine whether sexual intentions as recorded in surveys were predictive of sexual activity, pregnancy, birth, or STI. Also, while there is literature devoted to the success of community partnerships as a whole, there have been fewer studies of individual partnerships or the success of those partnerships in comparison to standalone community or school programming. Finally, more research is needed to better understand why teens are choosing to lessen their risk of birth in adolescence. Finer and Zolna (2013) showed that, among adolescents, nearly *all* of the decrease in birth rate from 2001 to 2008 came from a decrease in unintended pregnancies, while the opposite was true of adult women. Identifying the causal mechanism through which teens have been able to improve their outcomes would allow for the development of even better services and could help improve the rate of unintended pregnancies for older women, as well.

Appendices

Appendix 1: Coding of Indices

The Perceptions of Risk Index is coded by converting the following questions into a numeric scale with answers spaced equally from 0 to 1, where 1 is the answer listed below next to each question – this answer corresponds to the participant perceiving large amounts of risk from sexual activity, pregnancy, and STDs. Each student's answers for the five questions are averaged, with "Don't Know" or blanks ignored.

Imagine that sometime soon you were to have sexual intercourse with someone just once, but were unable to use any method of birth control for some reason. What is the chance that you (or your partner) would get pregnant?
ALMOST CERTAIN
Suppose that sometime soon you had sexual intercourse for a whole month, as often as you wanted to, without using any protection. What is the chance that you would get the AIDS virus?
ALMOST CERTAIN
Getting pregnant or getting someone pregnant at this time in your life is one of the worst things that could happen to you.
STRONGLY AGREE
If you got the AIDS virus, you would suffer a great deal.
STRONGLY AGREE
It would be a big hassle to do the things necessary to completely protect yourself from getting a sexually transmitted disease.
STRONGLY DISAGREE

The Abstinence Motivations Index is coded by converting the following questions into a numeric scale with answers spaced equally from 0 to 1, where 1 is the answer listed below next to each question – this answer corresponds to the participant being unmotivated to engage in sexual activity. Each student's answers for the five questions are averaged, with "Don't Know" or blanks ignored.

If you had sexual intercourse, your friends would respect you more.
STRONGLY DISAGREE
If you had sexual intercourse, your partner would lose respect for you.
STRONGLY AGREE
If you had sexual intercourse, afterward, you would feel guilty.
STRONGLY AGREE
If you had sexual intercourse, it would upset your mother (or main caretaker).

STRONGLY AGREE
If you had sexual intercourse, it would give you a great deal of physical pleasure.
STRONGLY DISAGREE
If you had sexual intercourse, it would relax you.
STRONGLY DISAGREE
If you had sexual intercourse, it would make you more attractive.
STRONGLY DISAGREE
If you had sexual intercourse, you would feel less lonely.
STRONGLY DISAGREE
If you got pregnant, or got someone pregnant, it would be embarrassing for your family.
STRONGLY AGREE
If you got pregnant, or got someone pregnant, it would be embarrassing for you.
STRONGLY AGREE
If you got pregnant, or got someone pregnant, you would have to quit school.
STRONGLY AGREE
If you got pregnant, or got someone pregnant, you might marry the wrong person just to get married.
STRONGLY AGREE
If you got pregnant, or got someone pregnant, you would be forced to grow up too fast.
STRONGLY AGREE
If you got pregnant, or got someone pregnant, you would have to decide (or help decide) whether or not to keep the baby, and that would be stressful and difficult.
STRONGLY AGREE

The Birth Control Motivations Index is coded by converting the following questions into a numeric scale with answers spaced equally from 0 to 1, where 1 is the answer listed below next to each question – this answer corresponds to the participant being motivated to use birth control. Each student's answers for the five questions are averaged, with "Don't Know" or blanks ignored. These questions were not administered to students in Middle School programming.

In general, birth control is too much of a hassle to use.
STRONGLY DISAGREE
In general, birth control is too expensive to buy.
STRONGLY DISAGREE
It takes too much planning ahead of time to have birth control on hand when you're going to have sex.
STRONGLY DISAGREE
It is, or would be, too hard to get a sexual partner to use birth control with you.
STRONGLY DISAGREE
For you, using birth control interferes, or would interfere, with sexual enjoyment.
STRONGLY DISAGREE
It is easy for you to get birth control.

STRONGLY AGREE
Using birth control is morally wrong.
STRONGLY DISAGREE
If you used birth control, your friends might think that you were looking for sex.
STRONGLY DISAGREE

Appendix 2: High School & Community Survey Instrument

The following survey was used for all settings except middle schools.

DELTA FUTURES
SUPPORTED BY DELTA HEALTH ALLIANCE

Student Survey
HIGH SCHOOL

This survey is designed to give us information about your health knowledge, attitudes and behaviors. None of your answers will be seen by parents, teachers or anyone at your school.

If you have any emotional discomfort or distress while taking this survey, please stop and ask a site administrator (leader) for help. You can also be connected to counseling or other services, if needed.

Please answer all of the questions honestly.

Part I		
Demographics		
1	Date of Birth:	
2	Grade:	
3	What school do you attend?	
4	What is your sex?	◯ Male ◯ Female
5	Which of the following best describes you?	◯ Heterosexual (straight) ◯ Gay or lesbian ◯ Bisexual ◯ Not sure
6	Who do you live with most of the time?	◯ Mother and father ◯ Only mother ◯ Mother and stepfather ◯ Only father ◯ Stepmother and father ◯ Some with mother/some with father ◯ Other relative ◯ Guardian or foster parent ◯ Alone or with friends ◯ Someone else

Part II

	On a scale from *"No chance"* to *"It will happen"* what do you think are the chances you will…	No chance	Some chance	About 50-50	Pretty likely	It will happen
7	Live to age 35?	○	○	○	○	○
8	Be married by age 25?	○	○	○	○	○
9	Graduate from college?	○	○	○	○	○
10	Get HIV or AIDS?	○	○	○	○	○
11	Get another sexually transmitted disease, such as gonorrhea or genital herpes?	○	○	○	○	○

Part III

These questions are about sexual intercourse. By sexual intercourse we mean a male putting his penis into a female's vagina. **Please mark only one response to each item.**

12	Have you ever had sexual intercourse?	○ Yes ○ No	*IF NO, GO DIRECTLY TO QUESTION NO 13.*
12a	Now please think about the past 3 months. <u>In the past 3 months</u>, have you had sexual intercourse, even once?	○ Yes ----------> ○ No	How many times have you had sexual intercourse in the past 3 months? ______
12b	To the best of your knowledge, have you ever been pregnant or gotten someone pregnant, even if no child was born?	○ Yes ----------> ○ No	How many times has this happened? ______
12c	How old were you when you had sexual intercourse for the first time?	○ 11 years or younger ○ 12 years ○ 13 years ○ 14 years	○ 15 years ○ 16 years ○ 17 years or older
12d	<u>During your life</u>, with how many people have you ever had sexual intercourse?	○ 1 person ○ 2 people ○ 3 people	○ 4 people ○ 5 people ○ 6 or more people
12e	<u>During the past 3 months</u>, with how many people did you have sexual intercourse?	○ No one ○ 1 people ○ 2 people ○ 3 people	○ 4 people ○ 5 people ○ 6 or more people
12f	In the past 3 months have you had sexual intercourse without you or your partner using a condom?	○ Yes ----------> ○ No ○ I don't know	How many times have you had sex without using a condom in the past 3 months? ______
12g	In the past 3 months, have you had sexual intercourse with you or your partner using any of these methods of birth control: • Condoms • Birth Control Pills • The Shot (Depo Provera) • The Patch • The Ring (NuvaRing) • IUD (Mirena or Paragard) • Implant (Implanon)?	○ Yes ----------> ○ No ○ I don't know	How many times have you had sexual intercourse without using any of these methods of birth control? ______
12h	Did you drink alcohol or use drugs before you had sexual intercourse the last time?	○ Yes ○ No	

13	Have you ever been physically forced to have sexual intercourse when you did not want to?	○ Yes ○ No
14	During the past 12 months, how many times did someone you were dating or going out with force you to do sexual things that you did not want to do? **(Count such things as kissing, touching, or being physically forced to have sexual intercourse.)**	○ I did not date or go out with anyone during the past 12 months ○ 0 times ○ 1 time ○ 2 or 3 times ○ 4 or 5 times ○ 6 or more times
15	During your life, with whom have you had sexual contact?	○ I have never had sexual contact ○ Females ○ Males ○ Female and males
16	Have you ever been tested for HIV, the virus that causes AIDS? **(Do not count tests done if you donated blood.)**	○ Yes ○ No ○ Not sure
17	Imagine that sometime soon you were to have sexual intercourse with someone just once, but were unable to use any method of birth control for some reason. What is the chance that you (or your partner) would get pregnant?	○ Almost no chance ○ Some chance, but probably not ○ A 50-50 chance ○ A good chance ○ Almost certain ○ Don't know
18	Suppose that sometime soon you had sexual intercourse for a whole month, as often as you wanted to, without using any protection. What is the chance that you would get the AIDS virus?	○ Almost no chance ○ Some chance, but probably not ○ A 50-50 chance ○ A good chance ○ Almost certain ○ Don't know
19	Do you intend to have sexual intercourse in the next year, if you have the chance?	○ Yes, definitely ○ Yes, probably ○ No, probably not ○ No, definitely not ○ Don't know
20	If you were to have sexual intercourse in the next year, do you intend to use (or have your partner use) any of these methods of birth control: • Condoms • Birth Control Pills • The Shot (Depo Provera) • The Patch • The Ring (NuvaRing) • IUD (Mirena or Paragard) • Implant (Implanon)?	○ Yes, definitely ○ Yes, probably ○ No, probably not ○ No, definitely not ○ Don't know

Part IV							
The next questions ask about your thoughts on pregnancy, sexually transmitted diseases, and AIDS. **Please mark your level of agreement with each of the following statements.**		Strongly agree	Agree	Neutral	Disagree	Strongly disagree	Don't know
21	Getting pregnant or getting someone pregnant at this time in your life is one of the worst things that could happen to you.	○	○	○	○	○	○
22	If you got the AIDS virus, you would suffer a great deal.	○	○	○	○	○	○
23	It would be a big hassle to do the things necessary to completely protect yourself from getting a sexually transmitted disease.	○	○	○	○	○	○
24	If you had sexual intercourse, your friends would respect you more.	○	○	○	○	○	○
25	If you had sexual intercourse, your partner would lose respect for you.	○	○	○	○	○	○
26	If you had sexual intercourse, afterward, you would feel guilty.	○	○	○	○	○	○
27	If you had sexual intercourse, it would upset your mother (or main caretaker).	○	○	○	○	○	○
28	If you had sexual intercourse, it would give you a great deal of physical pleasure.	○	○	○	○	○	○
29	If you had sexual intercourse, it would relax you.	○	○	○	○	○	○
30	If you had sexual intercourse, it would make you more attractive.	○	○	○	○	○	○
31	If you had sexual intercourse, you would feel less lonely.	○	○	○	○	○	○
32	If you got pregnant, or got someone pregnant, it would be embarrassing for your family.	○	○	○	○	○	○
33	If you got pregnant, or got someone pregnant, it would be embarrassing for you.	○	○	○	○	○	○
34	If you got pregnant, or got someone pregnant, you would have to quit school.	○	○	○	○	○	○

	Please mark your level of agreement with each of the following statements.	Strongly agree	Agree	Neutral	Disagree	Strongly disagree	Don't know
35	If you got pregnant, or got someone pregnant, you might marry the wrong person just to get married.	○	○	○	○	○	○
36	If you got pregnant, or got someone pregnant, you would be forced to grow up too fast.	○	○	○	○	○	○
37	If you got pregnant, or got someone pregnant, you would have to decide (or help decide) whether or not to keep the baby, and that would be stressful and difficult.	○	○	○	○	○	○
38	In general, birth control is too much of a hassle to use.	○	○	○	○	○	○
39	In general, birth control is too expensive to buy.	○	○	○	○	○	○
40	It takes too much planning ahead of time to have birth control on hand when you're going to have sex.	○	○	○	○	○	○
41	It is, or would be, too hard to get a sexual partner to use birth control with you.	○	○	○	○	○	○
42	For you, using birth control interferes, or would interfere, with sexual enjoyment.	○	○	○	○	○	○
43	It is easy for you to get birth control.	○	○	○	○	○	○
44	Using birth control is morally wrong.	○	○	○	○	○	○
45	If you used birth control, your friends might think that you were looking for sex.	○	○	○	○	○	○

<u>This concludes the survey, thank you for your participation!</u>

Appendix 3: Middle School Survey Instrument

The following survey was used for middle schools only.

DELTA FUTURES
SUPPORTED BY DELTA HEALTH ALLIANCE

Student Survey

MIDDLE SCHOOL/JUNIOR HIGH

This survey is designed to give us information about your health knowledge, attitudes and behaviors. None of your answers will be seen by parents, teachers or anyone at your school.

If you have any emotional discomfort or distress while taking this survey, please stop and ask a site administrator (leader) for help. You can also be connected to counseling or other services, if needed.

<u>Please answer all of the questions honestly.</u>

Part I

Demographics

1	Date of Birth:	
2	Grade:	
3	What school do you attend?	
4	What is your sex?	○ Male ○ Female
5	Which of the following best describes you?	○ Heterosexual (straight) ○ Gay or lesbian ○ Bisexual ○ Not sure
6	Who do you live with most of the time?	○ Mother and father ○ Only mother ○ Mother and stepfather ○ Only father ○ Stepmother and father ○ Some with mother/some with father ○ Other relative ○ Guardian or foster parent ○ Alone or with friends ○ Someone else

Part II

	On a scale from *'No chance'* to *'It will happen'* what do you think are the chances you will...	No chance	Some chance	About 50-50	Pretty likely	It will happen
7	Live to age 35?	○	○	○	○	○
8	Be married by age 25?	○	○	○	○	○
9	Graduate from college?	○	○	○	○	○
10	Get HIV or AIDS?	○	○	○	○	○
11	Get another sexually transmitted disease, such as gonorrhea or genital herpes?	○	○	○	○	○

Part III

These questions are about sexual intercourse. By sexual intercourse we mean a male putting his penis into a female's vagina. Please mark only one response to each item.

12	Have you ever had sexual intercourse?	○ Yes ○ No	*IF NO, GO DIRECTLY TO QUESTION NO 13.*
12a	Now please think about the past 3 months. In the past 3 months, have you had sexual intercourse, even once?	○ Yes ---------> ○ No	How many times have you had sexual intercourse in the past 3 months? ______
12b	To the best of your knowledge, have you ever been pregnant or gotten someone pregnant, even if no child was born?	○ Yes ---------> ○ No	How many times has this happened? ______
12c	How old were you when you had sexual intercourse for the first time?	○ 8 years or younger ○ 9 years ○ 10 years ○ 11 years ○ 12 years ○ 13 years or older	
12d	With how many people have you ever had sexual intercourse?	○ 1 person ○ 2 people ○ 3 people ○ 4 people ○ 5 people ○ 6 or more people	
12e	In the past 3 months have you had sexual intercourse without you or your partner using a condom?	○ Yes ---------> ○ No ○ I don't know	How many times have you had sex without using a condom in the past 3 months? ______
12f	In the past 3 months, have you had sexual intercourse with you or your partner using any of these methods of birth control: • Condoms • Birth Control Pills • The Shot (Depo Provera) • The Patch • The Ring (NuvaRing) • IUD (Mirena or Paragard) • Implant (Implanon)?	○ Yes ---------> ○ No ○ I don't know	How many times have you had sexual intercourse without using any of these methods of birth control? ______

13	Have you ever been physically forced to have sexual intercourse when you did not want to?	◯ Yes ◯ No
14	During the past 12 months, how many times did someone you were dating or going out with force you to do sexual things that you did not want to do? **(Count such things as kissing, touching, or being physically forced to have sexual intercourse.)**	◯ I did not date or go out with anyone during the past 12 months ◯ 0 times ◯ 1 time ◯ 2 or 3 times ◯ 4 or 5 times ◯ 6 or more times
15	Imagine that sometime soon you were to have sexual intercourse with someone just once, but were unable to use any method of birth control for some reason. What is the chance that you (or your partner) would get pregnant?	◯ Almost no chance ◯ Some chance, but probably not ◯ A 50-50 chance ◯ A good chance ◯ Almost certain ◯ Don't know
16	Suppose that sometime soon you had sexual intercourse for a whole month, as often as you wanted to, without using any protection. What is the chance that you would get the AIDS virus?	◯ Almost no chance ◯ Some chance, but probably not ◯ A 50-50 chance ◯ A good chance ◯ Almost certain ◯ Don't know
17	Do you intend to have sexual intercourse in the next year, if you have the chance?	◯ Yes, definitely ◯ Yes, probably ◯ No, probably not ◯ No, definitely not ◯ Don't know
18	If you were to have sexual intercourse in the next year, do you intend to use (or have your partner use) any of these methods of birth control: • Condoms • Birth Control Pills • The Shot (Depo Provera) • The Patch • The Ring (NuvaRing) • IUD (Mirena or Paragard) • Implant (Implanon)?	◯ Yes, definitely ◯ Yes, probably ◯ No, probably not ◯ No, definitely not ◯ Don't know

Part IV

	The next questions ask about your thoughts on pregnancy, sexually transmitted diseases, and AIDS. **Please mark your level of agreement with each of the following statements.**	Strongly agree	Agree	Neutral	Disagree	Strongly disagree	Don't know
19	Getting pregnant or getting someone pregnant at this time in your life is one of the worst things that could happen to you.	◯	◯	◯	◯	◯	◯
20	If you got the AIDS virus, you would suffer a great deal.	◯	◯	◯	◯	◯	◯
21	It would be a big hassle to do the things necessary to completely protect yourself from getting a sexually transmitted disease.	◯	◯	◯	◯	◯	◯
22	If you had sexual intercourse, your friends would respect you more.	◯	◯	◯	◯	◯	◯

	Please mark your level of agreement with each of the following statements.	Strongly agree	Agree	Neutral	Disagree	Strongly disagree	Don't know
23	If you had sexual intercourse, your partner would lose respect for you.	○	○	○	○	○	○
24	If you had sexual intercourse, afterward, you would feel guilty.	○	○	○	○	○	○
25	If you had sexual intercourse, it would upset your mother (or main caretaker).	○	○	○	○	○	○
26	If you had sexual intercourse, it would give you a great deal of physical pleasure.	○	○	○	○	○	○
27	If you had sexual intercourse, it would relax you.	○	○	○	○	○	○
28	If you had sexual intercourse, it would make you more attractive.	○	○	○	○	○	○
29	If you had sexual intercourse, you would feel less lonely.	○	○	○	○	○	○
30	If you got pregnant, or got someone pregnant, it would be embarrassing for your family.	○	○	○	○	○	○
31	If you got pregnant, or got someone pregnant, it would be embarrassing for you.	○	○	○	○	○	○
32	If you got pregnant, or got someone pregnant, you would have to quit school.	○	○	○	○	○	○
33	If you got pregnant, or got someone pregnant, you might marry the wrong person just to get married.	○	○	○	○	○	○
34	If you got pregnant, or got someone pregnant, you would be forced to grow up too fast.	○	○	○	○	○	○
35	If you got pregnant, or got someone pregnant, you would have to decide (or help decide) whether or not to keep the baby, and that would be stressful and difficult.	○	○	○	○	○	○

<u>This concludes the survey, thank you for your participation!</u>

Funding Disclosure This publication was supported by Award No. TP1AH000095-04-00 from the Office of Population Affairs (OPA). Its contents are solely the responsibility of the authors and do not necessarily represent the official views of OPA or HHS.

References

Add Health. (2019). *Add Health codebook explorer*. National Longitudinal Study of Adolescent to Adult Health. Retrieved from: https://www.cpc.unc.edu/projects/addhealth/documentation/ace

Boonstra, H. D. (2019). What is behind the declines in teen pregnancy rates? *Guttmacher Policy Review, 17*(3), 15–21.

Campa, M. I., & Eckenrode, J. J. (2006). Pathways to intergenerational adolescent childbearing in a high-risk sample. *Journal of Marriage and Family, 68*, 558–572.

Centers for Disease Control and Prevention (CDC). (2017). *Standard High School questionnaire, Youth Risk Behavior Survey, Standard High School*. Retrieved from: ftp://ftp.cdc.gov/pub/data/yrbs/2013/2013_hs_questionnaire.pdf

Centers for Disease Control and Prevention (CDC). (2019). *Reproductive health: Teen pregnancy*. Retrieved from: https://www.cdc.gov/teenpregnancy/about/

Coyle, K. K., Kirby, D. B., Marin, B. V., Gomez, C. A., & Gregorich, S. E. (2004). Draw the line/respect the line: A randomized trial of a middle school intervention to reduce sexual risk behaviors. *American Journal of Public Health, 94*, 843–851.

Driscoll, A. K., Sugland, B. W., Manlove, J., & Papillo, A. R. (2005). Community opportunity, perceptions of opportunity, and the odds of an adolescent birth. *Youth Sociology, 37*, 33–61.

Finer, L. B., & Zolna, M. R. (2013). Shifts in unintended pregnancies in the United States, 2001-2008. *American Journal of Public Health, 104*(S1), S43–S48.

Fuller, T. R., White, P., Chu, J., Dean, D., Clemmons, N., Chaparro, C., Thames, J. L., Henderson, A. B., & King, P. (2018). Social determinants and teen pregnancy prevention: Exploring the role of nontraditional partnerships. *Health Promotion Practice, 19*(1), 23–30.

Garney, W., Wilson, K., Nelon, J., Muraleetharan, D., McLeroy, K., & Baletka, D. M. (2019). Ecological approaches to teen pregnancy prevention: An examination of evidence-based interventions. *Health Promotion Practice, 4*, 494–501.

Guttmacher Institute. (2019a). *Emergency contraception*. Retrieved from: https://www.guttmacher.org/state-policy/explore/emergency-contraception

Guttmacher Institute. (2019b). *Insurance coverage of contraceptives*. Retrieved from: https://www.guttmacher.org/state-policy/explore/insurance-coverage-contraceptives

Guttmacher Institute. (2019c). *Minors' access to contraceptive services*. Retrieved from: https://www.guttmacher.org/state-policy/explore/minors-access-contraceptive-services

Hamilton, B. E., & Matthews, T. J. (2016, September). Continued declines in teen births in the united states, 2015. *NCHS Data Brief, No. 259*.

Iacus, S. M., King, G., & Porro, G. (2012). Causal inference without balance checking: Coarsened exact matching. *Political Analysis, 20*(1), 1–24.

Jemmott, J. B., III, Jemmott, L. S., & Fong, G. T. (2010). Efficacy of a theory-based abstinence-only intervention over 24 months: A randomized controlled trial with young adolescents. *Archives of Pediatrics & Adolescent Medicine, 164*(2), 152–159.

Jemmott, J. B., Jemmott, L. S., & Fong, G. T. (1998). Abstinence and safer sex HIV risk-reduction interventions for African American adolescents: A randomized controlled trial. *JAMA : The Journal of the American Medical Association, 279*(19), 1529–1536.

Kirby, D., Barth, R. P., Leland, N., & Fetro, J. V. (1991). Reducing the risk: Impact of a new curriculum on sexual risk-taking. *Family Planning Perspectives, 23*(6), 253–263.

Lindberg, L., Santelli, J., & Desai, S. (2016). Understanding the decline in adolescent fertility in the United States, 2007-2012. *Journal of Adolescent Health, 59*, 577–583.

Lindberg, L. D., Santelli, J. S., & Desai, S. (2018). Changing patterns of contraceptive use and the decline in rates of pregnancy and birth among U.S. adolescents, 2007-2014. *Journal of Adolescent Health, 63*, 253–256.

Manaseri, H., Roberts, K. D., Barker, L. T., & Tom, T. (2019). Pono choices: Lessons for school leaders from the evaluation of a teen pregnancy prevention program. *Journal of School Health, 89*(4), 246–256.

Maness, S. B., Buhi, E. R., Daley, E. M., Baldwin, J. A., & Kromrey, J. D. (2016). Social determinants of health and adolescent pregnancy: An analysis from the National Longitudinal Study of Adolescent to Adult Health. *Journal of Adolescent Health, 58*, 636–643.

MSTAHRS (Mississippi Statistically Automated Health Resource System). (2019). *Pregnancy* [Data file]. Retrieved from: http://mstahrs.msdh.ms.gov/forms/pregtable.html

Pearson, J. (2006). Personal control, self-efficacy in sexual negotiation, and contraceptive risk among adolescents: The role of gender. *Sex Roles, 54*, 615–625.

Penman-Aguilar, A., Carter, M., Snead, M. C., & Kourtis, A. P. (2013). Socioeconomic disadvantage as a social determinant of teen childbearing in the U.S. *Public Health Reports, 128*(S1), 5–22.

Santelli, J. S., Lindberg, L. D., Finer, L. B., & Singh, S. (2007). Explaining recent declines in adolescent pregnancy in the United States: The contribution of abstinence and improved contraceptive use. *American Journal of Public Health, 97*, 150–156.

Sedgh, G., Finer, L. B., Bankole, A., Eilers, M. A., & Singh, S. (2015). Adolescent pregnancy, birth, and abortion rates across countries: Levels and recent trends. *Elsevier: Journal of Adolescent Health, 56*, 223–230.

Skatrud, J. D., Bennett, T. A., & Loda, F. A. (1998). An overview of adolescent pregnancy in rural areas. *The Journal of Rural Health, 14*, 17–27.

The World Bank. (2017). *Adolescent fertility rate (births per 1,000 women ages 15–19)* [Data file]. Retrieved from https://data.worldbank.org/indicator/SP.ADO.TFRT

Trenholm, C., Devaney, B., Fortson, K., Quay, L., Wheeler, J., & Clark, M. (2007). *Impacts of four Title V, Section 510 abstinence education programs*. Princeton: Mathematica Policy Research, Inc.

Chapter 2
Trends and Determinants of Unmet Need for Contraception Among Married Women in Bangladesh: Rural Urban-Comparison

Md. Mostaured Ali Khan, Masud Karim, Md. Rafiqul Islam, Md. Nazrul Hoque, Md. Nurul Islam, Sumaiya Abedin, and Md. Mosharaf Hossain

Abbreviations

AAR	Average Annual Rate
BDHS	Bangladesh Demographic and Health Survey
CI	Confidence Interval
FP	Family Planning
OR	Odds Ratio
UNC	Unmet need for contraception

Md. M. A. Khan · M. Karim · Md. R. Islam (✉) · S. Abedin
Department of Population Science and Human Resource Development, University of Rajshahi, Rajshahi, Bangladesh

Md. N. Hoque
Hobby School of Public Affairs, University of Houston, Houston, TX, USA
e-mail: mnhoque@central.uh.edu

Md. N. Islam
Department of Statistics, University of Rajshahi, Rajshahi, Bangladesh

Md. M. Hossain
Faculty of Business, Economic and Social Development,, University Malaysia Terengganu, Kuala Nerus, Terenggan, Malaysia

B. Jivetti, Md. N. Hoque (eds.), *Population Change and Public Policy*, Applied Demography Series 11, https://doi.org/10.1007/978-3-030-57069-9_2

Introduction

Unmet Need for Contraception (UNC) is one of the important indicators for monitoring the progress of family planning program and has important implications on future population growth and wellbeing of maternal health. UNC refers to an unsatisfactory demand for contraception of women who are either currently married or in union and unwilling to become pregnant by limiting or postponing pregnancy but do not use or access consistent use of contraceptive methods (Bongaarts and Bruce 1995; Westoff 1992; Islam et al. 2013; Mills et al. 2010). According to Singh et al. (2017) an estimated 214 million women had an unmet need for modern contraception in the developing countries and if all need for contraception were met, 67 million unintended pregnancies, 32 million abortion, 76,000 maternal deaths, and 1.8 million infant deaths could be averted each year (Singh et al. 2017). Family planning program can help reduce the unmet need by providing contraceptive methods to the women of reproductive age. The measure of unmet need has become increasingly important in the context of the United Nations (UN) Millennium Development Goals (MDGs), Sustainable Development Goals (SDGs) and Family Planning (FP) 2020. FP 2020 is considered to be a key factor to eliminate the unmet need for contraception. Research indicates that addressing unmet need has served to mediate between the concerns of governments and social scientists focused primarily on controlling population growth and those of public health professionals and human rights activists who advocate for a focus on women's reproductive health and rights. Addressing unmet need result in increasing contraceptive prevalence rates and eliminate health risks associated with unwanted and unsafe pregnancies, thereby reducing population pressures (Sedgh et al. 2007b).

Since 1984 the Demographic and Health Surveys (DHS) program has been working with developing country governments to conduct household surveys of women of reproductive age in more than 90 countries to collect data on fertility, family planning, reproductive, maternal and child health and other health related topics. As an essential indicator of FP programs, unmet need for contraception, as well as the right to decide about contraceptive methods and it's uses, were included in millennium development goal (MDG) to promote maternal health (Alkema et al. 2013; Letamo and Navaneetham 2015) and to reduce fertility. A study estimated that the UNC varied between 5% and 33% for Asian countries, in Latin American and Caribbean countries it varied between 6% and 40% whereas it is 13% and 38% in sub-Saharan Africa (Moreland et al. 2010). Bangladesh is one of the countries where DHS surveys are conducted in every 3–4 years since 1984. Government of Bangladesh has committed to achieving the Sustainable Development Goals (SDGs) and Family Planning (FP) 2020 to ensure universal access to sexual and reproductive health-care services, including family planning program. A substantial increase in contraceptive use has occurred in Bangladesh since 1981 (Larson and Mitra 1992). Results from the 2014 DHS show that the proportion of currently married under age 50 using contraception has reached to 62.4%, a considerable increase from 18.6% in 1981 (NIPORT 2016; Hossain et al. 2018; Huda et al. 2017). DHS data also show that total fertility rate has declined substantially in the past 40 years, with women now having around 2.3 live births over their reproductive life spans as opposed to seven children during the 1970s. The total fertility rate has been stagnant at 2.3 since 2011. Bangladesh government wanted to achieve the replacement level of fertility, i.e., 2.1 children per woman by 2015 (MR 2018). In order to achieve that goal contraceptive prevalence rate would have to rise over 70.0%. According to 2014 BDHS data, 12.0% of currently married women in Bangladesh have an unmet need for family planning services, 6.7% for limiting and 5.3% for spacing of births (NIPORT 2016). The contraceptive prevalence rate could be raised by 12.0% if family planning program is able to convince to use the contraceptive method to the respondent described as having an unmet need for contraception. However, Bangladesh government is captivating initiatives in order to obtain an inclusive use of family planning up to the rate of 80% by 2020 (NIPORT 2016; FP 2021).

Reviewing various studies, many diverse factors have been found as predictors of UNC. In developing countries including Bangladesh, it is closely related to women's and husband's education, age, poverty, religious barrier, access to health service, parity, desire for more children etc. (Kabagenyi et al. 2014; Mosha et al. 2013; Bongaarts and Bruce 1995; Casterline et al. 1997; Casterline and Sinding 2000; Cleland et al. 2014; Fang et al. 2018; Khan 1997; Korra 2002). In Asia and Africa, exposure to family planning information, the price of contraceptive, lack of contraception knowledge, side effect and health risk, socio-cultural norms etc. were also found as the causes influencing UNC (Beguy et al. 2017; Sedgh and Hussain 2014; Yadav and Dhillon 2015; Cleland et al. 2014; Muanda et al. 2016). Few studies have investigated unmet need for contraception among Bangladeshi women, but none has focused on the differentials in unmet need for contraception in rural and urban areas in Bangladesh. The purpose of this study is to examine the trends and differentials and identifying the determining factors of unmet need for contraception among married women of reproductive age and the extent to which these effects vary among rural and urban areas in Bangladesh. Knowledge of rural-urban differentials in unmet need for contraception is important for policy purposes. By comparing the unmet need for contraception of rural women to those of urban women, we are able discern whether these differences can be explained by differing compositional characteristics with respect to demographic and socioeconomic characteristics of rural and urban women in Bangladesh. We are also able to establish whether the differences in unmet need for contraception of rural and urban women are diverging, converging or remaining the same.

Data and Methodology

Study Population and Sampling Procedure

The data analyzed for this study comes from Bangladesh Demographic and Health Surveys (BDHS) conducted in 2004, 2007, 2011, and 2014. Demographic and Health Surveys (DHS) are conducted in more than 90 developing countries using face-to-face interviews and typically collect nationally representative data on demographic and health indicators for women in the age group of 15–49. Bangladesh is one of the countries where DHS surveys are conducted in every 3–4 years since 1984.

Sponsored by the U.S. Agency for International Development (USAID) and conducted by Mitra Associates (Research Firm) under the authority of the National Institute of Population Research and Training (NIPORT) of the Ministry of Health and Family Welfare of Bangladesh. BDHS is a nationally representative cross-sectional survey of currently married women of 15–49 years of age living in households. ICF International of USA provided technical support to the project as part of its international Demographic and Health Surveys Program (2014). These nationally representative cross-sectional household surveys were conducted in seven administrative divisions of Bangladesh: namely, Dhaka, Chittagong, Sylhet, Rajshahi, Khulna, Rangpur, and Barisal. Every division is parted into districts (Zilas) and each district divided into sub-districts (Upazilas), which are further subdivided into urban and rural areas. The survey was based on a two-phase stratified sample of households. In the first phase, 600 primary sampling units were selected, the selection being probable in relation to the size of the unit. In the second phase, 30 households were selected in each primary sampling unit by means of systematic random sampling (BDHS 2004, 2007, 2013, 2015). The BDHS is one of the important sources of demographic data in Bangladesh. The BDHSs data are well suited for this study given its large and nationally representative samples of women of reproductive age (15–49) that collect data on fertility, family planning, reproductive, maternal and child health, and other health issues and unmet need for family planning. The overall response rate was around 98%. The total sample size was 11,440 in 2004, 10,996 in 2007, 17,842 in 2011, and 17,863 in 2014. In this study, women who were not currently married, were excluded from the analysis.

Outcome Variable

To accomplish the aim of this study, the *unmet need for contraception* (UNC) is used as the outcome variable. The information regarding our outcome variable was gathered by asking currently married women at reproductive age (15–49 years) about their contraceptive use status. The contraceptive use status of women further categorized in to categories such as, *"No unmet need"* defined as they have no UNC because they have desire for children right then, so contraception method is required; "*Unmet need for spacing birth*" and "*Unmet need for limiting birth*", defined as they do not want more children but are not using contraception either for limiting birth or for spacing birth; "*Using for spacing birth*" and "*Using for limiting birth*" refers to the women who do not want more children and using contraception either for limiting birth or for spacing birth, and the last category included the women who were "*Infecund or menopausal*", include those women who were infecund or in menopausal state. In this analysis we created a new variable by combining "*Unmet need for spacing birth*" and "*Unmet need for limiting birth*" into "*Unmet need for contraception*" and categories as (1, yes and 0, no). The occurrence of UNC is the primary outcome variable of the study, coded 1 for unmet need 0 otherwise.

Explanatory Variables

In this study, various socio-demographic variables like age, respondent's and their husband's education, economic status, employment status etc., and background characteristics like children ever born, age at marriage etc. were used as explanatory variables to identify the influencing factors of the UNC in rural and urban areas of Bangladesh. A brief explanation of the explanatory variable is represented in Table 2.1. These variables were selected in accordance with their importance based on previous research (Ahmed et al. 2012; Casterline and Sinding 2000; Islam et al. 2013, 2016; Letamo and Navaneetham 2015).

Statistical Analysis

First, we examine the trend of unmet need for contraception among married women of reproductive age in rural and urban Bangladesh. In order to examine the trend, we calculated the overall decline in UNC rate using the formula: $\frac{Y_t - Y_o}{Y_o} \times 100$ and we also calculated the average annual rate (ARR) of change in UNC among currently married women using four BDHSs data sets conducted in the year of 2004, 2007, 2011 and 2014. To perform this analysis following formula has been utilized: $r = \left(\sqrt[n]{\frac{Y_t}{Y_0}} - 1\right) \times 100$; where, Y_0 = prevalence of unmet need for contraception of any given year, Y_t = prevalence of unmet need for contraception of *tth* year, r = average annual rate of change, n = no. of years between two surveys. The method was obtained from UNICEF technical note and revised according to the provided information (UNICEF 2007). Second, an association between unmet need for

Table 2.1 A complete list of explanatory variables

Variables	Category
Age group	1 = 15–24 years; 2 = 25–34 years; 3 = 35–49 years
Education	1 = No education; 2 = Primary; 3 = Secondary; 4 = Higher
Wealth index	1 = Poor; 2 = Middle; 3 = Rich;
Employment	1 = Unemployed; 2 = Employed
Husband's education	1 = No education; 2 = Primary; 3 = Secondary; 4 = Higher
Children ever born	1 = ≤ 2 Children; 2 = 2 >Children
Ever had terminated pregnancy	0 = No; 1 = Yes
Visited by FP workers in past 6 months	0 = No; 1 = Yes
Watching television	0 = Not at all; 1 = Frequently
Recent sexual activity	0 = Almost inactive; 1 = Active (last 4 weeks)
Age at marriage	1 = ≤ 16 years; 2 = 16 < years

contraception and different explanatory variables were assessed by χ^2 tests, setting at $p < 0.05$ level of significance. Third, binary logistic regression model was fitted to measure the impact of selected explanatory variables on outcome variable. Since the outcome variable is dichotomous, binary logistic regression is an appropriate technique for the analysis. 95% confidence intervals (CIs) for each ORs were calculated to inspect the level of significance. All the analysis in this study was done by considering complex survey design and sample weights. Datasets were analyzed using Stata windows version 13.0 (Stata Corp., College Station, TX, USA) and MS excel version 2016.

Results

The trends of contraceptive use status among currently married women in rural and urban Bangladesh is presented in Table 2.2 and Fig. 2.1. Overall, the prevalence of UNC is following a gradual decreasing trend after the year 2007 (15.6%) and stand at 6.9% in year 2014. The UNC for spacing birth has increased from 4.7% in 2004 to 6.2% in 2007 and then declined to 5.4% in 2011 and further decrement to 3.1% in 2014. Similarly, UNC for limiting birth has increased from 5.7 in 2004 to 9.5 in 2007, then gradually decreased to 3.8% in 2014. A higher unmet need for limiting birth has observed than spacing birth across different time period. Likewise, the overall UNC in rural area has increased from 11.0% in 2004 to 16.4% in 2007, then decreased to 14.3% in 2011 and 8.3% in 2014 (Fig. 2.1). Besides, in urban settings, the overall UNC has increased from 8.4% in 2004 to 13.3% in 2007, then decreased to 11.1% in 2011 and 4.5% in 2014 (Fig. 2.1). This trend indicates that rural areas are more vulnerable in terms of UNC among married women in Bangladesh.

Table 2.2 The trends of contraceptive use status among currently married women in Bangladesh over the period 2004–2014

Contraception use	2004 (%, 95% CI)	2007 (%, 95% CI)	2011 (%, 95% CI)	2014 (%, 95% CI)
Unmet need for Contraception	10.4 (9.7–11.2)	15.6 (14.6–16.7)	13.5 (12.8–14.2)	6.9 (5.8–8.3)
No unmet need	21.9 (20.9–22.8)	18.1 (17.3–19.0)	13.7 (13.1–14.4)	13.8 (13.0–14.8)
Unmet need for spacing	4.7 (4.2–5.2)	6.2 (5.6–6.8)	5.4 (5.0–5.8)	3.1 (2.6–3.7)
Unmet need for limiting	5.7 (5.2–6.3)	9.5 (8.7–10.2)	8.1 (7.6–8.5)	3.8 (3.2–4.6)
Using for spacing	16.1 (15.4–16.8)	13.9 (13.1–14.7)	16.1 (15.4–16.8)	18.8 (17.9–19.7)
Using for limiting	39.9 (38.6–41.2)	38.2 (36.9–39.6)	45.1 (44.2–46.1)	46.5 (45.2–47.9)
Infecund or menopausal	11.8 (11.1–12.4)	14.1 (13.4–14.9)	11.5 (10.9–12.1)	13.9 (13.1–14.6)

Note: All percentages are weighted

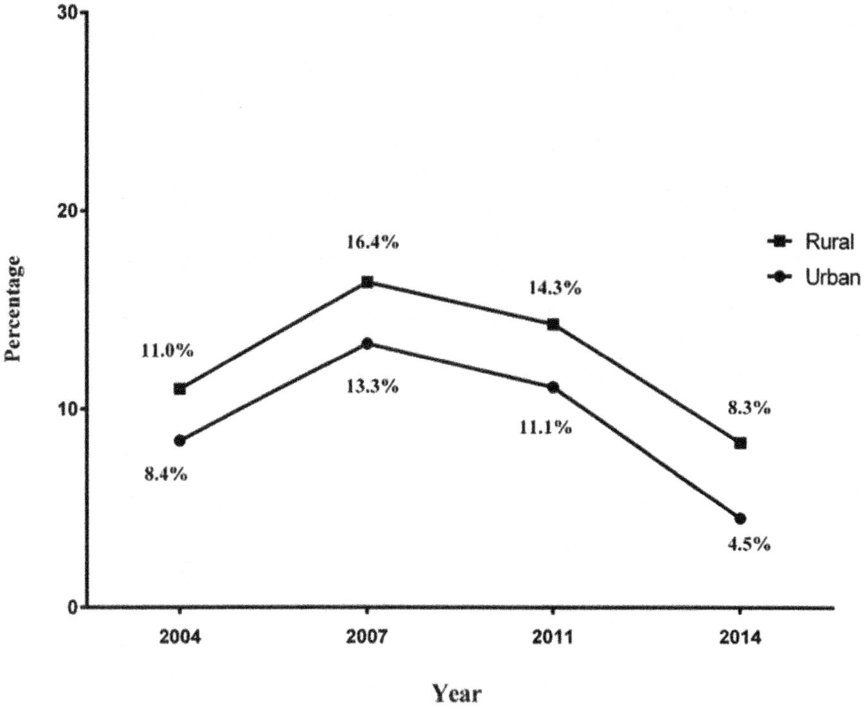

Fig. 2.1 Rural-urban trend of unmet need for contraception (weight %) among currently married women in Bangladesh over the period 2004–2014

In Table 2.3, Urban-rural differential trends of UNC during 2004–2014 among currently married women according to several socio-economic and demographic characteristics is represented. In this period, UNC was higher for women of each age group in rural areas than the urban areas in each of the year except for year 2004 (0.1% lower in rural than urban areas). The highest overall decrement (urban vs rural: 68.1% vs 26.1%) and ARR of decrement (urban vs rural: 10.8% vs 3.0%) was observed in young-adult cluster of women (24–35 years) for both of areas. The prevalence of UNC was higher in rural areas than urban areas among women according to their educational status (illiterate to higher educated). The highest rate of overall decrement (urban vs rural: 55.4% vs 51.2%) and AAR of decrement (urban vs rural: 7.8% vs 6.9%) was noticed among secondary educated women but among rural illiterate women, an uprising trend was observed. In urban areas, highest overall decrement (47.7%) and ARR of decrement (6.3%) was observed among poor community but in rural areas, middles class community had experienced the highest overall (29.7%) and ARR (3.5%) of decrement. The prevalence of UNC was always higher among unemployed rural women and the situation has not improved yet. There was a huge difference was observed in overall and ARR of decrement between urban and rural areas (overall: 49.5% vs 19.5% and ARR: 6.6% vs 2.1%, respectively in urban and rural areas). However, in both urban and rural areas, noticeable

Table 2.3 Trends and differential of unmet need for contraception among currently married women in rural and urban Bangladesh according to selected socio-economic and demographic characteristics

Background characteristics	Unmet need for contraception (%)															
	2004 (%)			2007 (%)			2011 (%)			2014 (%)			Change during 2004–2014 (%)			
													Overall change		ARR of change	
	Urban	Rural	Diff.	Urban	Rural	Diff.	Urban	Rural	Diff.	Urban	Rural	Diff.	Urban	Rural	Urban	Rural
Age group:																
15–24	9.7	14.4	4.7	15.4	17.9	2.5	12.8	16.9	4.1	5.3	11.1	5.8	−45.4	−22.9	−5.9	−2.6
25–34	11.6	11.5	−0.1	12.9	18.0	5.1	11.0	15.7	4.7	3.7	8.5	4.8	−68.1	−26.1	−10.8	−3.0
35–49	3.8	6.6	2.8	11.3	13.2	1.9	9.4	10.3	0.9	4.8	5.6	0.8	26.3	−15.2	2.4	−1.6
Education:																
No education	8.7	8.9	0.2	10.1	15.7	5.6	8.9	12.7	3.8	4.6	11.3	6.7	−47.1	27.0	−6.2	2.4
Primary	8.9	11.9	3.0	12.9	15.6	2.7	12.0	13.2	1.2	4.8	9.5	4.7	−46.1	−20.2	−6.0	−2.2
Secondary	9.2	12.1	2.9	16.5	18.1	1.6	11.6	16.7	5.1	4.1	5.9	1.8	−55.4	−51.2	−7.8	−6.9
Higher	4.5	7.9	3.4	10.3	16.6	6.3	10.7	14.6	3.9	5.2	6.9	1.7	15.6	−12.7	1.5	−1.3
Wealth index:																
Poor	11.1	11.2	0.1	16.5	16.0	−0.5	12.5	13.2	0.7	5.8	8.2	2.4	−47.7	−26.8	−6.3	−3.1
Middle	9.4	11.1	1.7	12.5	16.4	3.9	12.4	13.5	1.1	6.0	7.8	1.8	−36.2	−29.7	−4.4	−3.5
Rich	7.5	10.5	3.0	12.6	16.7	4.1	10.7	16.8	6.1	4.2	8.8	4.6	−44.0	−16.2	−5.6	−1.8
Employment:																
Unemployed	9.5	11.8	2.3	14.9	18.2	3.3	11.4	14.7	3.3	4.8	9.5	4.7	−49.5	−19.5	−6.6	−2.1
Employed	5.1	8.0	2.9	8.5	12.5	4.0	9.4	9.9	0.5	3.9	5.6	1.7	−23.5	−30.0	−2.6	−3.5
Husband's education:																
No education	7.9	9.7	1.8	12.1	14.9	2.8	10.3	13.0	2.7	5.1	11.4	6.3	−35.4	17.5	−4.3	1.6
Primary	10.3	12.2	1.9	14.2	17.4	3.2	9.9	14.6	4.7	4.6	9.4	4.8	−55.3	−23.0	−7.7	−2.6
Secondary	9.3	12.9	3.6	14.5	18.4	3.9	12.5	16.1	3.6	3.7	5.9	2.2	−60.2	−54.3	−8.8	−7.5
Higher	5.8	7.0	1.2	11.1	13.6	2.5	10.9	13.2	2.3	5.1	5.8	0.7	−12.1	−17.1	−1.3	−1.9

Division																
Barisal	8.6	12.2	3.6	14.3	19.4	5.1	11.7	12.2	0.5	8.8	14.8	6.0	2.3	21.3	0.2	2.0
Chittagong	11.4	17.1	5.7	16.3	22.8	6.5	15.9	22.3	6.4	4.7	10.5	5.8	−58.8	−38.6	−8.5	−4.8
Dhaka	8.1	10.7	2.6	12.8	17.5	4.7	10.3	14.8	4.5	2.6	4.2	1.6	−67.9	−60.7	−10.7	−8.9
Khulna	8.0	7.5	−0.5	9.1	11.0	1.9	7.3	10.1	2.8	5.0	5.3	0.3	−37.5	−29.3	−4.6	−3.4
Rajshahi	5.1	6.9	1.8	11.4	11.0	−0.4	10.6	11.1	0.5	6.5	8.9	2.4	27.5	29.0	2.5	2.6
Rangpur	–	–	–	–	–		8.5	9.8	1.3	8.2	5.0	−3.2	–	–	–	–
Sylhet	8.4	19.4	11	16.5	23.4	6.9	14.2	17.9	3.7	7.6	16.8	9.2	−9.5	−13.4	−1.0	−1.4

Note: All percentages are weighted. Diff. refers to absolute difference between rural and urban. "Plus (+)" sign indicates increment and "minus (−)" sign indicates decrement. *ARR* Average Annual Rate

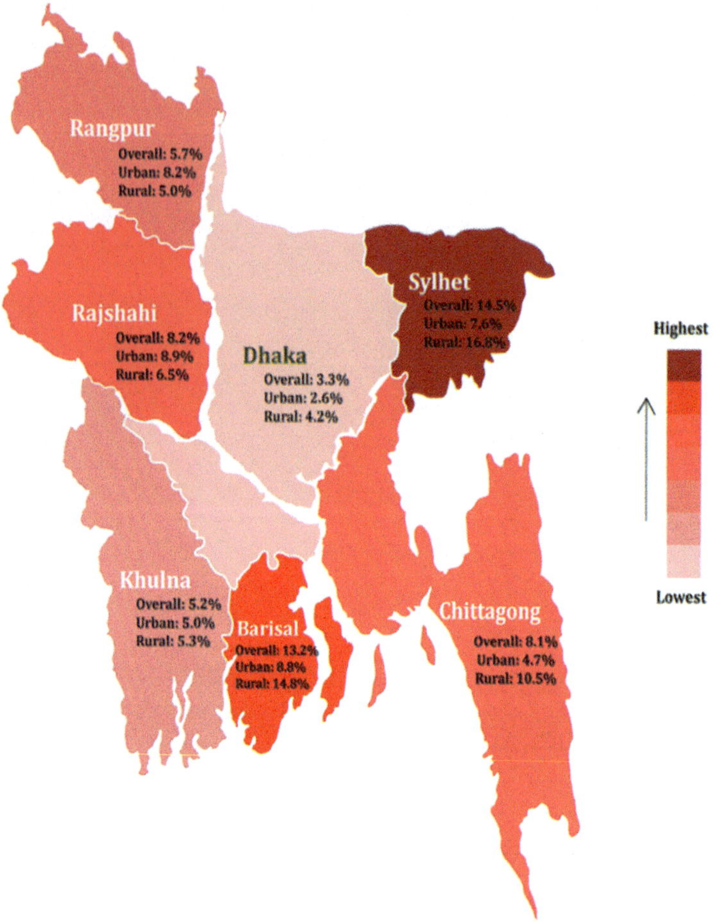

Fig. 2.2 Spatial distribution (weighted %) of unmet need for contraception among currently married women, using Bangladesh Demographic and Health Survey data, 2014

deterioration of UNC was observed among women who has secondary education level completed husband (overall decline: 60.2% vs 54.3%, respectively and ARR of decline: 8.8% vs 7.5%, respectively in urbans and rural areas). But UNC was showing an uprising trend among rural women (17.5% of overall and 1.6% of ARR of increment) who had illiterate husband. Regional variation shows significant deterioration of UNC in every regions of Bangladesh in both urban and rural areas except Barisal region (overall increase: 2.3% vs 21.3%, respectively and ARR of increment: 0.2% vs 2.0%, respectively in urban and rural areas). Highest decrement in UNC rate was found in Dhaka region (overall decrement: 67.9%% vs 60.7%, respectively and ARR of increment: 10.7% vs 8.9%, respectively in urban and rural areas).

The regional variation in UNC among currently married women is depicted in Fig. 2.2. In 2014, the lowest percentage of UNC was reported among married women in Dhaka region (overall: 3.3%; urban: 2.6%; and rural: 4.2%). Khulna region (overall: 5.2%; urban: 5.0% and rural: 5.3%). stood at the 2nd position and consequently Rangpur region stood at 3rd position (overall: 5.7%; urban: 8.2% and rural: 5.0%). The highest percentage of unmet need for contraception was found among women of reproductive age in Sylhet region (overall: 14.5%; urban: 7.6% and rural: 16.8%) followed by Barisal region (overall 13.2%; urban 8.8% and rural 14.8%).

Table 2.4 illustrates the association between UNC and explanatory variables. Chi-square test was used to observe the level of significance. Women younger cluster (aged 15–24 years) had the highest UNC prevalence in both urban (11.8%) and rural areas (16.9%). Also, educational status of both respondent's and their husband's played a significant role on UNC. A higher percentage of UNC was observed among women with no education both in urban and rural areas, 11.5% and 17.2%, respectively. A similar scenario was observed among women with illiterate husband, 11.7% in urban and 16.7% in rural area. In rural areas, respondent's wealth index was showing significant association with UNC and surprisingly higher unmet was deemed among rich families (15.1%). Unemployed urban and rural women were in higher UNC than employed women, 10.2% vs 14.5%, respectively. A large number of women with less than or equal of 2 children (10.1% in urban and 13.4% in rural areas) and women who were sexually inactive (33.3% in urban and 34.4% in rural areas) have/had desire for UNC. Besides, in rural area, 14.4% of women had a terminated pregnancy and also had desire for UNC. Respondent's frequency of watching television, visit by FP workers and age at marriage had also a good impact on UNC, especially in rural areas. There were 10.7% of urban and 14.6% of rural women with desire for UNC were not watching TV at all. Expectedly, a large number of women with UNC were not visited by FP workers in the past 6 months prior to the survey, 10.9% in urban area and 17% in rural area. Majority of women who got married at the age of 16 or above had desire for UNC (13.2% in urban and 15.9% in rural area).

Table 2.5 demonstrates the effect size of different risk factors of UNC among currently married women in rural and urban Bangladesh. The likelihood of UNC was decreased for rural women aged 25–34 years (AOR: 0.66, 95% CI: 0.52–0.84) and aged 35–49 years (AOR: 0.37, 95% CI: 0.27–0.51), and for urban women aged 35–49 years (OR = 0.549, 95% CI: 0.335–0.899) compared to the women of younger cluster (15–24 years) of both areas. However, the primary and secondary educated urban women had 0.47 (95% CI: 0.29–0.78) and 0.38 (95% CI: 0.22–0.65) times less likely to had UNC than illiterate urban women, respectively and the primary and secondary educated rural women had 0.543 (95% CI: 0.419–0.703) and 0.567 (95% CI: 0.428–0.753) times less likely to had UNC than their counterparts, respectively. Similarly, in urban area, respondents whose husbands had completed primary education had 0.55 (95% CI: 0.33–0.91) times fewer risk of UNC than the respondents with illiterate husband. Furthermore, in rural area, the likelihood of UNC was decreased than for rural women who husband had completed

Table 2.4 Bivariate distribution of unmet need for contraception among currently married women by selected background characteristics in Bangladesh, using Bangladesh Demographic and Health Survey data, 2014

	Unmet need for contraception (%)					
	Urban			Rural		
Background characteristics	No	Yes	χ^2_{cal} *p-value*	No	Yes	χ^2_{cal} *p-value*
Age group:						
15–24	1473 (88.2%)	198 (11.8%)	14.25 (0.001)	2825 (83.1%)	576 (16.9%)	102.24 (<0.001)
25–34	1937 (91.6%)	178 (8.4%)		3470 (86.9%)	521 (13.1%)	
35–49	1779 (91.1%)	174 (8.9%)		3367 (91.1%)	329 (8.9%)	
Education:						
No education	1144 (88.5%)	148 (11.5%)	9.46 (0.024)	2157 (82.8%)	447 (17.2%)	78.78 (<0.001)
Primary	1494 (90.1%)	165 (9.9%)		2841 (87.1%)	419 (12.9%)	
Secondary	2057 (91.6%)	188 (8.4%)		3736 (90.2%)	407 (9.8%)	
Higher	494 (91.0%)	49 (9.0%)		928 (85.8%)	153 (14.2%)	
Wealth index:						
Poor	797 (91.3%)	76 (8.7%)	1.10 (0.576)	4695 (88.2%)	626 (11.8%)	19.68 (<0.001)
Middle	691 (90.7%)	71 (9.3%)		2331 (87.5%)	332 (12.5%)	
Rich	3701 (90.2%)	403 (9.6%)		2636 (84.9%)	468 (15.1%)	
Employment:						
Unemployed	3699 (89.8%)	422 (10.2%)	7.27 (0.007)	6512 (85.5%)	1108 (14.5%)	61.35 (<0.001)
Employed	1490 (92.1%)	128 (7.9%)		3150 (90.8%)	318 (9.2%)	
Husband's education:						
No education	1444 (88.3%)	192 (11.7%)	16.31 (0.001)	2573 (83.3%)	516 (16.7%)	67.28 (<0.001)
Primary	1430 (90.4%)	151 (9.6%)		2585 (87.5%)	370 (12.5%)	
Secondary	1543 (92.4%)	127 (7.6%)		3012 (90.1%)	332 (9.9%)	
Higher	769 (90.7%)	80 (9.4%)		1492 (87.8%)	208 (12.2%)	

Children ever born:						
≤2 Children	3395 (89.9%)	383 (10.1%)	3.92 (0.048)	5312 (86.6%)	825 (13.4%)	4.16 (0.041)
>2 Children	1794 (91.5%)	1679 (8.5%)		4183 (87.9%)	601 (12.1%)	
Ever had a terminated pregnancy						
No	4196 (90.5%)	439 (9.5%)	0.35 (0.554)	7836 (87.5%)	1120 (12.5%)	5.24 (0.022)
Yes	993 (89.9%)	111 (10.1%)		1826 (86.6%)	306 (14.4%)	
Visited by FP worker in last 6 months:						
No	1449 (89.1%)	177 (10.9%)	4.02 (0.045)	2755 (83.0%)	564 (17.0%)	7.23 (0.007)
Yes	256 (93.1%)	19 (6.9%)		668 (87.0%)	100 (13.0%)	
Watching television:						
Not at all	2580 (89.3%)	309 (10.7%)	8.28 (0.004)	4468 (85.4%)	766 (14.6%)	27.85 (<0.001)
Yes	2608 (91.5%)	241 (8.5%)		5194 (88.7%)	660 (11.3%)	
Recent sexual activity:						
Almost inactive	591 (66.7%)	295 (33.3%)	679.87 (<0.001)	1362 (65.6%)	713 (34.4%)	1052.99 (<0.001)
Active (last 4 weeks)	4598 (94.7%)	25 5(5.3%)		8300 (92.1%)	713 (7.9%)	
Age at marriage:						
≤16 years	3587 (92.1%)	307 (7.9%)	40.38 (<0.001)	6479 (88.7%)	823 (11.3%)	48.2 (<0.001)
>16 years	1602 (86.8%)	243 (13.2%)		3183 (84.1%)	603 (15.9%)	

Note: Sample was unweighted for this analysis. The analysis was calculated for row percentage

Table 2.5 Adjusted odds ratio to measure the effect size of different risk factors of unmet need for contraception among currently married women in Bangladesh, using Bangladesh Demographic and Health Survey data, 2014

	Urban		Rural	
Variables	AOR[a]	95% CI	AOR[a]	95% CI
Age group:				
15–24[(RC)]	1.00	……	1.00	……
25–34	0.82	(0.54–1.23)	0.66***	(0.52–0.84)
35–49	0.55*	(0.34–0.89)	0.37***	(0.27–0.51)
Education:				
No education [(RC)]	1.00	……	1.00	……
Primary	0.47**	(0.27–0.78)	0.54***	(0.42–0.70)
Secondary	0.38***	(0.22–0.65)	0.57***	(0.43–0.75)
Higher	0.65	(0.30–1.40)	1.09	(0.72–1.68)
Employment:				
Unemployed [(RC)]	1.00	……	1.00	……
Employed	1.11	(0.73–1.68)	0.89	(0.71–1.12)
Wealth index:				
Poor [(RC)]	1.00	……	1.00	……
Middle	0.77	(0.43–1.38)	1.24	(0.99–1.54)
Rich	1.14	(0.75–1.73)	1.07	(0.87–1.31)
Husband's educational:				
No education [(RC)]	1.00	……	1.00	……
Primary	0.55*	(0.33–0.91)	0.75*	(0.58–0.98)
Secondary	0.62	(0.36–1.06)	0.71**	(0.53–0.94)
Higher	1.05	(0.55–2.01)	1.09	(0.75–1.59)
Children ever born:				
≤2 Children [(RC)]	1.00	……	1.00	……
>2 Children	1.02	(0.66–1.52)	1.59***	(1.23–2.05)
Ever had a terminated pregnancy:				
No [(RC)]	1.00	……	1.00	……
Yes	1.18	(0.77–1.83)	1.12	(0.89–1.41)
Visited by FP workers in last 6 months:				
No [(RC)]	1.00	……	1.00	……
Yes	0.69	(0.40–1.19)	0.76*	(0.59–0.98)
Watching television:				
Not at all [(RC)]	1.00	……	1.00	……
Yes	2.16***	(1.45–3.18)	2.14***	(1.76–2.63)
Recent sexual activity:				
Almost inactive [(RC)]	1.00	……	1.00	……
Active (last 4 weeks)	0.08***	(0.06–0.12)	0.14***	(0.12–0.17)
Age at marriage:				
≤16 years [(RC)]	1.00	……	1.00	……
>16 years	1.24	(0.86–1.79)	0.70***	(0.57–0.87)

(continued)

Table 2.5 (continued)

Variables	Urban		Rural	
	AOR[a]	95% CI	AOR[a]	95% CI
Model summary:				
Model Chi-Square (ρ-value)	1673.194 (<0.001)		2717.628 (<0.001)	
-2Log Likelihood	957.993		2948.157	
Cox & Snell R-square	0.586		0.486	
Nagelkerke R-square	0.781		0.648	

Note: [a]Model was adjusted for all the predictors included in this table. Values with superscript asterisks *, **, and *** indicate $p < 0.05$, $p < 0.01$, and $p < 0.001$, respectively. *RC* Reference category, *AOR* adjusted odds ratio, *CI* confidence interval

primary level (OR = 0.75, 95% CI: 0.58–0.99) and secondary level (OR = 0.71, 95% CI: 0.53–0.94) of education than the respondents with illiterate husband. Bangladeshi rural women with more than two children had 1.59 (95% CI: 1.23–2.05) times more at risk to had UNC than the women with 2 children or less. If the women in rural area were visited by FP workers, the likelihood of having UNC (OR = 0.764, 95% CI: 0.595–0.981) was reduced than their counterparts. The chance of UNC was 2.16 times (95% CI: 1.45–3.18) and 2.14 times (95% CI: 1.76–2.63) higher among urban and rural women who were watching television frequently than the women who were not watching television at all. The chance of having UNC among sexually active urban women (AOR: 0.08, 95% CI: 0.06–0.12) and rural women (AOR: 0.14, 95% CI: 0.12–0.17) was decreased than their counterparts. Rural women whose age at marriage was > 16 years, had 0.70 times (95% CI: 0.566–0.869) reduced risk of having UNC than their counterparts.

Discussion

Using nationally representative survey datasets of Bangladesh, this study focused on observing the trend of UNC, the recent variation in the UNC change rate, and also identified the risk factors associated with UNC in both urban and rural areas. The finding of this study concerning the higher rate of UNC for limiting birth than spacing birth among both urban and rural women which are identical to the findings of other studies that conducted in different time across different countries (Shah et al. 2004). For example, in India, Yadav and Dhillon (2015) found a very similar result. In general, the rate of UNC in Bangladesh is much lower than many African countries like Ethiopia, Sudan etc. (Ali and Okud 2013) and the trend is gradually decreasing after increasing at a high rate in the year 2007 in both urban and rural area. A systematic and comprehensive study has found a decreasing trend of UNC for 111 countries around the world over the year 1990–2015 (Alkema et al. 2013). But very few studies have described the rural-urban trend of UNC. If we focus on the overall change of UNC over the period 2004–2014 then the UNC will show an increasing trend. The results regarding higher UNC in Sylhet and Barisal region compared to other divisions of Bangladesh is similar to the study result of Barkat-e-Khuda et al. (Roy and Rahman 1999) and Islam et al. (2013). So, these regions can be termed as most vulnerable to UNC in Bangladesh.

Among Bangladeshi women, various factors were found to be responsible for having UNC. Reasons may vary in urban and rural area. Comparatively, rural women are troubling with UNC more than urban women because of different factors (Roy and Rahman 1999; Islam et al. 2013). Though the highest AAR of increase in UNC with a decreasing trend after 2007 was observed among women aged 35–49 years in the urban and rural area but comparatively higher for urban women. Besides, our study result regarding the higher percentage and odds for UNC among young women (15–24 years) both in urban and rural area than the mid-aged adult women (35–49 years) (Petry 2002) is showing consistency with other studies of developing countries (Mawajdeh 2007), though the percentage was higher among rural women than urban women. So, it is very clear that young women are at high risk of UNC in Bangladesh, especially in rural area.

In Bangladesh, the UNC has been increased at the highest annual rate among higher educated urban women and illiterate rural women and the same trend were observed in case of their husband's education between years 2004 and 2014. In this study, the respondent's and their husband's education have a significant effect on UNC. Expectedly, both urban and rural educated women had a lower chance to have UNC. Evidence from previous studies showed couples education is a vital predictor of low contraception use in Bangladesh (Khan 1997; Islam et al. 2017). Moreover, Emina et al. (2014) identified women's education as a key factor of contraceptive use. But another study found a negative correlation between UNC and couple's education in Jordan (Mawajdeh 2007). As expected, a greater portion of contraceptive nonuser is always illiterate Bangladeshi women (Islam et al. 2017). So, education is found to have a strong effect in fetching social change in Bangladesh.

Findings of this study showed a high percentage of UNC among rich people in the rural area shows irrelevance with earlier endeavors of August et al. (2015) and Uddin et al. (2016), as they stated that low-income women were at high risk of having UNC. The results found a greater AAR of increment in UNC among unemployed women than employed women in rural area but the scenario is different for the urban area; UNC was increased at a higher AAR among employed women. In Bangladesh, a greater part of unemployed women has an unmet need than employed women in both urban and rural area. Similarly, Khan et al. (1997) observed women's work experience has a little effect on contraceptive use in both urban and rural area.

However, a higher AAR of increase in UNC was observed among women with children ever-born 2 or less in both urban and rural women in Bangladesh. This study result regarding a higher percentage and more likelihood of having UNC among rural women with children more than 2 is supported by Imasiku et al. (2013). Thereafter, Khan et al. (1997) explained, parity of women has a significant positive effect on contraceptive use in the contrast of Bangladesh. So, women with more children ever born have to more careful about family planning. Besides, according to our observation, pregnancy termination has shown an increasing trend among women with UNC in both urban and rural area. Similarly, a study on trend analysis observed a very increasing trend of abortion in developing countries and eastern Europe (105 for every 100 live birth in eastern Europe) and pregnancy termination is significantly associated with UNC (Sedgh et al. 2007a). Moreover, our result regarding a higher UNC among women who ever had a terminated pregnancy is also supported by Westhoff's (2005) findings as he found that the unmet need for family planning contributes most of the abortions in one-third of the countries within his study sample. Hopefully, there is no further clarification is needed to clarify the importance of reducing UNC to lessen the rate of pregnancy termination.

In Bangladesh, FP programs have a vital impact on reducing UNC among women (Bongaarts 2014). A very rising trend is observed among rural women who have not visited by FP workers in last 6 months prior to the survey, especially in rural area. Based on this study result, visits from FP workers is found to be a significant predictor of UNC in both rural and urban area though it is most important for rural women. Showing consistency with our study, Bugey et al. (2017) and Casterline and Sinding (2000) noticed family planning program and policy as a vital predictor of contraceptive use. So, particular attention is needed to increase the visit of FP workers to women to reduce UNC, especially in the rural area of Bangladesh. The present study demonstrated a greater part of women with UNC were watching television frequently in both areas and the trend is growing. Mass media can play an emergent role in increasing contraceptive use and on reducing UNC (Beguy et al. 2017) especially among women living in rural areas. Islam et al. (2016) found a significant role of newspaper/magazine on UNC in Bangladesh.

Our study result concerning a higher percentage of UNC among sexually inactive women showed inconsistency with other studies (Bell and Bishai 2017; Bradley and Casterline 2014). Most of the sexually almost inactive women are not much conscious about contraceptive use in Bangladesh. Similarly, a previous study on developing countries found a greater share of contraceptive nonusers in developing countries had infrequent sexual activity (Sedgh and Hussain 2014). So, these people are at high risk of UNC. In rural setting, age at marriage is an important indicator of UNC when unmet need for spacing and limiting are taken into account in itself (Korra 2002). This study found women's age at marriage of 16 years or above has lower chance to have UNC in the rural area. Therefore, more attention is needed to pay to aware the women whose age at marriage is more than 16 years about the use of contraception.

Limitations and Strengths of the Study

This study have few limitations and strengths that need to be mentioned. In terms of strengths, we used nationally representative data to perform the study analysis with appropriate statistical methods by taking into account complex survey design and sample weights. This large dataset of several surveys also gives us strengths to make a reasonable conclusion. In terms of limitations, these data are cross-sectional, non-response error and recall bias is a common problem for such kind of datasets.

Moreover, few important factors that could also be an influencing factor of UNC, were missing from these datasets. Nevertheless, we have tried to depict a compact illustration of the current situation, trend and risk factors of UNC comparing the differences of the urban and rural women in Bangladesh. To our knowledge, only a few studies were conducted in this area of study in Bangladesh. Hopefully, this research will contribute to developing appropriate interventions for decreasing UNC among currently married women especially in the rural areas of Bangladesh.

Conclusions

This study observed an overall rising trend of unmet need for contraception among women in Bangladesh and significantly, Sylhet and Chittagong are the most perilous region. Age of women, education, husband's education, watching television, and respondent's sexual activity were the dominating factors of unmet need for contraception among urban women, over against age, couple's education, children ever born, visit by FP workers, watching television, sexual activity, and age at marriage were the most influencing factors of unmet need for contraception among rural Bangladeshi women.

This study considers that the unmet need for contraception can be met by enhancing the light of education to women and men, especially in the rural area. Proper knowledge and importance of contraceptive use need to spread through family planning workers, especially in rural areas and among women who were almost sexually inactive. Mass media, especially television can play as a momentous part in increasing knowledge and consciousness about family planning by which unmet need for contraception can be eradicated.

Acknowledgments The data used in this study has been obtained from MEASURE DHS Archive. The authors are thankful to MEASURE DHS for granting data use permission. The authors also acknowledge the support of the Department of Population Science and Human Resource Development, University of Rajshahi, Bangladesh, where this study was conducted.

Ethics Statement Ethical approval was obtained from the Ministry of Health and Family Welfare, Dhaka, Bangladesh. The international credit finance (ICF) supported financial and technical assistance for the survey via USAID/Bangladesh. Data collection process was also approved by the ORC Macro Institutional Review Board. The BDHS is part of the worldwide Demographic and Health Surveys program, which is designed to collect data on fertility, family planning, and maternal and child health. Informed written consent was obtained from all the participants prior to the study.

Competing Interest The authors have declared that there are no competing interests exist.

Funding This research did not receive any specific grant either from the University or from any funding agencies in the public, commercial, or not-for-profit sectors.

References

Ahmed, S., Li, Q., Liu, L., & Tsui, A. O. (2012). Maternal deaths averted by contraceptive use: An analysis of 172 countries. *The Lancet, 380*(9837), 111–125.

Ali, A. A. A., & Okud, A. (2013). Factors affecting unmet need for family planning in Eastern Sudan. *BMC Public Health, 13*(1), 102.

Alkema, L., Kantorova, V., Menozzi, C., & Biddlecom, A. (2013). National, regional, and global rates and trends in contraceptive prevalence and unmet need for family planning between 1990 and 2015: A systematic and comprehensive analysis. *The Lancet, 381*(9878), 1642–1652.

August, E. M., Steinmetz, E., Gavin, L., Rivera, M. I., Pazol, K., Moskosky, S., et al. (2015). Projecting the unmet need and costs for contraception services after the affordable care act. *American Journal of Public Health, 106*(2), 334–341. https://doi.org/10.2105/AJPH.2015.302928.

BDHS. (2004). *Bangladesh Demographic and Health Survey 2004*. National Institute of Population Research and Training (NIPROT), Mitra and Associates & Macro International, Dhaka, Bangladesh and Calverton, Maryland, USA.

BDHS. (2007). *Bangladesh demographic and Health Survey 2007*. National Institute of Population Research and Training (NIPROT), Mitra and Associates & Macro International, Dhaka, Bangladesh and Calverton, Maryland, USA.

BDHS. (2013). *Bangladesh Demographic and Health Survey 2011*. National Institute of Population Research and Training (NIPROT), Mitra and Associates & Macro International, Dhaka, Bangladesh and Calverton, Maryland, USA.

BDHS. (2015). *Bangladesh Demographic and Health Survey 2014*. National Institute of Population Research and Training (NIPROT), Mitra and Associates & Macro International, Dhaka, Bangladesh and Calverton, Maryland, USA.

Beguy, D., Ezeh, A. C., Mberu, B. U., & Emina, J. B. (2017). Changes in use of family planning among the urban poor: Evidence from Nairobi slums. *Population and Development Review, 43* (S1), 216–234.

Bell, S. O., & Bishai, D. (2017). Unmet need and sex: Investigating the role of coital frequency in fertility control. *Studies in Family Planning, 48*(1), 39–53. https://doi.org/10.1111/sifp.12012.

Bongaarts, J. (2014). The impact of family planning programs on unmet need and demand for contraception. *Studies in Family Planning, 45*(2), 247–262.

Bongaarts, J., & Bruce, J. (1995). The causes of unmet need for contraception and the social content of services. *Studies in Family Planning, 22*(6), 57–75.

Bradley, S. E., & Casterline, J. B. (2014). Understanding unmet need: History, theory, and measurement. *Studies in Family Planning, 45*(2), 123–150.

Casterline, J. B., & Sinding, S. W. (2000). Unmet need for family planning in developing countries and implications for population policy. *Population and Development Review, 26*(4), 691–723.

Casterline, J. B., Perez, A. E., & Biddlecom, A. E. (1997). Factors underlying unmet need for family planning in the Philippines. *Studies in Family Planning, 28*(3), 173–191.

Cleland, J., Harbison, S., & Shah, I. H. (2014). Unmet need for contraception: Issues and challenges. *Studies in Family Planning, 45*(2), 105–122.

Emina, J. B., Chirwa, T., & Kandala, N.-B. (2014). Trend in the use of modern contraception in sub-Saharan Africa: Does women's education matter? *Contraception, 90*(2), 154–161.

Fang, D., Sun, R., & Wilson, J. R. (2018). Joint modeling of correlated binary outcomes: The case of contraceptive use and HIV knowledge in Bangladesh. *PLoS One, 13*(1), e0190917.

FP. (2021). *Family planning 2020*. Available at: http://www.familyplanning2020.org/entities/70. Accessed 21st Aug 2018.

Hossain, M., Khan, M., Ababneh, F., & Shaw, J. (2018). Identifying factors influencing contraceptive use in Bangladesh: Evidence from BDHS 2014 data. *BMC Public Health, 18*(1), 192.

Huda, F. A., Robertson, Y., Chowdhuri, S., Sarker, B. K., Reichenbach, L., & Somrongthong, R. (2017). Contraceptive practices among married women of reproductive age in Bangladesh: A review of the evidence. *Reproductive Health, 14*(1), 69.

Imasiku, E. N. S., Odimegwu, C. O., Adedini, S. A., & Ononokpono, D. N. (2013). Variations in unmet need for contraception in zambia: Does ethnicity play a role? *Journal of Biosocial Science, 46*(3), 294–315. https://doi.org/10.1017/S0021932013000357.

Islam, R., Islam, A. Z., & Rahman, M. (2013). Unmet need for family planning: Experience from urban and rural areas in Bangladesh. *Public Health Research, 3*(3), 37–42.

Islam, A. Z., Mostofa, M. G., & Islam, M. A. (2016). Factors affecting unmet need for contraception among currently married fecund young women in Bangladesh. *The European Journal of Contraception & Reproductive Health Care, 21*(6), 443–448. https://doi.org/10.1080/13625187.2016.1234034.

Islam, M. R., Islam, M. N., Rahman, M. M., & Islam, M. R. (2017). Influencing socio-demographic factors on contraceptive use in Bangladesh: Multivariate approach. *International Journal of Mathematical Sciences, Technology and Humanities, 7*(3), 23–36.

Kabagenyi, A., Jennings, L., Reid, A., Nalwadda, G., Ntozi, J., & Atuyambe, L. (2014). Barriers to male involvement in contraceptive uptake and reproductive health services: A qualitative study of men and women's perceptions in two rural districts in Uganda. *Reproductive Health, 11*(1), 21.

Khan, H. A. (1997). A hierarchical model of contraceptive use in urban and rural Bangladesh. *Contraception, 55*(2), 91–96.

Korra, A. (2002). *Attitudes toward family planning and reasons for nonuse among women with unmet need for family planning in Ethiopia*. Calverton: Citeseer.

Larson, A., & Mitra, S. N. (1992). Family planning in Bangladesh: An unlikely success story. *International Family Planning Perspectives, 18*(4), 123–144. https://doi.org/10.2307/2133539.

Letamo, G., & Navaneetham, K. (2015). Levels, trends and reasons for unmet need for family planning among married women in Botswana: A cross-sectional study. *BMJ Open, 5*(3), e006603.

Mawajdeh, S. (2007). Demographic profile and predictors of unmet need for family planning among Jordanian women. *BMJ Sexual & Reproductive Health, 33*(1), 53–56.

Mills, S., Bos, E., & Suzuki, E. (2010). *Unmet need for contraception*. Washington, DC: World Bank.

Moreland, S., Smith, E., & Sharma, S. (2010). *World population prospects and unmet need for family planning* (p. 62). Washington, DC: Futures Group.

Mosha, I., Ruben, R., & Kakoko, D. (2013). Family planning decisions, perceptions and gender dynamics among couples in Mwanza, Tanzania: A qualitative study. *BMC Public Health, 13*(1), 523.

MR, M. (2018). *Fertility stalls in Bangladesh*. Dunedin: University of Otago.

Muanda, M., Ndongo, P. G., Taub, L. D., & Bertrand, J. T. (2016). Barriers to modern contraceptive use in Kinshasa, DRC. *PLoS One, 11*(12), e0167560.

NIPORT. (2016). *Bangladesh Demographic and Health Survey 2014*. Dhaka: National Institute of Population Research and Training (NIPORT).

Petry, N. M. (2002). A comparison of young, middle-aged, and older adult treatment-seeking pathological gamblers. *The Gerontologist, 42*(1), 92–99.

Roy, N. C., & Rahman, D. M. (1999). Unmet contraceptive need in Bangladesh: Evidence from the 1993/94 and 1996/97 Demographic and Health Surveys. *Asia-Pacific Population Journal, 14*(2), 37.

Sedgh, G., & Hussain, R. (2014). Reasons for contraceptive nonuse among women having unmet need for contraception in developing countries. *Studies in Family Planning, 45*(2), 151–169.

Sedgh, G., Henshaw, S., Singh, S., Åhman, E., & Shah, I. H. (2007a). Induced abortion: Estimated rates and trends worldwide. *The Lancet, 370*(9595), 1338–1345. https://doi.org/10.1016/S0140-6736(07)61575-X.

Sedgh, G., Hussain, R., Bankole, A., & Singh, S. (2007b). *Women with an unmet need for contraception in developing countries and their reasons for not using a method* (Occasional report, Vol 37, pp. 5–40). New York: Guttmacher Institute.

Shah, M. A., Shah, N. M., Chowdhury, R. I., & Menon, I. (2004). Unmet need for contraception in Kuwait: Issues for health care providers. *Social Science & Medicine, 59*(8), 1573–1580. https://doi.org/10.1016/j.socscimed.2004.01.033.

Singh, S., Darroch, J. E., & Ashford, L. S. (2017). *Adding it up: Investing in contraception and maternal and newborn health, 2017*. New York, USA: Guttmacher Institute.

Uddin, J., Pulok, M. H., & Sabah, M. N.-U. (2016). Correlates of unmet need for contraception in Bangladesh: Does couples' concordance in household decision making matter? *Contraception, 94*(1), 18–26. https://doi.org/10.1016/j.contraception.2016.02.026.

UNICEF. (2007). How to calculate average annual rate of reduction of underweight prevalence. *Technical note*. Available at: https://data.unicef.org/wp-content/uploads/2015/12/Technical_Note_AARR_185.pdf

Westoff, C. F. (1992). Measuring the unmet need for contraception: Comment on Bongaarts. *Population and Development Review, 18*(1), 123–125.

Westoff, C. F. (2005). *Recent trends in abortion and contraception in 12 countries* (DHS analytical studies no. 8). Calverton: ORC.

Yadav, D., & Dhillon, P. (2015). Assessing the impact of family planning advice on unmet need and contraceptive use among currently married women in Uttar Pradesh, India. *PLoS One, 10*(3), e0118584.

Chapter 3
Changes in Ethnic Composition and Fertility of the Australian Population

Jo. M. Martins and Farhat Yusuf

Socioeconomic and Demographic Change

Since World War II, Australia has gone through a major socioeconomic and demographic transformation associated with a change in its economic structure, employment and migration policies, since World War II. These have been associated with a general decline in fertility after the baby boom of the 1950s and early 1960s.

One driving factor has been government policies regarding migration, origin of migrants and their level of skill. The promotion of immigration led to a substantial increase in the proportion of residents born outside Australia from 9.8% in 1947 to 23.8% in 2011. The ethnic composition of residents born outside Australia remained mostly European at almost 90% in the period 1947–1971, in accordance with the White Australia Migration Policy that prevailed until the 1970s and discriminated against the migration of non-Europeans. This policy was changed in the 1970s and migration discrimination on the basis of colour or race ended (APIC 1977). Although, some shift in the origin of Australian migrants had taken place by the time of the 1981 Census, more significant change took place in the period between the 1981 and 2011 censuses. In that period, the proportion of residents born in Europe declined from 75% of the total born outside Australia in 1981 to 41.5% in 2011,

Disclaimer: This article is an original work by the named authors and is being submitted exclusively to the International Association of Applied Demography for publication consideration into the edited book on Applied Demography by springer Publishing Company.

J. M. Martins (✉)
Macquarie University, Sydney, Australia

F. Yusuf
Menzies Centre for Health Policy, School of Public Health, Faculty of Medicine & Health, The University of Sydney, Sydney, NSW, Australia

Macquarie Business School, Macquarie University, Macquarie Park, NSW, Australia

B. Jivetti, Md. N. Hoque (eds.), *Population Change and Public Policy*, Applied Demography Series 11, https://doi.org/10.1007/978-3-030-57069-9_3

Table 3.1 Country of birth of the Australian population, 1947–2011

Country of birth	1947	1954	1961	1971	1981	2011	% change 1947–2011
	Percentage						
Australia	90.2	85.7	83.1	80.0	79.7	76.2	−24.0
Other	9.8	14.3	6.9	20.0	20.3	23.8	+24.0
All residents	100.0	100.0	100.0	100.0	100.0	100.0	
Born outside Australia:							
Europe	87.6	89.8	89.7	85.7	75.0	41.5	−46.1
Asia	3.2	4.0	4.4	6.4	12.3	38.0	+34.8
Africa	1.0	1.2	1.6	2.4	3.1	5.4	+4.4
America	1.6	1.1	1.1	2.0	3.0	3.0	+1.4
New Zealand & other Pacific	6.6	3.8	3.1	3.6	6.6	12.1	+5.5
All born outside Aust.	100.0	100.0	100.0	100.0	100.0	100.0	

Sources: CBCS (1960, 1963) and ABS (2004a, b, undated). Computations by the authors

Table 3.2 Labour force employment by industry, 1954–2011

Industry	1954	1971	1981	2011	% Change 1954–2011
	Percentage				
Primary	15.0	8.8	7.4	4.2	−10.8
Manufacturing	27.8	23.2	17.7	9.0	−18.8
Other	57.2	68.0	74.9	86.8	+29.6
All industries	100.0	100.0	100.0	100.0	

Sources: CBCS (1960, 1972) and ABS (1983a, 2012). Computations by the authors

while the proportion of those born in Asia rose more than threefold from 12.3% to 38% (Table 3.1).

Another factor has been the evolution of the Australian economy, related employment and changes in migrant selection criteria. The importance of employment in manufacturing diminished from 27.8% in 1954 to 17.7% in 1981 and then to only 9.0% in 2011 (Table 3.2). Thus, migration selection policies that encouraged less skilled labour employed in manufacturing were changed to favour more skilled migrants, at the time that discrimination against non-European migrants was lifted (APIC 1977).

Previous studies of fertility of females born in and outside Australia showed that fertility rates of females born outside Australia in some European countries tended to be higher than those of Australian born (Yusuf and Rockett 1981; Yusuf 1986).

The following analysis will test the hypothesis that the fertility of female residents born outside Australia moved from 1981 to 2011 to be lower on average than that of those born in Australia, in view of the abandonment of the White Australia Policy and changed migrant skill selection criteria since the 1970s. This is in spite of the large change in the origin of migrants into Australia from Europe to Asia. It also provides new knowledge on the evolution of fertility of resident females born in and outside Australia, during a period of significant change in the origin of Australian migrants from 1981 to 2011. It uses data not previously available in the public domain sourced by the authors from the records of the Australian Bureau of Statistics.

Data Sources and Methods

The data for the following analysis of Australian fertility by ethnic group (country of birth of the mother) were obtained from Australian Bureau of Statistics (ABS). As the relevant data were not in the public domain, the authors requested tabulations from the ABS that met the Bureau's privacy requirements but allowed for high quality data. The tabulations provided data on the number of births in 5-year groups for mothers born in Australia and Overseas in the years 1981 and 2011 (ABS 2016a, b, c, d). When the numbers for some countries of birth were relatively small the figures were aggregated into groups. The estimated resident populations for 1981 and 2011 were also tabulated in 5-year age groups by age, sex and country of birth. For 1981, there were some population and birth counts that had "not stated" country of birth. They constituted 0.02% of all births and 1.2% of the total resident population in 1981. The estimated resident population for 2011 by ABS was adjusted by the Bureau for "not stated" country of birth. The number of births in 2011 in the "not stated" category amounted to 0.05% of all births. These features of the tabulations are too small to be of consequence to the results of the analysis carried out.

The raw data on births and population obtained from ABS were used by the authors to estimate the total fertility rates (TFRs) of Overseas-born females and females born in Australia by the calculation of age-specific fertility rates (excluding "not stated"), and related resident population TFRs for 1981 and 2011. As per the usual definition, the TFR refers to the average total number of births that females have during their lifetime. In this case, these estimates are cross-sectional for the given years resulting in TFRs that are of a synthetic rather than longitudinal nature. Cumulative fertility rates were derived for the 5-year age groups for the average of three aggregates in 1981 and 2011 to assess differences in the cumulative effects during the reproductive period of females (15–49 years of age). Further analysis of age-specific fertility rates was also carried out for females born in a number of countries that made up a large proportion of the population born Overseas. The selection of the countries of birth was based on their relatively importance in size in 1981 and 2011, and the availability of adequate birth and population figures for both years.

An indirect standardisation procedure (Yusuf et al. 2014) was used to estimate the standardised fertility ratios (SFRs), as the ratio of actual number of births, of the specific female population group, to the expected number of births if the particular group was subjected to the age-specific fertility rates of the females born in Australia in 1981.

Previous Findings

Some earlier studies of the fertility of ethnic groups in the Australian population were based on the number of children ever-born to females from population census data. Day (1969) used a set of cross-classifications of the number of children ever-born by age, country of birth and religion of mothers in the 1961 Australian population census to assess the fertility of Australian female residents aged 40 years and over born in the United Kingdom (UK), Italy, Netherlands, Germany and Greece. These were then the major countries of birth outside Australia. He found that the average number of children ever-born to mothers born in the United Kingdom and Germany to be lower while those of mothers born in the Netherlands, Greece and Italy to be higher than their Australia-born counterparts.

As indicated earlier, the ABS made data available in 1972 on births by age and country of birth of the mother. Yusuf and Eckstein (1980) used these data to estimate the post-arrival age-specific fertility rates and some other indicators of the fertility of females resident in Australia who were born in a number of countries. Their findings showed that females born in Southern European countries had higher fertility than Australia-born. Those born in the UK had similar fertility while females born in Poland had lower fertility than the Australia-born. They also found that females born in Greece, Italy and Yugoslavia had higher fertility in Australia than females living in those countries, while the reverse applied to those born in Poland.

Yusuf and Rockett (1981) carried out analyses of changes in ethnic composition and fertility of the Australian resident population during the period 1971–1976 and Yusuf (1986) extended this study to cover the period 1976–1981. Both studies used ABS cross-tabulations of births by age, country of birth and marital status of the mother. The authors also sourced the age distribution, country of birth and marital status of females resident in Australia from the Australian population censuses of 1971, 1976 and 1981. A finding from their analyses showed that the TFR of females resident in Australia fell during the 10-year period 1971–1981. However, the rate of decline varied with a decrease of 31% among Australia-born and those born in the UK, Ireland, Netherlands, Malta and New Zealand, but females born in Italy, Greece, Yugoslavia and Eastern European countries experienced even larger decreases, while females born in the Middle East and Poland had the smallest declines.

Gunn (1986) used a 1% sample file of the 1981 Australian population census to estimate the complete family size of Australian female residents aged 45–54 years. He also found that females born in the UK, Ireland and other European countries, with the exception of those born in the Netherlands and Malta, had a smaller average number of children than Australia-born females, while females born in the Middle East and other Asia, with the exception of South Asia, had an average higher number. The same applied to females born in New Zealand. However, some of these estimates relied on small numbers and the differences may not be statistically significant.

Abbasi-Shavazi and McDonald (2000) used 1% samples of the Australian population censuses of 1986 and 1991 to apply a reverse survival technique – "the own-children approach" (Cho et al. 1986) – to estimate the fertility rates during the period 1977–1991. They found that the fertility of all earlier European female immigrants (except those from Malta) had declined to levels below those of the Australia-born. This was also the case for most other immigrants with the exception of those born in Lebanon, Turkey, Cyprus and Egypt. Again, some of these estimates relied on small numbers and the differences may not be statistically significant.

Population Growth and Fertility Changes of Australia-Born and Overseas-Born Females 1981–2011

As indicated earlier, Australia has experienced a substantial change in the source of its migrants from 1981 to 2011. This was also accompanied by fast population growth from 14.6 to 22.3 million people (Table 3.3). During that period, the average annual population growth rate of 1.4% was high compared with the average of high-income countries, which was about half of the Australian growth rate (UN 2017).

This high growth in population took place in spite of a fall in the average fertility rate from 3.5 children per woman in 1961 (CBSC 1968) to just below replacement level of about 2.0 in 1981 and 1.9 in 2011 (ABS 2018). The emergence of migration as the major factor in sustained population growth is reflected in the contributions of natural increase and net migration to population growth. In 1981 net migration made up 49% of population growth (ABS 1983b), by 2011 it made up 57% of the increase in the Australian population (ABS 2013). Thus, adding to the ethnic diversity of the Australian population (Table 3.3).

Estimates of the TFRs of females resident in Australia in 1981 and 2011 show that the TFR of both those born in Australia and Overseas declined during the 30-year period of large ethnic change in the origin Australian migrants. However,

Table 3.3 Australian population by country of birth, 1981 and 2011

	1981	2011	Population growth rate (% per year)
Population (millions)	14.6	22.3	1.4
Place of birth:	% total		% Difference 2011–1981
Australia	78.2	73.2	−5.0
Europe	15.3	10.8	−4.5
Asia	2.5	10.3	+7.8
Africa	0.6	1.7	+1.1
America	0.7	1.2	+0.5
New Zealand & other Pacific	1.4	2.8	+1.4
Other & not stated	1.3	na	−1.3
Total place of birth	100.0	100.0	

Note: *na* not applicable
Source: ABS (2004b, 2012). Computations by the authors

Table 3.4 Total fertility rates of Australia-born and Overseas-born females, Australia, 1981 and 2011

	TFR		
Year	Australia-born	Overseas-born	All
1981	1.993	2.140	2.027
2011	1.929	1.864	1.916
Difference 2011–1981	−0.064	−0.279	−0.111

Sources: ABS (2016a, b, c, d). Computations by the authors

Table 3.5 Standardised fertility ratio of Australia and Overseas born females, 1981 and 2011

	Ratio of actual to expected number of births (percent)* 1981 female Australia-born age fertility rates as standard	
Year	Australia-born	Overseas-born
1981	100.0	107.1*
2011	94.5*	98.7*
Difference 2011–1981	−5.5	−8.4

Note: (*) Ratio of actual expected births (times 100) using 1981 age-specific fertility rates of Australian born females as the standard
Sources: ABS (2016a, b, c, d). Computations by the authors

as hypothesised, in view of the migrant selection criteria for more skilled people (APIC 1977), while in 1981 the TFR of Overseas-born females was larger (TFR 2.140) than that of females born in Australia (TFR 1.993), in 2011 the TFR of Overseas-born was lower (1.864) than those born in Australia (1.929) (Table 3.4).

The estimation of expected age-standardised number of births for 1981 and 2011 of Overseas and Australian born females confirms that in 1981 the fertility of Overseas-born females was 7.1% higher than that of Australia-born. The estimation of expected births for females born Overseas and Australia for 2011 using the same standard confirms the greater decline in fertility of Overseas-born females than those of born in Australia (Table 3.5).

The cumulative fertility from 15–19 to 45–49 years of age indicates that in 1981 Overseas-born females contributed to a higher average age-specific fertility rates with the exception of those 25–29 years of age, when fertility reached its peak for both groups. The pattern was different in 2011, the cumulative age-specific rates indicate that the younger groups of Overseas-born females – including the peak in fertility at 30–34 years of age – pushed average fertility rate downwards but it pulled it upwards at older ages of 35–49, as in 1981.

The different pattern in the role of Overseas-born female fertility in 1981 and 2011 in comparison with those born in Australia, in the same years, points to early higher fertility of Overseas-born females at most ages in 1981 but lower fertility at younger ages with possible catch up after the later peak period of 30–34 years of age in 2011 (Fig. 3.1).

This analysis provides evidence of the changes that took place, on average, in the fertility of Overseas-born and Australia-born females in the 30-year period 1981–2011. In fact, because of the range of the reproductive age of females, the

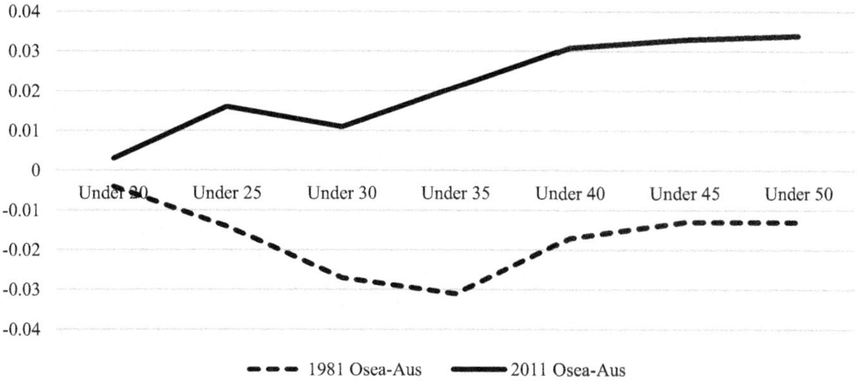

Fig. 3.1 Cumulative age-specific fertility rates differences Overseas & Australia-born resident females 1981 & 2011. Note: (Aus) Australia-born females, (Osea) Overseas-born females. (Under x) means under x years of age. The differences are expressed cumulative age-specific fertility rates up to the given age. (Source: ABS (2016a, b, c, d). Computations by the authors)

analysis relates to two different groups of females in 1981 and 2011, with a small overlap at the age of low fertility (45–49 years of age). They are also females who might have migrated to Australia at two different time periods that followed different migration policies: not just in terms of country of origin but also under different social and economic circumstances (Tables 3.1 and 3.2).

Fertility of Major Ethnic Groups

An analysis of the major groups of Overseas-born females resident in Australia, for which there is adequate data, indicates that in most cases there was a decline in fertility from 1981 to 2011 as shown previously (Tables 3.4 and 3.5), with some exceptions. However, there was no consistent pattern. Females born in some countries had fertility rates below replacement level while others had fertility well above it. Accordingly, their impact on the average Australian fertility was dependent on the relative weights of each country of origin that made the average of Overseas-born females. The following analysis illustrates the fertility changes from 1981 to 2011 of the largest four groups of females born Overseas in comparison to the changes of the larger Australian-born group.

The TFR of females born in the UK and China declined substantially from 1981 to 2011, especially of those born in China from above replacement level in 1981 (TFR 2.2) to well below it in 2011 (TFR 1.5). The TFR of females born in India remained about the same, just above replacement level (TFR 2.2). While the TFR of females born in New Zealand, the other major country-of-birth group, actually rose from below replacement level in 1981 (TFR 1.8) to replacement level in 2011 (TFR 2.1) (Table 3.6).

Table 3.6 Total fertility rates of females born in United Kingdom, China, India, New Zealand, all Overseas-born, Australia-born and all residents, 1981 and 2011

Female country of birth	TFR 1981	TFR 2011	TFR 2011 less 1981
United Kingdom	1.990	1.858	−0.132
China	2.224	1.535	−0.689
India	2.176	2.179	0.003
New Zealand	1.790	2.113	0.323
All Overseas-born	2.140	1.864	−0.279
Australia-born	1.993	1.929	−0.064
All Australia residents	2.027	1.916	−0.111

Sources: ABS (2016a, b, c, d). Computations by the authors

An examination of the fertility of females born in countries that constituted a considerable proportion of the Australian population does not show a consistent pattern in relation of the fertility of Australia-born. In the case of females born in the United Kingdom, the largest group of female migrants in Australia, their fertility in 2011 (TFR 1.89) fell below that of Australia-born resident females, but it was similar to that of females born and resident in the United Kingdom (TFR 1.89) (Zumpe et al. 2012). Australia resident females born in China showed a major decline in fertility from 1981 (TFR 2.2) to 2011 (TFR 1.5). The latter was close to that prevailing in China in 2011 (TFR 1.5) (WB 2016). The fertility of females born in India and resident in Australia remained about the same (TFR 2.2) just above replacement level in 1981 and 2011, above the fertility rate Australian-born females, but somewhat lower than that of females in India in 2011 (TFR 2.4) (GI undated). A notable exception to the downward trend in fertility was that of females born in New Zealand that rose from 1981 to 2011. In 2011, the fertility of New Zealand born females resident in Australia (TFR 2.1) was about that for the average in New Zealand (TFR 2.1). A possibility is that the characteristics of the females born in New Zealand and resident in Australia changed with a higher proportion of Maori females in 2011 than in 1981, as Maori females have a higher fertility rate (TFR 2.7) than the average in New Zealand (SNZ 2016). However, this possibility cannot be tested from Australian data that lack that specificity.

Conclusion

The population of Australia experienced substantial growth in the three decades from 1981 to 2011, well above the average for high-income countries, in spite of a small drop in fertility. During the same period the relative contribution of natural increase fell while that of net migration rose to more than half of the rate of population growth. An outcome was an increment in the proportion of residents born outside Australia. There was also a major shift in the source of immigrants from Europe to Asia. This reflected a change in Australia's immigration policies that discriminated against non-European immigrants until the 1970s. The social and economic circumstances leading to migration also changed considerably leading to a change in selection to more skilled migrants and related characteristics that have affected the fertility of female migrants. A feature of the changes that took place was that the fertility of females born outside Australia fell, on average, substantially more than that of Australia-born females between 1981 and 2011. As hypothesized, the outcome was an average decline in the fertility rate of Overseas-born females that was higher than that of Australia-born females in 1981 but somewhat lower in 2011.

References

Abbasi-Shavazi, M. J., & McDonald, P. (2000). Fertility and multiculturalism: Immigrant fertility in Australia, 1977-1991. *International Migration Review, 34*, 215–242.

Australian Bureau of Statistics. (ABS). (1983a). *Census of population and housing, 30 June 1981. Summary of characteristics of persons and dwellings*. Canberra: Australian Bureau of Statistics.

Australian Bureau of Statistics. (ABS). (1983b). *Australian demographic statistics, June quarter 1983* (Catalogue No. 3101.0). Canberra: Australian Bureau of Statistics.

Australian Bureau of Statistics. (ABS). (2004a). *Australian historical population statistics. Population, sex and country of birth, states and territories, 1971 census*. Canberra: Australian Bureau of Statistics.

Australian Bureau of Statistics. (ABS). (2004b). *Australian historical population statistics. Population, sex and country of birth, states and territories, 1981 census*. Canberra: Australian Bureau of Statistics.

Australian Bureau of Statistics. (ABS). (2012). *2011 Census of population and housing. Basic community profile*. Canberra: Australian Bureau of Statistics.

Australian Bureau of Statistics. (ABS). (2013). *Australian demographic statistics, June quarter 2013* (Catalogue No. 3101.0). Canberra: Australian Bureau of Statistics.

Australian Bureau of Statistics. (ABS). (2016a). *Australian historical population statistics, 2014* (Catalogue No. 3105.65.001). Canberra: Australian Bureau of Statistics.

Australian Bureau of Statistics. (ABS). (2016b). *1981 Census customised table. Count of persons (excluding Overseas visitors) by sex and age group by birthplace of individual, Australia*. Canberra: Australian Bureau of Statistics.

Australian Bureau of Statistics. (ABS). (2016c). *Births, Australia, customised table. Table 1: Number of births registered in 1981 by selected country of birth of mother by age group of mother, Australia*. Canberra: Australian Bureau of Statistics.

Australian Bureau of Statistics. (ABS). (2016d). *Births, Australia, customised table. Table 1: Number of births country of birth of mother by age group of mother, Australia, reference period 2011*. Canberra: Australian Bureau of Statistics.

Australian Bureau of Statistics. (ABS). (2018). *Births, Australia, 2018* (Catalogue No. 3301.0). Canberra: Australian Bureau of Statistics.

Australian Bureau of Statistics. (ABS). (undated). *Census of population and housing 2011. Country of birth of persons by age and sex*. Canberra: Australian Bureau of Statistics.

Australian Population and Immigration Council. (APIC). (1977). *Immigration policies and Australia's Population: Green Paper*. Canberra: Australian Government Publishing Service.

Cho, L.-J., Retherford, R. D., & Choe, M. K. (1986). *The own-children method of fertility estimation*. Honolulu: University of Hawaii Press.

Commonwealth Bureau of Census and Statistics. (CBCS). (1960). *Year book of the Commonwealth of Australia No. 46-1960*. Canberra: Commonwealth Bureau of Census and Statistics.

Commonwealth Bureau of Census and Statistics. (CBCS). (1963). *Census of the Commonwealth of Australia, 30th June 1961. Birthplaces of the population of Australia by states and territories* (Census Bulletin No. 28). Canberra: Commonwealth Bureau of Census and Statistics.

Commonwealth Bureau of Census and Statistics. (CBCS). (1968). *Year book of the Commonwealth of Australia No. 54-1968.* Canberra: Commonwealth Bureau of Census and Statistics.

Commonwealth Bureau of Census and Statistics. (CBCS). (1972). *Census of population and housing, 30 June 1971.The labour force. Part 9.* Canberra: Commonwealth Bureau of Census and Statistics.

Day, L. H. (1969). Differential fertility in Australia. In *Proceedings of the international population conference*, London, Vol. 3, pp. 2043–2052.

Government of India (GI). (undated). *Total fertility trends.* https://data.gov.in/resources/trends-total-fertility-rate. Accessed 2 Mar 2016.

Gunn, P. A. (1986). Birthplace differentials in age-specific fertility of Australian women. *Journal of the Australian Population Association, 2*, 40–51.

Statistics New Zealand (SNZ). (2016). *Infoshare: Population, births, total fertility rate.* www.stats.govt.nz.Default.aspx. Accessed 2 Mar 2016.

United Nations (UN). (2017). *World population prospects – The 2017 revision. Volume I: Comprehensive tables.* New York: United Nations (UN).

World Bank (WB). (2016). *Fertility rate, total (births per woman).* Washington, DC. https://data.worldbank.org/indicator/SP.DYN.TFRT.IN?location=CN. Accessed 2 Mar 2016.

Yusuf, F. (1986). Ethnic differences in Australian fertility. *Clinical Reproduction and Fertility, 4*, 107–116.

Yusuf, F., & Eckstein, G. (1980). Fertility of migrant women in Australia. *Journal of Biosocial Science, 12*, 179–190.

Yusuf, F., & Rockett, I. (1981). Immigrant fertility patterns and differentials in Australia. *Population Studies, 35*(3), 413–424.

Yusuf, F., Martins, J. M., & Swanson, D. A. (2014). *Methods of demographic analysis.* Dordrecht: Springer.

Zumpe, J., Dormon, O., & Jeffries, J. (2012). *Childideraring among UK born and non-UK born women living in the UK.* Kew: Office of National Statistics.

Chapter 4
Contraceptive Use Dynamics Among Urban Poor Women in India

Dewaram Abhiman Nagdeve

The largest increases in urban population are expected in India, China, Nigeria, the United States of America and Indonesia. The urbanisation of India is taking place at a faster rate than in the rest of the world. Over the next four decades (2010–2050), India will add another 497 million to its urban population (UN 2012). India is the second most populous country in the world after China, having population of 1210 million. Urbanisation is taking place at a faster rate in India. The urban population in India has been increased from 28.53% in 2001 to 31.16% in 2011 (Census of India 2011). India is the first country in the world to launch a national family planning program soon after its independence (in 1952) with the objective of reducing the birth rate. Thereafter, the promotion of maternal and child health through safe motherhood programmes within the wider context of reproductive health. In the National Population Policy, 2000, the Government of India set as its immediate objective to reduce fertility and the task of addressing unmet need for contraception to achieve the medium-range objective of bringing the total fertility rate down to replacement level by 2010 (Ministry of Health and Family Welfare 2000). The rise in the population is thought to be lack of effective contraceptive methods in spite of women desire to control their family and to space births. Fertility and contraception use in developing countries is associated with socioeconomic status and other related factors (Gribble 2012). This demographic transformation will have profound implications for health.

Further, the levels of unwanted fertility too have been quite high in India among all and particularly among married young women. The total fertility rate for the country as a whole was 2.18 children per woman in 2015–2016 compared to 2.85 children in 1998–1999. Contraception has been the single most important intervention to contraceptive use among young married women in India reduce burden of

D. A. Nagdeve (✉)
Fertility Studies Department, International Institute for Population Sciences, Mumbai, India
e-mail: dnagdeve@iips.net

B. Jivetti, Md. N. Hoque (eds.), *Population Change and Public Policy*, Applied Demography Series 11, https://doi.org/10.1007/978-3-030-57069-9_4

unwanted pregnancy as well as to promote healthy living among young adults. However, a considerable proportion of the fertility in India is reported as unwanted fertility, according to NFHS – 4, about 27% of the fertility in India was unwanted, perhaps indicating towards a huge gap between the demand and supply of family planning. As per the NFHS 4 estimates, the unmet need for family planning for the country as a whole is about 13%. The unmet need for family planning is alarmingly high 19.9% for spacing and 2.3% for limiting among married adolescent women aged 15–19 and 15.7% for spacing and over 6.5% for limiting among those aged 20–24 years (IIPS and Macro International 2017).

Childbearing among the urban poor is often mistimed or unwanted. These high rates of unintended births suggest that the urban poor may have high unmet need for family planning or lack access to effective family planning methods and services (Ezeh et al. 2010). Sulthana et al. (2015), found high unmet need for contraception in Puducherry, India indicating a necessity to address user perspective to meet the contraception needs Meeting the reproductive health needs of the urban poor may thus not only improve their health and wellbeing, but could also contribute to reduced rates of urban growth in the region (Chen et al. 1998). Improving access to voluntary family planning services among slum dwellers may play a key role in reducing the growth of slums in the region.

As the world becomes more urbanized, urbanites enjoy an advantage in health relative to rural villagers but the health needs of urban populations differ among all urban residents. The social and economic diversity of urban populations include large groups of the poor whose health environments differ little from those of rural villagers (Montgomery 2009). The urban health advantages are not reaching to the urban poor because supply side of the urban health system is just as diverse as the urban population. The private sector is a far more important presence in cities than in rural areas, and urban health care is consequently more monetized (UN 2008).

Most of the developing countries are trying to reduce poverty through implementing programs that address the Millennium Development Goals (MDG) and emphasis is given on family planning (FP) programs for the attainment of the MDG. Contraception has widespread positive impacts for population health such as reduced burden of unintended pregnancies, morbidity, infant and maternal mortality and well-being – addressing MDG 4 and MDG 5. Due to relatively higher growth rate of slum population compared to the overall growth rate of the urban population and a strong interest in urban health, as well as the MDG on improving the lives of slum dwellers, all the programmes and services relating to health of urban poor need to adopt a synergistic approach. The increasing concentration of the population in slums and urban poverty has drawn a strong interest in urban health in general, and the health of slum dwellers and the urban poor in particular in the recent years.

As the urbanization increases, contraceptive use among urban populations is increases but differential exists in slum and non-slum areas. Consideration of the family planning and reproductive health needs of women, men, and couples in urban India is increasingly important, given the rapid growth in the number of people living in urban poor settings throughout India (Census of India 2001). There have been limited efforts to study the contraceptive use dynamics among urban poor especially women living in slums in India. Most of the few studies that exist have reported considerable differences in the use of family planning methods between slum and non-slum areas. However, the major limitation of most of these studies is that they have been confined to specific cities and based on a small sample size and the findings cannot be generalized (Hazarika 2010; Speizer et al. 2012). Although many focused and detailed studies of the contraceptive use of slum-dwellers have been undertaken, systematic studies and comparisons with other urbanites are still scarce. It is difficult to divide the overall contraceptive use among slum-dwellers into those attributable to household poverty. In view of the above, the present paper is undertaken with the following objectives.

Objectives

(i) To study the levels and patterns of contraceptive use among urban poor women in slums and non-slums of three cities in India.
(ii) To examine the unmet need for family planning among urban poor women in the slum and non-slum areas in three cities.
(iii) Factors affecting contraceptive use among urban poor women in three cities of India.

Hypotheses

1. The use of family planning services varies by socio-economic and demographic characteristics of women within the urban poor population.
2. Women living in slum areas are more likely to have an unmet need than women living in non-slum areas.

Data and Methods

The data have been analysed from survey on Health of the Urban Poor (HUP) in three cities: Bhubaneswar, Jaipur and Pune conducted in 2011 (IIPS 2011) for the present paper. A sample of 4351 women (1322 women in Bhubaneswar, 1613 in Jaipur and 1416 in Pune) in the age group of 15–49 years has been analysed to assess the relationship between contraceptive use and socio-economic characteristics. The analysis has been done by using bivariate and multivariate logistic regression analysis. We have also tried to identify the socio-demographic factors that influence the contraceptive use among urban poor women. The bivariate and multivariate logistic regression were used to analyse the strength of association between contraceptive use and relevant socio-demographic factors. An attempt has been made with the help of multivariate logistic regression analysis to explore the factors influencing the contraceptive use in Bhubaneswar, Jaipur and Pune.

Results

Profile of Respondents

The percentage distributions of women by selected background characteristics in Bhubaneswar, Jaipur and Pune are presented in Table 4.1. The age distributions of

Table 4.1 Percentage distribution of women by selected background characteristics in three cities of India, 2011

Background characteristics	Bhubaneswar	Jaipur	Pune
Age of women			
15–24	8.1	16.5	12.5
25–34	37.3	39.3	42.4
35+	54.6	44.2	45.1
Education of women			
< 5 years	9.2	36.4	14.4
5–9 years	21.2	31.6	24.7
10 years & above	69.6	32.0	60.9
Caste of household			
SC/ST	11.9	30.5	16.8
OBC	25.4	23.5	26.5
Others	62.7	46.0	56.7
Religion of head of HH			
Hindu	96.1	84.0	82.4
Muslim	2.8	15.0	9.4
Christian	0.9	0.1	2.4
Others	0.2	0.9	5.8
Media exposure			
No	5.7	7.3	2.9
Partial	92.8	87.7	92.1
Full	1.5	5.0	5.0
Standard of living index			
Low	28.7	38.4	32.2
Medium	35.7	33.7	32.1
High	35.6	27.9	35.7
Type of locality			
Slum	19.0	25.5	24.0
Non-slum	81.0	74.5	76.0
Number of women	**1322**	**1613**	**1416**

women in all the three cities are very similar with increasing age. The percentage of women who are in the youngest age group is higher in the Jaipur and Pune than in the Bhubaneswar. The distribution of women by completed number of years of education reveals a low overall educational attainment among women in three cities of India. The higher percentage of women in Jaipur has less than 5 years of education than the percentage of women in Bhubaneswar and Pune. Only 32% of women in Jaipur have 10 or more years of education, compared with 61% of women in Pune and 70% of women in Bhubaneswar. The highest percentage of women belonging to the scheduled castes/scheduled tribes is found in Jaipur, OBC in Pune and women who do not belong to any of these two groups in Bhubaneswar. In keeping religious mix of the population in three cities, the majority of women in Bhubaneswar, Jaipur and Pune are Hindu and a minority are Muslim, followed by Christians. Most of the women are partially exposed to mass media and this percentage is lower in case of women who are not having regular exposure to any mass media and women who are fully exposed to mass media in all the three cities. The considerable proportion of women in Bhubaneswar, Jaipur and Pune are belonging to low standard of living index (29%, 38% and 32% respectively). A considerable proportion of women are living in slums (19%, 26% and 24%) as compared to non-slums (81%, 75% and 76%) in Bhubaneswar, Jaipur and Pune respectively.

Number of Living Children at First Use of Contraception

The fertility impact of contraception depends not only on its prevalence but also on the stage of family planning when it is used. Initiation of contraceptive use when women have a small number of children can result in a greater fertility reduction. The percent distribution of currently married women who have ever used any method of contraception by number of living children at the time of first use of contraception by selected background characteristics in three cities of India is depicted in Table 4.2. More than four-fifths of ever users in Bhubaneswar, 64% ever users in Jaipur and 82% of ever users in Pune used contraception for the first time when they had two or fewer children. This percentage decreases with increasing age in all the three cities. The percentage of currently married women who have ever used any method of contraception when they had two and less than two children increases with increasing education in Bhubaneswar, Jaipur and Pune. The use of contraception when women had two and less than two children is highest in case of women who have full exposure to mass media and lowest in case of women who was not having exposure to mass media in Bhubaneswar, Jaipur and Pune. The use of contraception is strongly and positively related to standard of living index. The percentage of currently married women who have ever used any method of contraception when they had two and less than two children increases from 72%, 56% to 70% of ever users among women belonging to low standard of living index to 87%, 74% and 90% of ever users among women belonging to high standard of living index in Bhubaneswar, Jaipur and Pune respectively. The use of contraception is

Table 4.2 Percentage of currently married women who have ever used any method by different methods of contraception according to selected background characteristics in three cities of India, 2011

	Bhubaneswar			Jaipur			Pune		
Background characteristic	Spacing method	Limiting method	Any method	Spacing method	Limiting method	Any method	Spacing method	Limiting method	Any method
Age of women									
15–24	28.9	4.2	31.9	36.7	8.7	45.4	32.8	10.4	43.4
25–34	40.4	13	53.4	46	30.6	75.7	33.7	39.1	72.7
35+	23.5	30.9	54.5	21.5	51.2	72.7	11.5	70.6	82.1
Education of women									
< 5 years	13.2	49.3	62.4	16.6	47.5	63.9	9.8	70.2	79.9
5–9 years	24.4	27.3	51.2	35.2	37.2	72.4	10.9	63.4	74.3
10 years & above	34	17.5	51.5	50.1	25.4	74.8	31.3	40.8	72.1
Caste of household									
SC/ST	27.7	27.3	55	24.3	41	64.9	19.7	55.3	74.8
OBC	25.5	26.2	51.6	33.3	38.8	71.9	20	49.6	69.3
Others	32.3	20.1	52.2	39.9	33.4	72.8	25.3	50.1	75.4
Media exposure									
No	34.5	15.1	50	29.2	13	43.5	3.8	53.8	57.7
Partial	29.9	23.1	52.9	32.8	39.6	72.1	22.7	51.3	74
Full	24.3	2.8	27.8	60.9	13.8	74.7	40	40	80
Standard of living index									
Low	24.4	24.9	48.9	26.4	34.6	60.9	14.5	56	70.6
Medium	34.6	23.6	58.3	36	38.4	73.3	23.5	49.7	73.2
High	29.8	19.6	49.4	39.5	38.6	78.3	30.2	47.1	77.2
Type of locality									
Slum	24.5	26.9	51.4	22.3	37.5	59.4	14.6	56.1	70.7
Non-slum	30.5	22.1	52.5	38.9	36.8	75.5	25.2	49.4	74.6
Total	**30**	**22.5**	**52.4**	**33.4**	**37.1**	**70.2**	**23**	**50.8**	**73.8**
Number (N)	**1278**	**1278**	**1278**	**1381**	**1381**	**1381**	**1347**	**1347**	**1347**

lower in slums than non-slums among currently married women in Bhubaneswar, Jaipur and Pune when they used contraception for the first time when they had two and fewer children.

Current Use of Contraceptive Methods

The percentage of currently married women according to current use of different methods of contraception by selected background characteristics in three cities of India is shown in Table 4.3. The contraceptive prevalence rate varies across the cities, from highest 74% in Pune, followed by 70% in Jaipur to lowest 52% in Bhubaneswar. The use of spacing method of family planning varies from highest 33% in Jaipur to lowest 23% in Pune. The use of limiting method of family planning is varying from highest 51% of women in Pune to lowest 23% of women in Bhubaneswar. The use of spacing, limiting and any method of family planning varies with broad age groups of the women in all the three cities. The use of spacing

Table 4.3 Percent of currently married women according to use of any modern method of contraception by selected background characteristics and locality in three cities of India, 2011

	Bhubaneswar		Jaipur		Pune	
Background characteristic	Slum	Non-slum	Slum	Non-slum	Slum	Non-slum
Age of women						
15–24	29.5	21.0	30.3	46.8	45.7	43.1
25–34	60.2	41.2	62.2	74.5	71.5	69.4
35+	48.2	42.8	58.7	71.5	87.0	81.8
Education of women						
< 5 years	52.1	61.4	56.1	69.1	83.5	84.3
5–9 years	40.7	47.7	49.3	74.8	72.6	76.2
10 years & above	60.9	37.1	59.0	67.1	63.0	69.7
Caste of household						
SC/ST	50.0	44.3	52.2	73.1	76.0	76.6
OBC	58.0	44.5	60.3	69.3	65.8	71.0
Others	35.1	38.8	54.5	69.0	75.9	71.9
Media exposure						
No	62.1	44.4	44.4	43.8	66.7	66.7
Partial	48.0	41.0	56.9	71.1	74.1	72.3
Full	0.0	23.5	76.9	64.9	50.0	75.4
Standard of living index						
Low	51.0	46.3	48.8	65.5	76.6	69.7
Medium	50.0	45.2	63.6	70.8	67.7	73.7
High	28.6	33.3	51.1	72.2	73.3	72.6
Total	**50.0**	**40.8**	**54.3**	**70.0**	**73.2**	**72.3**
Number (N)	**220**	**1102**	**311**	**1302**	**358**	**1058**

method is higher among younger women than older women whereas the use of limiting and any method of family planning are higher among older women in all the three cities except Bhubaneswar. The use of limiting and any method of family planning by education are highest among less than 5 years educated women as compare to educated groups in all the three cities. The use of spacing method of family planning increases with education in all the three cities. The use of any method of family planning varies among caste groups in all the three cities. Overall, the use of spacing method of family planning is highest among women who do not belong to SC/ST and OBC as compared to SC/ST. The use of limiting method of family planning is highest among SC/ST women as compared to the women who do not belong to SC/ST and OBC in all the three cities. Overall, the use of any method of family planning, the use of spacing and limiting method of family planning by mass media exposure varies among women in all the three cities. The percentage of currently married women who are using spacing method, limiting method and any method of family planning varies by standard of living in all the three cities. The use of spacing method of family planning is high among women belonging to high standard of living index and low among women belonging to low standard of living index in all the three cities except Bhubaneswar. The use of limiting method of family planning is negatively related to standard of living index in Bhubaneswar and Pune and positively related to standard of living index in Jaipur. The use of spacing method and any method of family planning is lower in slums than non-slums among currently married women in all the three cities. However, the use of limiting is higher in slums as compared to non-slums among currently married women in all the three cities.

Current Use of Any Modern Method of Contraception

Differentials in any modern method of contraception among currently married women age 15–49 by their selected background characteristics and type of locality in three cities is presented in Table 4.4. The use of any modern method varies by background characteristics and type of locality in three cities. The use of any modern method of contraception is slightly higher in slum areas (50% and 73%) than in non-slum areas (41% and 72%) of Bhubaneswar and Pune respectively. However, the use of any modern method of contraception is higher in non-slum areas (70%) than slum areas (54%) of Jaipur. This percentage increases with increasing age of the women both in slum and non-slum areas of Bhubaneswar, Jaipur and Pune respectively. The use of any modern method of contraception increases with education in slum areas of Bhubaneswar and Jaipur whereas it decreases in Pune. However, use of any modern method of conception in non-slum areas decreases with education in all the three cities. The use any modern method of contraception varies by caste groups both in slum and non-slum areas in Bhubaneswar, Jaipur and Pune. The use of any modern method of contraception by mass media exposure varies both in slum and non-slum areas in all the three cities. The use of any modern method of

Table 4.4 Percent distribution of currently married women who have ever used any method of contraception by number of living children at the time of first use by selected background characteristics in the three cities of India, 2011

	Bhubaneswar				Jaipur				Pune			
Background characteristic	0	1	2	3+	0	1	2	3+	0	1	2	3+
Age												
15–24	16.0	60.0	22.0	2.0	16.7	54.8	16.7	11.9	20.5	54.2	24.1	1.2
25–34	5.8	44.2	40.9	9.1	2.4	39.7	31.0	27.0	15.0	38.0	35.4	11.7
35 +	2.7	20.9	49.2	27.2	1.3	20.5	30.6	47.6	4.4	23.9	44.1	27.6
Education												
< 5	0.0	5.3	36.8	57.9	1.5	16.8	21.3	60.3	4.0	8.7	37.3	50.0
5–9	5.6	24.7	46.5	23.2	3.7	30.3	32.0	34.0	4.2	19.4	46.4	30.0
10 years & above	5.0	39.5	44.7	10.8	4.5	47.7	33.7	14.2	14.1	43.2	35.9	6.8
Exposure to mass media[1]												
No	0.0	23.1	46.2	30.8	0.0	41.2	17.6	41.2	8.3	0.0	50.0	41.7
Partial	4.8	32.6	44.4	18.2	3.3	30.7	29.0	37.0	9.1	32.5	39.0	19.4
Full	0.0	50.0	33.3	16.7	4.6	57.5	33.3	4.6	29.6	37.0	33.3	0.0
Standard of living index[2]												
Low	3.6	25.7	42.4	28.3	3.7	30.9	21.5	43.9	6.5	24.0	39.6	29.9
Medium	5.5	33.5	45.5	15.5	2.9	31.2	30.2	35.7	9.2	31.5	40.7	18.7
High	3.7	38.2	44.9	13.1	3.5	34.7	35.6	26.2	13.8	40.0	36.5	9.6
Type of locality												
Slum	0.7	22.6	43.8	32.9	3.4	23.8	26.3	46.5	7.0	19.4	34.8	38.8
Non-slum	5.4	34.3	44.4	15.9	3.3	34.6	30.0	32.0	11.1	36.0	39.8	13.1
Total	**5.0**	**33.4**	**44.4**	**17.2**	**3.3**	**31.5**	**28.9**	**36.3**	**10.3**	**32.9**	**38.9**	**17.8**
Number of women	**51**	**266**	**360**	**163**	**53**	**375**	**344**	**446**	**111**	**331**	**411**	**195**

contraception varies with standard of living index in both slum and non-slum areas in Bhubaneswar, Jaipur and Pune.

Contraceptive Use Status

The percentage of currently married women by their contraceptive use status by selected background characteristics in three cities of India is depicted in Table 4.5. Among the users, more than half of the women (51%) in Pune, 37% of women in Jaipur and lowest 23% of women in Bhubaneswar reported that they or their husband undergone sterilization. The use of other modern spacing method is highest 27% of women in Jaipur as compared to 20% of women in Pune and 19% of women in Bhubaneswar. The use of any traditional method is highest 12% in Bhubaneswar, followed by 7% in Jaipur and lowest 3% in Pune. The use of sterilization, other modern spacing method and any traditional method varies with broad age groups of the women showed that sterilization is lowest in younger age group and maximum for older age groups. The use of other modern spacing method and any traditional method is higher among older women than younger women in all the three cities. The use of sterilization with respect to educational level indicates that it is highest among less than 5 years educated women as compare to educated groups in all the three cities. The use of other modern spacing method and any traditional method increases with education in all the three cities. The use of sterilization is highest among SC/ST women, followed by OBC women and lowest among women who do not belong to any of these two groups in all the three cities except Pune. The use of other modern spacing method and any traditional method is highest among women who do not belong to SC/ST and OBC as compared to OBC and SC/ST women in all the three cities. The use of sterilization, any traditional method and other modern spacing method varies by mass media exposure in all the three cities. Overall, the use of other modern spacing method is highest among women who were having full exposure to mass media as compared to women who were not having exposure to mass media and women who were having partial exposure to mass media in all the three cities except Bhubaneswar. The use of sterilization is highest among women belonging to low standard of living in all the three cities except Jaipur where it is highest among women belonging to high standard of living. Overall, the use of other modern spacing method and any traditional method is positively related to standard of living index. The use of sterilization is higher in slums than non-slums among currently married women in all the three cities. However, the use of other modern spacing method and any traditional method among currently married women is higher in non-slums as compared to slums in all the three cities except Bhubaneswar in case of other modern spacing method.

Multinomial regression is used in case of contraceptive use as it has more than two categories, i.e., sterilization, other modern spacing method and any traditional method. In order to avoid the complexity of interpretation, the results obtained from the multinomial logit regression are presented in terms of adjusted percentages using

Table 4.5 Percent of currently married women by their contraceptive use status by selected background characteristics in three cities of India, 2011

Background characteristic	Bhubaneswar			Jaipur			Pune		
	Sterilization[a]	Any traditional method[b]	Other modern spacing method[c]	Sterilization[a]	Any traditional method[b]	Other modern spacing method[c]	Sterilization[a]	Any traditional method[b]	Other modern spacing method[c]
Age of women									
15–24	4.2	9.0	19.9	8.7	7.3	29.8	10.4	0.0	32.8
25–34	13.0	11.6	29.4	30.6	6.7	39.5	39.1	4.2	29.4
35+	30.9	12.0	12.0	51.2	6.3	15.7	70.6	2.1	9.4
Education of women									
< 5 years	49.3	5.0	9.0	47.5	2.8	13.8	70.2	0.4	8.9
5–9 years	27.3	6.5	18.8	37.2	7.5	27.8	63.4	1.3	9.5
10 years & above	17.5	14.1	20.3	25.4	10.2	40.9	40.8	3.9	27.4
Caste of household									
SC/ST	27.3	9.7	18.7	41.0	4.7	19.8	55.3	3.1	16.5
OBC	26.2	5.8	20.0	38.8	6.1	27.5	49.6	0.7	19.2
Others	20.1	14.3	18.4	33.4	8.6	31.8	50.1	3.8	21.5
Media exposure									
No	15.1	7.0	32.2	13.0	4.2	26.1	53.8	0.0	3.8
Partial	23.1	11.9	18.4	39.6	7.0	26.3	51.3	2.6	20.1
Full	2.8	2.8	21.6	13.8	6.9	52.9	40.0	8.0	32.0
Standard of living index									
Low	24.9	3.6	21.1	34.6	4.8	21.6	56.0	0.8	13.7
Medium	23.6	13.3	21.5	38.4	6.3	29.9	49.7	1.6	21.7
High	19.6	16.4	14.4	38.6	9.9	30.7	47.1	5.3	24.8

(continued)

Table 4.5 (continued)

Background characteristic	Bhubaneswar			Jaipur			Pune		
	Sterilization[a]	Any traditional method[b]	Other modern spacing method[c]	Sterilization[a]	Any traditional method[b]	Other modern spacing method[c]	Sterilization[a]	Any traditional method[b]	Other modern spacing method[c]
Type of locality									
Slum	26.9	2.5	21.7	37.5	5.4	16.9	56.1	0.9	13.7
Non-slum	22.1	12.4	18.6	36.8	7.4	32.1	49.4	3.3	22.0
Total	22.5	11.6	18.9	37.1	6.8	27.0	50.8	2.8	20.3
Number (N)	**1278**	**1278**	**1278**	**1381**	**1381**	**1381**	**1347**	**1347**	**1347**

Note: [a]Sterilization includes male and female sterilization [b]Traditional method includes Rhythm and withdrawal method [c]Other modern spacing method includes pill, IUD, Injectable, implants, condom, female condom, Diaphragm and form/jelly a multiple classification application (MCA). The effects of the predictor variables differ somewhat for specific family planning methods. Table 4.6 shows the adjusted effects of the predictor variables on use of the sterilization, other modern spacing method and any traditional method. The effects of the predictor variables are quite similar for sterilization. This is not surprising, given that sterilization accounts for 77% of all modern method use in India. Age of the women and education, however, have strong adjusted effects in the expected direction – younger women and more educated women are much more likely to use other modern spacing method and any traditional method than older women and women with less education in all the three cities. Table 4.6 also shows that, neither media exposure and standard of living index nor type of locality has a significant adjusted effect on use of the other modern spacing method and any traditional method.

Not Using Any Method of Contraception

The women who are not using any method of contraception, varies across the cities, from a low of 19% in Jaipur and 20% in Pune to a high of 33% in Bhubaneswar. The women who are not using any contraceptive method is higher among currently married women in non-slum areas (33% and 21%) than in slum areas (29% and 17%) of Bhubaneswar and Pune respectively whereas currently married women who are not using any contraceptive method is higher in slum areas (29%) than in non-slum areas (16%) of Jaipur (Fig. 4.1).

Unmet Need for Family Planning

Unmet need for family planning is an important indicator for assessing the potential demand for family planning services. Currently married women who are not using any method of contraception but who do not want any more children are defined as having an unmet need for limiting and those who are not using contraception but want to wait 2 or more years before having another child are defined as having an unmet need for spacing. The sum of the unmet need for limiting and the unmet need for spacing is the unmet need for family planning. Current contraceptive users are said to have a met need for contraception. According to these definitions, about 13% currently married women in Bhubaneswar have unmet need for family planning, while proportion of such women is 4% in each of the Jaipur and Pune city. As expected the percentage of unmet need for family planning is relatively higher among women living in non-slums than slums. These figures portraits that women living in non-slum area are more articulate of their desired number of children and family preferences (Fig. 4.2).

Table 4.6 Adjusted percentages based on multinomial logistic regression model showing the effect of selected background characteristics on contraceptive use in three cities of India, 2011

Background characteristic	Bhubaneswar			Jaipur			Pune		
	Sterilization[a]	Other modern method[b]	Any traditional method[c]	Sterilization[a]	Other modern method[b]	Any traditional method[c]	Sterilization[a]	Other modern method[b]	Any traditional method[c]
Age of women									
15–24	17.4	59.2***	23.4**	11.5	73.5***	15.0***	36.0	64.0***	0.0
25–34	25.3	57.0***	17.8***	31.2	59.8***	9.0**	68.3	31.7***	0.0***
35 + ®	60.8	22.7	16.5	61.0	29.3	9.7	94.2	5.8	0.0
Education of women									
< 5 years	67.9	18.1***	14.0**	60.3	33.8***	5.9***	91.7	8.3***	0.0
5–9 years	55.0	32.6***	12.4***	40.8	45.9***	13.3**	92.3	7.7***	0.0*
10 years & above®	34.4	43.1	22.5	23.0	62.6	14.4	71.7	28.3	0.0
Media exposure									
No	43.2	46.1	10.7	44.8	45.6	9.6	85.1	14.9	0.0
Partial	43.7	36.6	19.7	39.6	49.6	10.9***	82.3	17.7	0.0**
Full®	13.1	64.5	22.4	36.5	34.6	28.9	80.6	19.4	0.0
Standard of living index									
Low	44.4	45.7	9.8**	38.0	39.8	22.1	93.8	06.2	0.0**
Medium	39.4	36.8	23.8	53.0	38.9	8.0	73.8	25.9	0.3*
High®	43.1	29.5	27.4	26.3	67.3	6.4	69.4	29.8	0.8
Type of locality									
Non-slum	41.1	37.6	21.3*	38.5	51.2	10.4*	81.3	18.7	0.0
Slum®	53.4	36.1	10.4	45.8	40.3	13.8	84.9	15.1	0.0

Note: [a]Sterilization includes male and female sterilization [b]Traditional method includes Rhythm and withdrawal method [c]Other modern method includes pill, IUD, Injectable, implants, condom, female condom, Diaphragm and form/jelly
®Reference category, ***$P < 0.01$, **$P < 0.05$ and *$P < 0.10$

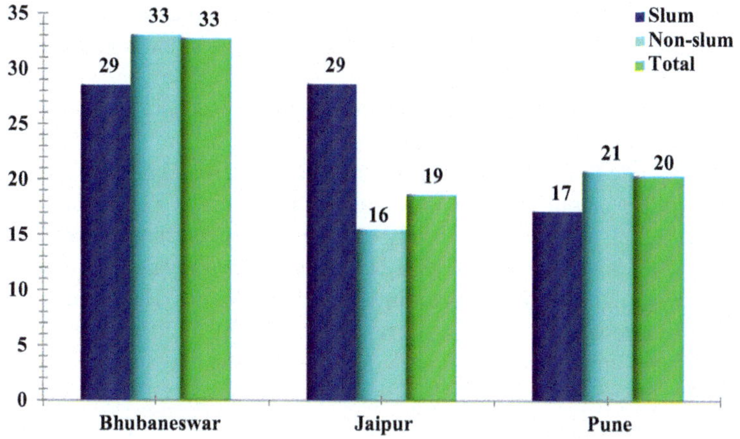

Fig. 4.1 Percentage of currently married women age 15–49 years, not using any method of contraception by type of locality in Bhubaneswar, Jaipur and Pune, 2011

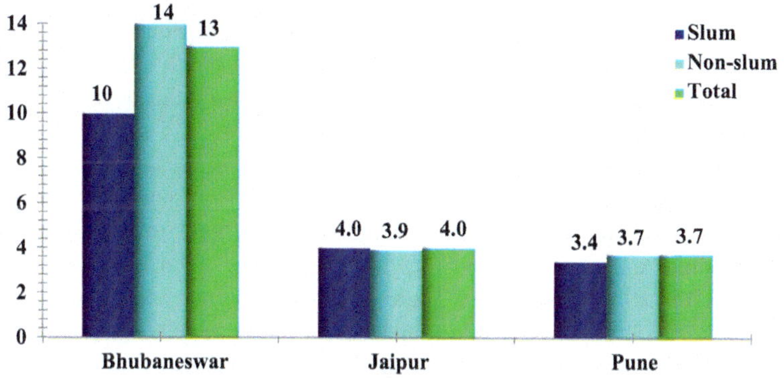

Fig. 4.2 Percent of currently married women age 15–49 years by unmet need for family planning services according to type of locality in three cities, 2011

Discussion

This study provides contraceptive use dynamics of three cities in India with a representative data from slum and non-slum areas in each of the three cities. Our findings revealed similarities and variations in contraceptive use across cities and between slum and non-slum areas within three cities. Most of the couples are using family planning to limit family size adopt contraception when they have already had the number of children they want. When contraception is used to space births, couples may start using family planning earlier, with the intention of delaying a possible pregnancy. This may be done before a couple has had their desired number of children or even if they have no living child. As seen earlier, a majority of ever users of family planning have been sterilized. Popularizing the concept of spacing births for the welfare of children and women and promoting spacing methods can lead to contraceptive use among women at lower parities and consequent fertility reduction (IIPS and ORC Macro 2007).

India has experienced a fertility transition due to the widespread contraceptive use among uneducated women. Smaller families are slowly becoming the norm in India. Total Fertility rate in India has declined to 2.2 children, mainly due to increased use of contraception among women. In spite of this, India is yet to achieve below replacement level fertility, with contraceptive prevalence rate for married women being only 54% and total unmet need being 13%. The prevalence rate of India is almost similar to global rates. According to National Family Health Survey – 4 the contraceptive prevalence of India during the year 2015–2016 was 57% (urban population) and unmet need for family planning in urban area is 12.1% consisting of 5.1% for spacing and 7% for limiting births (IIPS and ORC Macro 2017). In particular, slum residents are known to have poorer sexual and reproductive health outcomes than their counterparts living in other parts of the city or in rural areas (African Population and Health Research Center 2002). Evidence also shows that women from the lowest wealth quintiles exhibit the greatest gap between wanted and observed fertility (Kenya National Bureau of Statistics (KNBS), ICF Macro 2010).

More than half of currently married women in Pune and Bhubaneswar and 7 out of 10 women in Jaipur are currently using any method of family planning but the use of limiting method is higher than spacing method in all the three cities. Across the cities, when there is a difference in method use by slum and non-slum areas, the common pattern is that the slum areas are more likely to be using sterilization than the non-slum areas. Multivariate analyses of method use found that the younger women and more educated women are much more likely to use other modern spacing method and any traditional method than older women and women with less education in all the three cities. There is a slightly higher unmet need for family planning in non-slum areas than slum areas in all the three cities.

In a study done in Vietnam the researcher describes the relationship between women's age and modern contraceptive methods as an inverted 'U-shape'. While the likelihood of contraceptive use was low among women aged 15–24, it was lower among the those 35 and above and highest among women aged 25–35 (Vu et al. 2016). However, Fotso and colleagues suggest that the gap between poor and rich in the use of modern contraception in urban Kenya is dwindling, and that the difference in the contraceptive prevalence rate for modern methods between the lowest wealth quintile and the highest quintile in 2008–2009 was insignificant (Fotso et al. 2013). This suggests that family planning programs may be increasingly reaching the urban poor, although unmet need for contraception was greater among the poorest women.

The contraceptive use differential exists in both slums and non-slums of urban areas. At the same time, there is a substantial unmet need for contraception. The contraceptive use is mostly dominated by non-reversible methods, limited use of spacing, male/couple dependent methods, substantial levels of discontinuation and negligible use of contraceptives among married adolescents in both slums and non-slums of urban areas. There is poor acceptance of contraceptive methods either due to ignorance or fear of complications using them. Sterilisation was the most commonly adopted method even among married adolescents in India. A review of data on contraceptive behaviour of adolescents in Asian countries shows that India is the only country where such a pattern prevails (Pachauri and Santhya 2002). The official family planning programmes began during the 1960s adopted a "cafeteria approach," whereby clients are provided with a choice of contraceptive methods. However, it is well documented that, until recently, the emphasis of the programme remained skewed towards promoting non-reversible methods, particularly female sterilisation. Hence, not surprisingly, female sterilisation continues to be used by the majority of contraceptive users in India. Nationally, data from NFHS–3 show that female sterilisation accounted for two thirds of contraceptive use and 77% of modern method use (IIPS and ORC Macro 2007).

A recent study using the National Family Health Survey (NFHS) conducted in 2015–2016 for India showed that women living in slums were significantly less likely to use modern family planning methods than women outside slums. Multivariate analyses demonstrate a significant difference between current modern method users and all others based on slum residence (less likely to use) and education (lower education, less likely to use) among other factors. This is consistent with our findings related to the role of education in both slum and non-slum areas (Hazarika 2010). But our findings show that the use of modern family planning methods is higher in slum areas as compared to non-slum areas of Bhubaneswar and Pune and lower in Jaipur. Our findings are also similar to those performed by (Speizer et al. 2012) that compared six cities in Uttar Pradesh, India to examine family planning use among the urban poor in slum and non-slum areas. Data were collected from about 3000 currently married women in each city (Allahabad, Agra, Varanasi, Aligarh, Gorakhpur, and Moradabad) for a total sample size of 17,643 women. In their study they found that women in slum areas generally report less family planning use and among those women who use, slum women are more likely to be sterilized than to use other methods, including condoms and hormonal methods. There is a higher unmet need for family planning to limit childbearing than for spacing births. Poorer women are more likely to have an unmet need than richer women in both the slum and non-slum samples; this effect is attenuated when education is included in the analysis (Speizer et al. 2012). But our findings show that there are fewer differentials in modern method of family planning use both in slum and non-slum areas except Jaipur. But our study also found that among those women who use, slum women are more likely to be sterilized than to use any traditional method and other method spacing method. There is a slightly higher unmet need for family planning in non-slum areas than slum areas in Bhubaneswar and Pune and lower unmet need for family planning in Jaipur. Sudha et al. (2017) concluded in their study that unmet need for family planning was found to be fairly high in urban area of Puducherry. Health care personnel, preferably field staff should be trained to give an informed choice of contraception to the eligible couples to reduce the unmet need for contraception.

Summary and Conclusions

The contraceptive use varies in all the three cities of India and there is scope of further improvement in the acceptance of family planning methods by the couples. The main method of contraception is sterilization in both slum and non-slum areas and particularly among the women who are less educated and belonging to low standard of living index. The analysis revealed that, only one-third of the women with one child use contraception in three cities of India. This proportion of using contraception for the first time when they had two or fewer children increases to more than four-fifths of ever users in Bhubaneswar, 64% of ever users in Jaipur and 82% of ever users in Pune. Clearly spacing is not very common but most of the contraceptive use is for limiting the family size. Age and educational levels of women have a significant effect on the use of contraception. Therefore, these aspects may be given due attention while framing family planning programme and efforts should be made to make people aware of the benefit of small family norm. If other modern spacing methods, such as the pill, are made more widely acceptable and available, women may choose to use them to space births and then sterilization can be used as a common limiting method. About 13% currently married women in Bhubaneswar have unmet need for family planning, while proportion of such women is 4% in each of the Jaipur and Pune city. Current use patterns and unmet need for family planning should be an indication of where the biggest gaps for services are. There is a need to promote modern contraceptive methods specially focusing at social marketing strategy with emphasis at brand promotion and product positioning. Availability and accessibility of contraception should be ensured within the framework of CBOs/NGOs functioning among socially marginalized and economically deprived group of people. To conclude, our study demonstrates that programs seeking to increase contraceptive use in three cities of India should consider the city context within which they are working.

References

African Population and Health Research Center (APHRC). (2002). *Population and health dynamics in Nairobi's informal settlements. Report of the Nairobi Cross-sectional Slums Survey (NCSS) 2000*. Nairobi: African Population and Health Research Center.

Census of India. (2001). *Metadata and brief highlights on slum population*. Available at: http://censusindia.gov.in/Data_Products/Data_Highlights/Data_Highlights_link/metadatahighlights.pdf. Accessed 11th Sept 2009.

Census of India. (2011). *Provisional population total, India Census, 2011*. Available at: http://www.censusindia.gov.in. Accessed 2nd Apr 2018.

Chen, N., Valente, P., & Zlotnik, H. (1998). What do we know about recent trends in urbanization? In R. E. Bilsborrow (Ed.), *Migration, urbanization, and development: New directions and issues* (pp. 59–88). Norwell: UNFPA-Kluwer Academic Publishers.

Ezeh, A. C., Kodzi, I., & Emina, J. (2010). Reaching the urban poor with family planning services. *Studies in Family Planning, 41*(2), 109–116.

Fotso, J., Speizer, I., Mukiira, C., Kizito, P., & Lumumba, V. (2013). Closing the poor-rich gap in contraceptive use in urban Kenya: Are family planning programs increasingly reaching the urban poor? *International Journal for Equity in Health, 12*(1), 71.

Gribble, J. N. (2012). *Unmet need for family planning. World population data sheet 2012, Population reference Bureau, USA*. Available at: http://www.prb.org/publications/Datasheets/2012/world-population-data-sheet/fact-sheet-unmet-need.aspx. Accessed 2nd Apr 2017.

Hazarika, I. (2010). Women's reproductive health in slum populations in India: Evidence from NFHS-3. *Journal of Urban Health, 87*(2), 264–277.

International Institute for Population Sciences (IIPS). (2011). *Health of the urban poor in three cities: Bhubaneswar, Jaipur and Pune*. Mumbai: IIPS.

International Institute for Population Sciences (IIPS) and ICF. (2017). *National Family Health Survey (NFHS-4), 2015–16: India*. Mumbai: IIPS.

International Institute for Population Sciences (IIPS) and Macro International. (2007). *National Family Health Survey (NFHS-3), 2005–06: India: Volume I*. Mumbai: IIPS.

Kenya National Bureau of Statistics (KNBS), ICF Macro. (2010). *Kenya demographic and health survey 2008–09*. Calverton: KNBS and ICF Macro.

Ministry of Health and Family Welfare (MOHFW) (2000). *National Population Policy,* New Delhi: Department of Family Welfare, MOHFW.

Montgomery, M. R. (2009). Urban poverty and health in developing countries. *Population Bulletin, 64*(2), 1–6.

Pachauri, S., & Santhya, K. G. (2002). Reproductive choices for Asian adolescents: A focus on contraceptive behaviour. *International Family Planning Perspectives, 28*, 186–195.

Speizer, I. S., Nanda, P., Achyut, P., Pillai, G., & Guilkey, D. K. (2012). Family planning use among urban poor women from six cities of Uttar Pradesh, India. *Journal of Urban Health, 89* (4), 639–658.

Sudha, V., Vrushabhendra, H. N., Srikanth, S., & Suganya, E. (2017). 2017. Unmet need for contraception among urban women: A cross sectional study in Puducherry. *International Journal of Community Medicine and Public Health, 4*(5), 1494–1499.

Sulthana, B., Shewade, H. D., Sunderamurthy, B., Manoharan, K., & Subramanian, M. (2015). Unmet need for contraception among married women in an urban area of Puducherry, India. *Indian Journal of Medical Research, 141*, 115–118.

United Nations. (2008). *World urbanization prospects: The 2007 revision*. New York: United Nations Population Division.

United Nations, Department of Economic and Social Affairs, Population Division. (2012). *World urbanization prospects, the 2011 revision*. Press Release. New York.

Vu, L. T. H., Oh, J., Bui, Q. T.-T., & Le, A. T.-K. (2016). Use of modern contraceptives among married women in Vietnam: A multilevel analysis using the multiple Indicator cluster survey (2011) and the Vietnam population and housing census (2009). *Global Health Action, 9*(1), 29574.

Chapter 5
Birth Weight Outcomes for Non-Hispanic Black Women in a Home Visiting Program in Rural Mississippi: Observations from the Field

Chris N. King, Anna C. Church, Wesley L. James, Rhonda G. Okoth, and Karen C. Matthews

Introduction

Preventing low (LBW; < 2500 grams) and very low (VLBW; < 1500 grams) weight births is a major public health concern, as LBW and VLBW are the leading cause of infant mortality (Shan and Austin 2014; Conley et al. 2003). Preterm birth, LBW and VLBW are also highly associated with increased risk of a range of co-morbidities, from neonatal breathing difficulties and cerebral palsy to cognitive and motor skills impairments that emerge later in childhood, and has even been associated with lower intelligence well into adulthood (Synnes 2016; Shan and Austin 2014; Conley et al. 2003). Improving these outcomes is difficult, because social and environmental factors independent of and antecedent to pregnancy, like race and socioeconomic status, have been shown to contribute a great deal of the risk to pregnancies resulting in LBW and VLBW (Shan and Austin 2014).

Contributing Authors: Anna C. Church, Wesley L. James, Rhonda G. Okoth, Karen C. Matthews

---

C. N. King (✉) · W. L. James
Center for Community Research and Evaluation, The University of Memphis, Memphis, TN, USA
e-mail: cnking@memphis.edu; wljames1@memphis.edu

A. C. Church
Ohio State University, Columbus, OH, USA
e-mail: church.213@buckeyemail.osu.edu

R. G. Okoth
St. Jude Children's Research Hospital, Memphis, TN, USA

K. C. Matthews
Delta Health Alliance, Stoneville, MS, USA
e-mail: kmatthews@deltahealthalliance.org

B. Jivetti, Md. N. Hoque (eds.), *Population Change and Public Policy*, Applied Demography Series 11, https://doi.org/10.1007/978-3-030-57069-9_5

In recent years, practitioners and policy makers have used home visiting programs as one method of intervening and attempting to improve low birth weight outcomes. Home visitors including nurses, social workers, early childhood educators, or other trained professionals typically conduct periodic visits to clients' homes to support preventive health and prenatal best practices and assist mothers in caring for themselves and their child(ren) (Health Resources Services Administration [HRSA] n.d.). Standards of care and practices vary by program, but participants are typically visited at least monthly and supported through education, connection to resource providers, provision of some resources in-kind, and social connections.

Despite promising evidence of home visiting's effects on birth outcomes, including increased birth weights (Kothari et al. 2014) and reduced risk of LBW and preterm birth in some studies (Salihu et al. 2009), major gaps in knowledge remain, most notably with regard to populations who are especially at-risk due to social and environmental factors like geographic region, racial discrimination, and resource deficiencies (Peacock et al. 2013). For example, we currently know of no study that specifically investigates home visiting programming at the intersection of rurality and minority racial or ethnic status, despite the fact that a full 10.0% of non-Hispanic Black Americans lived in nonmetro areas in 2010 (compared to 14.9% of the total population and 18.8% of the non-Hispanic white population) and exhibited some of the worst health outcomes of any group (Lawrence 2015; USDA Economic Research Service [USDA] n.d.). In 2015, race and place compounded to produce especially risky conditions for the pregnancies of Black women in Mississippi (MS) (Thomson et al. 2016), leading to VLBW and LBW rates that were the highest for any racial group in any state in the country (3.5%, 16.5%). In fact, the rate of LBW births to Black women in Mississippi was three times higher than the rate for births to white women in Alaska (5.2%) or Hispanic women in South Dakota (5.5%) (Takeuchi et al. 2010; Mississippi State Department of Health [MSDH] n.d.). The pathways through which race and place affect birth weights are not fully understood, but explanations have ranged from socio-structural and discriminatory to psychological, cultural, and behavioral (Conley et al. 2003; Synnes 2016).

Here, we begin to bridge this gap in the literature by reporting preliminary outcomes of a home visiting program in a rural, majority minority, poor community in Mississippi. We assert that if home visiting is effective here, its ability to reduce a major public health burden is more portable and possibly more potent than previously understood. This study investigates the association between birth weight outcomes and enrollment in a HRSA Healthy Start program, the Delta Healthy Start Collaborative (DHSC), among non-Hispanic Black women in Leflore County, MS. With a VLBW rate of 3.4% and a LBW rate of 15.6%, Leflore County, population 32,317, has had rates more comparable to areas of the developing world than those of the United States (Bhati 2017; Li et al. 2017; Taywade and Pisudde 2017; Afriyie et al. 2016; UNICEF 2015; MSDH n.d.). DHSC's position in a county that is classified nonmetro, non-adjacent to a metro area, and also 72.8% Black and non-Hispanic, makes it a novel case (MSDH n.d.; USDA n.d.). However, it is also a case that is generalizable to underserved, understudied Black Belt areas throughout the US Deep South, and to the Black population in rural areas across the country.

Materials and Methods

Study Population

We test differences in observations of LBW and VLBW rates between births to non-Hispanic Black prenatal enrollees in the DHSC program (n = 109) and births to all other non-Hispanic Black women in Leflore County, MS, in 2015 (n = 281). Program participation included monthly visits to women's homes by licensed social workers, who primarily used the Parents as Teachers "Foundational Curriculum" (2011) in their visits, supplemented with educational materials from Florida State University Partners for a Healthy Baby "Before Baby Arrives" curriculum (2017), and collected assessments at specified time intervals to gather information about the women's living conditions and identify participant needs (which were subsequently addressed through education and referrals to resources). Participation was purely voluntary and without cost to participants. Participants became aware of and entered the program either through word-of-mouth or active recruitment on the part of home visitors, who were encouraged to seek out the most at-risk, but also the most interested, women for their services.

Programmatic rates are derived from birth weights observed by home visitors within the first month subsequent to the child's birth. The weights were recorded from maternal recall, as birth certificates were typically not available at the time of the first postnatal visit. Rates only include women whose dates of enrollment into the program predate the child's date of birth. We include only births to non-Hispanic Black women to ease comparisons, since they made up nearly the entirety of DHSC's enrollment in 2015 (93.2%) and the majority of births in Leflore County, as well (78.9%). Leflore County is chosen because, as of the beginning of 2015, it was the only county in which DHSC was active. The year 2015 is chosen for simplicity, because it is the first full year of the program's existence and prior to its expansion into multiple counties. Multiple births were included in all samples due to the inability to remove multiples from state and county aggregates.

Data and Analysis

Programmatic LBW and VLBW rates are derived from birth weights (in pounds) taken in home visits in the month following birth. These program birth weights were coded into binaries for low (< 5.5 pounds, or 2500 grams) or very low (< 3.3 pounds, or 1500 grams) weight. Population aggregates come from the publicly-available

Mississippi Statistically Automated Health Resource System (MSTAHRS), published by the Mississippi Department of Health (MSDH n.d.).

Since MSTAHRS only publishes data in aggregate form, we were unable to control for covariates in our analysis, but we include aggregate rates for comparisons to participants for all the variables they provide, including age, prenatal care, education level, marital status, and smoking. We also include comparisons to state and national rates on other variables not provided by MSTAHRS, including insurance status, obesity, and WIC enrollment, from the CDC public use natality files (CDC n.d.).

We test differences in association on confounders between program participants and county, state, or national aggregates using one-sample binomial tests against participant rates. To test the association between the independent variable, program participation, and the dependent variable, LBW and VLBW incidence, we built a dataset of all births to non-Hispanic Black women in Leflore County, with the observed proportion of healthy and LBW or VLBW births, and then assigned the appropriate number of those cases in each category as program participants. We then tested for significance using chi-square, Fisher's exact, and one-sample binomial tests of association. All testing was done with SPSS 22 and base R (R Core Team n.d.).

Results

Table 5.1 displays, for context, the descriptive statistics of DHSC participants and comparison populations, including county non-participants. Program participants were on average younger, more obese, less likely to be privately insured, and less likely to be married than non-Hispanic Black mothers at the county, state, and national levels (MSDH n.d.; CDC n.d.). Further, almost 3 out of 4 participants (73.1%; data not provided by MSTAHRS) reported household incomes *below $10,000*, and more than 9 out of 10 (90.7%) reported incomes below $20,000, indicating that nearly all of the participant cases lived on incomes below the federal poverty level. However, participants were slightly more educated than comparison populations on average, indicating possible positive selection effects, and were also less likely to have smoked tobacco and more likely to have been enrolled in WIC and/or to have started prenatal care in the first trimester, activities emphasized in education by home visitors to encourage healthier pregnancies.

Table 5.2 presents overall rates of very low and low birth weight incidence for births to non-Hispanic Black women from the level of the United States down to the participant and non-participant groups, along with confidence intervals for DHSC participants [Martin et al. 2017; MSDH n.d.). The VLBW rate was 4.1% in Leflore County, compared to 3.5% in Mississippi and 2.9% in the US. Among the 16 very low birth weight births to non-Hispanic Black women in Leflore County, only one was born to a DHSC participant, leaving rates of 0.9% for participants and 5.3% for non-participants (OR = 0.16, 95% CI = 0.02–1.25). The LBW rate in Leflore

Table 5.1 Demographic proportions

	DHSC (95% CI) (n = 109)	Leflore Co. non-DHSC[a] (n = 281)	Mississippi 2015[a] (n = 16,364)	United States 2014[b] (n = 547,424)
Mother's age at birth				
<18	5.5% (1.2–9.8)	4.3%	3.8%	2.7%
18–25	46.8% (37.4–56.2)	48.6%	45.8%	***37.6%***
25–30	25.7% (17.5–33.9)	28.4%	27.2%	27.4%
30–35	17.4% (10.3–24.5)	12.9%	15.3%	19.9%
35+	4.6% (0.7–8.5)	6.8%	7.8%	***12.3%***
Prenatal care in first trimester	87.5% (77.3–97.7)	***53.4%***	***69.8%***	***65.8%***
Insured	93.3% (88.1–98.5)	–	–	93.0%
Receives Medicaid	86.5% (79.4–93.6)	–	–	***66.7%***
Insured privately	6.7% (1.5–11.9)	–	–	***25.5%***
Obese BMI classification	38.6% (24.2–53.0)	–	–	34.8%
Enrolled in WIC at birth	73.8% (64.2–83.4)	–	–	64.5%
Some college or more	54.5% (44.1–64.9)	***43.0%***	48.9%	50.0%
Unmarried at birth	90.3% (84.2–96.3)	88.5%	***80.1%***	***71.1%***
Non-smoker	100%	99.5%	93.6%	93.3%

[a]MSDH (n.d.); [b]CDC (n.d.). *Bold italics denote difference from DHSC significant at p < .05*

Table 5.2 Low and very low weight birth rates: 2015 Births to non-Hispanic Black Women

	DHSC (95% CI) (n = 109)	Leflore Co. non-DHSC[a] (n = 281)	Mississippi 2015[a] (n = 16,364)	United States 2015[b] (n = 589,047)
LBW rate	14.7% (8.99–22.2)	19.9%	16.5%	13.3%
VLBW rate	0.92% (0.10–4.21)	***5.34%***	3.45%	2.89%

[a]MSDH (n.d.); [b]Martin et al. (2017); *Bold italics denote difference from DHSC significant at p < .05*

County in 2015, including participants and non-participants, was 18.5%, compared to 16.5% for the state of Mississippi and 13.3% for the US. Within the county, LBW rates were 14.7% for DHSC participants and 19.9% for non-participants (OR = 0.69,

95% CI = 0.38–1.27). The participant VLBW rate was well below both US and Mississippi rates, and was significantly lower than the rate for Leflore County non-participants in a one-sample binomial test ($p = .033$) and Fisher's exact test ($p = .035$). The LBW rate for participants was 26% lower than the rate for Leflore County non-participants, a large difference but not significant in the one-sample binomial test ($p = .105$) or Fisher's exact test ($p = .146$).

Discussion

We find that non-Hispanic Black women who were not prenatal participants in DHSC were 6.1 times more likely to have given birth to a VLBW child (significant at $p < .05$), and 1.4 times more likely to have given birth to a LBW child (n.s.), versus those who participated in the program. This follows similar findings from Kothari et al. (2014), Salihu et al. (2009), and Wells et al. (2008) regarding improved birth weights and reduced morbidities among births to non-Hispanic Black participants in urban home visiting programs.

Isolating the county non-participants highlights their alarming VLBW and LBW rates. About 1 in 5 non-participant births to a non-Hispanic Black woman in Leflore County was LBW in 2015, and more than 1 in 20 was VLBW. This is compared to about 1 in 7 LBW and 1 in 100 VLBW births among DHSC participants, respectively, and 1 in 14 LBW and 1 in 100 VLBW births to non-Hispanic white women across the country.

DHSC participant rates are especially encouraging in light of the low socioeconomic status of program participants, with nearly all women reporting household incomes below the federal poverty threshold. Other demographic differences in Table 5.1 demonstrate pathways through which DHSC services may have affected birth outcomes. None of the DHSC participants smoked, almost three-fourths were enrolled in WIC, and about 83% received SNAP benefits (not shown for lack of comparison data). Although almost all of the insured had healthcare access through Medicaid, consistent with their socioeconomic status, the rate of basic coverage was high at 93%. And more than 4 in 5 DHSC women reported starting prenatal care in the first trimester, compared to about 3 in 5 non-participants ($OR = 1.433$; $P < .001$). Prenatal care and social and economic support have been called the "most likely known targets for prenatal interventions to prevent low birth weight" (Alexander and Korenbrot 1995).

This approach has various opportunities for improvement in future research. At least three are related to our dependence upon publicly available aggregate data at the county and state levels, which resulted in the inability to include control variables in regression models, to account for or eliminate multiple births (which are heavily associated with LBW and have been accounted for in most other analyses (*e.g.*, Kothari et al. 2014), and to include multiple years of data because of the lag on state availability, which would have allowed higher testing power and possibly a trend analysis. Other limitations are associated with the research design, including: the program's mandate to reach and serve as many participants as economically feasible; the program's targeted, snowball recruitment strategy; the lack of an experimental design, placebo condition, or randomization of any sort; limited generalizability from focusing on one county and racial group; validity issues associated with using maternal recall for participant birth weights; the choice not to account for the quality and duration of program enrollment for simplicity; and the inability to determine whether the participants received insurance and prenatal care as a result of enrollment in the program. Since these and other factors are unaccounted for, we cannot mitigate any selection effects or isolate the mechanisms through which the program would affect birth weight. Finally, data were not collected by research personnel but by trained home visiting staff. While this is an advantage in that it reduces the artificiality of the research encounter, data collection was not the primary concern of home visitors, and participants were not anonymous to program staff (though they were anonymous to external evaluators and researchers).

Implications

To our knowledge, there have been no reports of the birth weight outcomes of a home visiting program in a rural, predominantly non-Hispanic Black community, despite the fact that these communities are situated at the intersection of two groups with heightened risk of adverse birth outcomes. Infants born VLBW are as likely to die as live (Lau et al. 2013), accounting for about half of all infant deaths and 30% of newborn health care costs nationwide despite making up less than 2% of all births (Johnson et al. 2013). The social and economic implications of reducing the occurrence of LBW, and especially VLBW, then, are of utmost importance. Through home visiting, interventionists can cost-effectively decrease this burden on household, community, and state economies and institutions, which in turn could alter generational disadvantages through lowering societal costs and increasing productivity and quality of life (Aizer and Currie 2014; Conley et al. 2003).

Acknowledgments The authors would like to thank Karin Scott Ewbank and Carolyn Willis, staff and leadership of the Delta Healthy Start Collaborative including Melanie Williams, and others who work directly with the program and its data collection and management.

Conflicts of Interest The principal author and three co-authors have worked as external evaluation consultants for Delta Health Alliance, the governing entity of the Delta Healthy Start Collaborative. One coauthor is employed by the organization directly. This research was performed in fulfillment of those roles.

Funding The DHSC project was supported by the Health Resources and Services Administration (HRSA) of the U.S. Department of Health and Human Services (HHS) under grant number H49MC27819, titled "Healthy Start Initiative-Eliminating Racial/Ethnic Disparities" (award amount: $3.51 M; 0% responsibility from nongovernmental sources). The project was also supported in part by a grant from the WAPACK Foundation. The information, content, and conclusions presented here are those of the authors and should not be construed as the official position or policy of, nor should any endorsements be inferred by HRSA, HHS, the U.S. Government, or the WAPACK Foundation.

References

"Before baby arrives". (2017). Tallahassee, FL: Florida State University Center for Prevention & Early Intervention Policy.

"Foundational curriculum". (2011). St. Louis, MO: Parents as Teachers National Center, Inc.

Afriyie, J., Bedu-Addo, K., Asiamah, E. A., & Boateng, S. T. (2016). Low birth weight among adolescents at Cape Coast metropolitan Hospital of Ghana. *International Journal of Reproduction, Contraception, Obstetrics and Gynecology, 5*(12), 4242–4247.

Aizer, A., & Currie, J. (2014). The intergenerational transmission of inequality: Maternal disadvantage and health at birth. *Science, 23*(344), 856–861.

Alexander, G. R., & Korenbrot, C. C. (1995). The role of prenatal care in preventing low birth weight. *The Future of Children, 5*(1), 103–120.

Bhati, D. K. (2017). Predictors of low birth weight infants in western desert region, Rajasthan. *International Journal of Child Health and Human Development, 10*(1), 31–38.

CDC. (n.d.). *The public use natality file – 2015 update* [database online]. Accessed 11 Nov 2017 from https://www.cdc.gov/nchs/data_access/vitalstatsonline.htm.

Conley, D., Strully, K. W., & Bennett, N. G. (2003). *The starting gate: Birth weight and life chances*. Berkeley: University of California Press.

Health Resources and Services Administration [HRSA]. (n.d.). *Home visiting program: Mississippi* [Web page]. Accessed 22 Jan 2018 from https://mchb.hrsa.gov/sites/default/files/mchb/MaternalChildHealthInitiatives/HomeVisiting/pdf/ms.pdf

Johnson, T. J., Patel, A. L., Jegier, B. J., Engstrom, J. L., & Meier, P. P. (2013). Cost of morbidities in very low birth weight infants. *The Journal of Pediatrics, 162*(2), 243–249. https://doi.org/10.1016/j.jpeds.2012.07.013.

Kothari, C. L., Zielinski, R., James, A., Charoth, R. M., & Sweezy, L. (2014). Improved birth weight for black infants: Outcomes of a Healthy Start program. *American Journal of Public Health, 104*(Suppl 1), S96–S104.

Lau, C., Ambalavanan, N., Chakraborty, H., Wingate, M. S., & Carlo, W. A. (2013). Extremely low birth weight and infant mortality rates in the United States. *Pediatrics, 131*(5), 855–860. https://doi.org/10.1542/peds.2012-2471.

Lawrence, D. T. (2015). The rural black nowhere: Invisibility, ubannormativity, and the geography of indifference. *The Journal of the Midwest Modern Language Association, 48*(1), 221–244.

Li, N., Zhiwen, L., Ye, R., Liu, J., & Ren, A. (2017). Impact of periconceptional folic acid supplementation on low birth weight and small-for-gestational-age infants in China: A large prospective cohort study. *The Journal of Pediatrics, 187*, 105–110.

Martin, J. A., Hamilton, B. E., Osterman, M. J. K., Driscoll, A. K., & Matthews, T. J. (2017). Births: Final data for 2015. *National Vital Statistics Reports, 66*(1), 58–59.

Mississippi State Department of Health [MSDH]. (n.d.). Mississippi Statistically Automated Health Resource System (MSTAHRS) [database online]. Accessed 27 Aug 2017 from http://mstahrs.msdh.ms.gov/

Peacock, S., Konrad, S., Watson, E., Nickel, D., & Muhajarine, N. (2013). Effectiveness of home visiting programs on child outcomes: A systematic review. *BMC Public Health, 13*, 17. https://doi.org/10.1186/1471-2458-13-17.

R Core Team. R: A language and environment for statistical computing (n.d.). *R foundation for statistical computing*. Vienna, Austria. Accessed from https://www.R-project.org/

Salihu, H. M., Mbah, A. K., Jeffers, D., Alio, A. P., & Berry, L. (2009). Healthy start program and feto-infant morbidity outcomes: Evaluation of program effectiveness. *Maternal and Child Health Journal, 13*(1), 56–65.

Shan, M. K., & Austin, K. R. (2014). Do home visiting services received during pregnancy improve birth outcomes? Findings from Virginia PRAMS 2007-2008. *Public Health Nursing, 31*(5), 405–413.

Synnes, A. R. (2016). Developmental outcome. In M. G. MacDonald & M. M. K. Seshia (Eds.), *Avery's neonatology* (7th ed.). Philadelphia: Wolters Kluwer.

Takeuchi, D. T., Walton, E., & Leung, M. (2010). Race, social contexts and health: Examining geographic spaces and places. In C. E. Bird, P. Conrad, A. M. Fremont, & S. Timmermans (Eds.), *Handbook of medical sociology* (6th ed., pp. 92–105). Nashville: Vanderbilt University Press.

Taywade, M. L., & Pisudde, P. M. (2017). Study of sociodemographic determinants of low birth weight in Wardha district, India. *Clinical Epidemiology and Global Health, 5*, 14–20.

Thomson, J. L., Tussing-Humphreys, L. M., Goodman, M. H., & Oldender, S. (2016). Baseline demographic, anthropometric, psychosocial, and behavioral characteristics of rural, Southern women in early pregnancy. *Maternal and Child Health Journal, 20*(9), 1980–1988. Accessed 18 Nov 2017 from https://doi.org/10.1007/s10995-016-2016-y.

UNICEF. (2015). *Low birthweight: Percentage of infants weighing less than 2,500 grams at birth* [database online]. https://data.unicef.org/wp-content/uploads/2015/12/Low-Birthweight-SOWC-5-November-20-14_68.xlsx. Accessed 22 Jan 2018.

USDA Economic Research Service [USDA]. (n.d.). *Atlas of rural and small-town America* [database online]. Accessed 11 Feb 2018 from https://www.ers.usda.gov/data-products/atlas-of-rural-and-small-town-america/download-the-data/.

Wells, N., Sbrocco, T., Hsiao, C. W., Hill, L., Vaughn, N. A., & Lockley, B. (2008). The impact of nurse case management home visitation on birth outcomes in African-American women. *Journal of the National Medical Association, 100*(5), 547–552.

Part II
Health Behaviors and Their Correlates

Chapter 6
Neighborhood Food Environment and Self-Rated Health: An Investigation with a Spatial Perspective

Danhong Chen and Tse-Chuan Yang

Introduction

Food Environment and Health

The relationship between the food environment and dietary behaviors has been well established in the literature. Since supermarkets and other large grocery stores are more likely to offer a variety of healthy food at relatively low costs, they are often found to be positively related to dietary healthfulness (Larson et al. 2009). For example, Morland, Wing, and Roux (2002) showed that each additional supermarket in a census tract was associated with increased fruit and vegetable intake for both Black and White Americans. In contrast, the promotion effect for a healthy diet was not significant in other forms of food outlets, including grocery stores, full-service restaurants, and fast-food restaurants. A positive impact of proximity to supermarkets on dietary healthfulness during pregnancy was also reported by Laraia, Siega-Riz, Kaufman, and Jones (2004). Similarly, compared to women shopped at independent grocery stores, women who shopped at supermarkets or specialty stores had higher consumption of fruits and vegetables (Zenk et al. 2005).

There is also evidence on the role of a healthy diet in disease prevention. After reviewing 206 human epidemiologic and 22 animal studies, Steinmetz and Potter

---

Disclaimer: This article is an original work by Danhong Chen and Tse-Chuan Yang, and it is being submitted exclusively to the International Association of Applied Demography for publication consideration into an edited book on Applied Demography by Springer Publishing Company.

D. Chen (✉)
Department of Agricultural Sciences, Sam Houston State University, Huntsville, TX, USA
e-mail: dchen@shsu.edu

T.-C. Yang
Department of Sociology, Center for Social and Demographic Analysis, University at Albany, SUNY, Albany, NY, USA

B. Jivetti, Md. N. Hoque (eds.), *Population Change and Public Policy*, Applied Demography Series 11, https://doi.org/10.1007/978-3-030-57069-9_6

(1996) concluded that consumption of fruits and vegetables was found to have protective effects against cancers of the stomach, esophagus, lung, oral cavity and pharynx, endometrium, pancreas, and colon. A diet rich in fruits and vegetables might also reduce the risk of weight gain, obesity (He et al. 2004), and other chronic diseases (E. S. Ford and Mokdad 2001).

There is a growing body of literature assessing the impact of the built food environment on health outcomes. The majority of the studies focused on the association between access to food stores and weight (Ford and Dzewaltowski 2011; Hanibuchi et al. 2011; Inagami, Cohen, Finch, and Asch, 2006; Morland, Diez Roux, and Wing, 2006; Morland and Evenson 2009; Powell, Auld, Chaloupka, O'Malley, and Johnston, 2007), and relatively few studies examined the relationship between food environment and self-rated overall health status (Brown, Vargas, Ang, and Pebley, 2008). In terms of research methods, multilevel models were commonly used among these studies (Brown et al. 2008; Ford and Dzewaltowski 2011; Hanibuchi et al. 2011; Inagami et al. 2006; Morland et al. 2006) to capture the influence of contextual-level food environment on individual-level health outcomes.

Previous studies primarily focused on analyzing the statistical correlations between health status and food environment, without considering the spatial autocorrelation of health outcomes. Failure to account for spatial dependence may result in less reliable statistical inference (Chi and Zhu 2008). This study aims to fill this research gap by utilizing spatial regression models to analyze the association between health status and the food environment.

Spatial Autocorrelation of Health Outcomes

Spatial autocorrelation or spatial dependence is often used to describe the similarity or dissimilarity between values of an attribute in neighboring geographical units (Chi and Zhu 2008). For example, spatial autocorrelation has been found in pneumonia and influenza hospitalizations (Crighton et al. 2007), and influenza-related mortalities (Greene et al. 2006). This observation of spatial autocorrelation led us to propose Hypothesis One: Health outcomes were spatially dependent.

Spatial autocorrelation is likely to be observed under the following circumstances (Voss et al. 2006): (1) Feedback: residential proximity tends to increase interactions between individuals and households, creating the feedback effect. (2) Grouping Forces: individuals or households with similar attributes tend to cluster together because of self-selection or combined external forces. (3) Grouping Responses: individuals or households with common characteristics may respond similarly to external forces. (4) Nuisance Autocorrelation: spatial patterns happen at a level higher than the unit of observation. In the case of health outcomes, there is likely a direct feedback or diffusion process between people living close to each other, especially for highly contagious diseases. Therefore, we hypothesized that the spatial autocorrelation of health outcomes could be partially attributed to the spatial diffusion process (Hypothesis Two).

There may also be clusters of individuals or ethnic groups that exhibit similar health behaviors or patterns (Schuurman et al. 2009), leading to similar health outcomes. More importantly and more specific to this study, the spatial similarity in the prevalence of poor or fair health and specific types of health outcomes might be attributed to the similarity in neighborhood characteristics (e.g., socio-demographic features and food environment) between neighboring places.

The food environment affects dietary behaviors, and consequently, health outcomes, a mechanism that is especially salient for disadvantaged communities. In contrast to wealthy and racially segregated neighborhoods, low-income, predominately minority and racially mixed neighborhoods were found to have significantly more small grocery stores, but fewer supermarkets (Moore and Diez Roux 2006). People living in these neighborhoods are limited by not only the availability of large retail services that provide healthy foods in the neighborhoods, but also insufficient budget to purchase relatively high-priced healthy foods and lack of mobility to travel to other parts of the area for provisions (Brown et al. 2008; Clifton 2004). These factors all contribute to worse diet and subsequently lower health outcomes (Moore and Diez Roux 2006). Therefore, we hypothesized that the spatial inequalities in health outcomes could be partially attributed to the geographical disparities in the neighborhood environment, especially the food environment (Hypothesis Three).

Relatively little research considered the spatial dependence observed in health outcomes due to the spillover effects from neighboring geographical units and the similarity in neighborhood characteristics across distances. Based on data from Southeastern Pennsylvania Household Health Surveys (see next section for details), this study first examines whether there is a spatial autocorrelation in self-rated health (SRH). It then evaluates the impact of the neighborhood food environment on SRH status with both non-spatial and spatial perspectives.

Data and Methods

Data

The data used in this paper were derived from the biennial Philadelphia Health Management Corporation's (PHMC) Southeastern Pennsylvania Household Health Surveys, which were focused on residents living in five Philadelphia metropolitan counties—Bucks, Chester, Delaware, Montgomery, and Philadelphia. The adult file of the PHMC data set contained information on the health status, health care utilization, health behaviors, as well as the socio-demographic characteristics of all respondents 18 years of age and older. There were 10,007 and 10,006 individual observations in the 2008 and 2010 surveys, respectively. The surveys were representative of the population in the area.

The 2008 and 2010 PHMC surveys were combined to increase the reliability of the aggregated sample. The tracts for respondents in 2008 and 2010 PHMC were originally defined by the 2000 Census and 2010 Census, respectively, and there was additional information on how respondents in 2010 were linked to 2000 tracts boundaries. Thus, we first concatenated the two data sets and then aggregated the pooled data set based on the 2000 Census tracts. The resulting data set contained 959 tracts, with an average of 21 respondents in one tract. All of our analyses in this study were conducted at the tract level.

In order to visualize the spatial patterns of health status and conduct spatial regressions, the shapefile for all the tracts in the Philadelphia metropolitan area corresponding to the 2000 Census boundaries was obtained from the census website. This shapefile included 987 tracts. Since the aggregated data set only included information on 959 tracts, there was missing information on 28 tracts in the shapefile.

A common approach to handle missing data in spatial econometrics is to replace missing values with zeros, which creates a problem when assessing the spatial dependence using global and local Moran's I, as well as when estimating the spatial lag model and spatial error model (SEM). While there is no agreement on how to deal with missing data, several options are available, such as excluding the areas with missing values or interpolating missing values. However, the latter will inevitably introduce spatial autocorrelation to data and may not be suitable for spatial analysis. Therefore, we adopted the former and deleted the 28 tracts without any PHMC respondents.

Measures

Broadly speaking, "food environment" refers to home, community, and media/information environments (Glanz et al. 2005). This paper studies the food environment pertaining to the neighborhood level. Various measurements of the community food environment have emerged in the literature to evaluate its effects on dietary behaviors (McKinnon et al. 2009). Based on a review of 137 articles that contain measures of food environment between 1990 and 2007, McKinnon et al. (2009) summarized a list of food outlets in the four types of community food environments, including food stores, restaurants, schools, and worksites (see Fig. 6.1). While food stores and restaurants are places where people get direct access to food, schools and worksite are also included because they provide students and workers with food through vending machines, cafeteria, etc.

Ideally, a good measure of the food environment would capture all the aspects illustrated in Fig. 6.1. However, to our knowledge, previous research has not been able to include multiple measures of the food environment due to limited data. This study is no exception. We employ three measures of the food environment based on data availability. The first variable, *difficulty in finding fruits and vegetables*, is a subjective measure of the community food environment in general. Specifically, respondents in both 2008 and 2010 PHMC were asked to assess how easy or difficult for them to find fruits and vegetables in the neighborhood on a scale of 1 (very easy) to 4 (very difficult). In contrast to measures of specific food outlets, *difficulty in finding fruits and vegetables* is more of measuring the overall access to healthy food

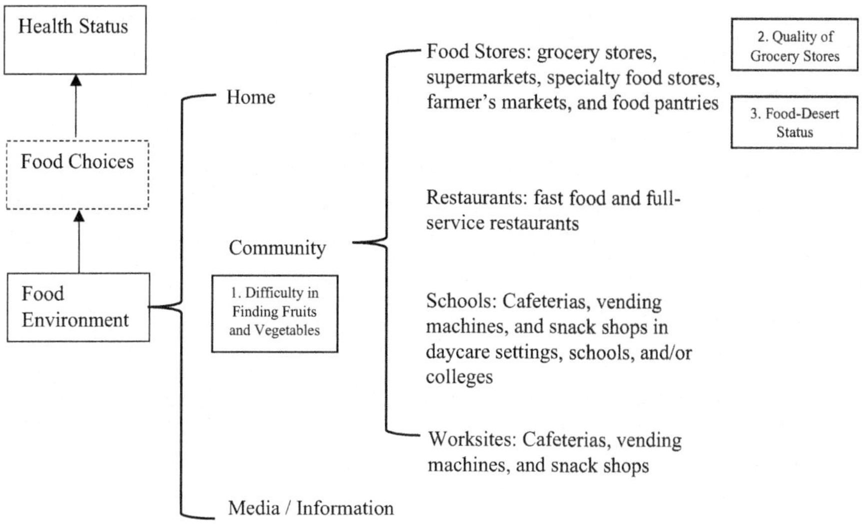

Fig. 6.1 Measures of the food environment (based on McKinnon et al. 2009) and the pathway from food environment to health status

in the neighborhood than measuring any particular category (e.g., food stores, restaurants) of the food environment.

The second measure of the neighborhood food environment, *quality of grocery stores*, largely pertains to the food store environment. Respondents were asked to rate the overall quality of groceries available in the neighborhood on a scale of 1 (excellent) to 4 (poor). It should be noted that there is no explicit definition for "neighborhood" in the questionnaire. This means that respondents answered these two questions according to their own perceptions of the neighborhood, which may or may not correspond to tract boundaries.

The third measure, *a dummy variable for food-desert tracts*, is also a measure for the food store environment. The term "food desert" was introduced by the United Kingdom Nutrition Task Force's Low Income Project Team to describe "areas of relative exclusion where people experience physical and economic barriers to accessing healthy food" (Reisig and Hobbiss 2000). Furthermore, strategic action plans for the government were put forward to mitigate the barriers.

A food-desert tract in this paper was defined by the Healthy Food Financing Initiative Working Group as "a low-income tract where a substantial number or share of residents has low access to a supermarket or large grocery store" in the USDA 2006 estimates for food desert census tracts in the United States. Although the 2010 estimates for food desert census tracts were also available, the 2006 data set was adopted because it used the 2000 census tract geography. The numbers and percentages of people who had low access to a supermarket or a large grocery store were only provided for those tracts that meet the food desert criteria. Therefore, we created a dummy variable indicating whether a census tract was a food desert.

Difficulty in finding fruit and vegetables and quality of grocery stores in the neighborhood are both subjective measurements of the food environment with vague neighborhood boundaries. In contrast, the food-desert dummy variable is based on a directory of supermarkets and large grocery stores within tract boundaries. As such, the food desert dummy variable is a relatively objective measure and is introduced to validate the results from perceived food environment. The food-desert variable also represents tracts with extreme food deficiency due to lack of access to supermarkets and large grocery stores, and thus it can be used to examine the impact of deprived neighborhoods.

The dependent variable, *self-rated health (SRH)* was defined as the percentage of participants reporting poor or fair health in a tract. Aside from the three measures of food environment above, all regressions in this paper included the following controls: neighborhood social capital, health behaviors, average age, educational attainment, percent of people married or living with a partner, racial composition, the percentage of males, the percentage of people who are living below the federal poverty line, and a dummy variable for the tracts in the county of Philadelphia. In particular, the "Willingness to Help" variable measured how likely people in the neighborhood were willing to help their neighbors with routine activities on a scale of 1 (always) to 5 (never). The "% Collective Improvement" was the percentage of respondents who indicated that people in their neighborhood had worked together to improve the neighborhood. "Sense of Belonging" and "Trust" variables were used to denote the degrees to which respondents agreed with the statements—"I feel that I belong and am a part of my neighborhood" and "Most people in my neighborhood can be trusted" respectively, on a scale of 1(strongly agree) to 4 (strongly disagree). Respondents were also asked to provide the number of servings of fruits and vegetables they consumed on a typical day. "Physical Activities" was a variable indicating the number of times per week respondents engaged in any form of physical activities that lasted for at least one-half hour in the month prior to the survey month, ranging from 0 (none) to 4 (more than 3 days per week). The "Smoking Intensity" variable was used to measure the respondents' intensity of smoking at the time of the survey, ranging from 1 (every day) to 4 (never smoked).

Statistical Methods

To examine if spatial autocorrelation exists in SRH (Hypothesis 1), we calculated the Moran's I for SRH in our study region (Anselin 1996). Local Moran statistics (Anselin 1995) were presented in the form of the Local Index of Spatial Association (LISA) map.

To test Hypotheses 2 and 3, we used three modeling approaches in this study: OLS, a spatial lag model, and a spatial error model (SEM). They are presented as follows.

(1) OLS model: $Y = X\beta + \varepsilon$
(2) Spatial Lag model: $Y = \rho WY + X\beta + \varepsilon$
(3) SEM model: $Y = X\beta + \varepsilon$ and $\varepsilon = \lambda W\varepsilon + u$

Y is an n-by-1 vector of dependent variables. X is an n-by-n design matrix of independent variables. β is an n-by-1 vector of regression coefficients. W is an n-by-n spatial weight matrix. WY denotes the spatially lagged variable, which is the weighted average of the neighboring response variables. $W\varepsilon$ is a spatial lag of the error term. ε and u are n-by-1 vectors of error terms. ρ and λ are the autoregressive coefficients for the lagged dependent variable and the lagged error term, respectively.

We employed a decision process recommended by (Anselin 2005) to determine the best model for analyzing the data. First, we conducted an OLS model and calculated the diagnostic statistics for spatial dependence. The Lagrange Multiplier (LM) test and robust LM test were used to diagnose spatial lag and/or spatial error dependence, based on the OLS model. If none of the diagnostic tests were significant at a significance level of 0.05, the OLS model would be chosen, and no spatial regressions would be performed. If one of the diagnostic tests was significant, the corresponding model would be used. That is, if only the LM and robust LM test for spatial lag dependence were significant, the spatial lag model would be used. In addition, we used R^2, log-likelihood value, Akaike information criterion (AIC), and Bayesian Information Criterion (BIC) to compare the fit of the three types of models. We also conducted likelihood ratio tests for spatial lag and/or spatial error dependence in respective models.

Results

Descriptive Statistics

Table 6.1 provides the descriptive statistics, as well as Moran's I, for all variables in the model. Except for the percentage of people who are Native Americans, the Moran's I values were statistically significant based on 999 random permutations for all the variables. Variables with a large Moran's I (greater than 0.5) included quality of grocery stores, neighborhood trust level, percentage of people married or living with a partner, percentages of White or Black population, percentages of people who are below 100% of the federal poverty line, and the dummy variable for tracts in Philadelphia County, suggesting spatial clustering in these aspects. Except for the dummy variable for tracts in Philadelphia County (which were spatially autocorrelated by definition), the percentages of White or Black population had the highest Moran's Is, with values equal to 0.813, and 0.804, respectively.

The Moran's I results (see Figs. 6.2, 6.3, and 6.4) suggested that there was a positive and statistically significant spatial autocorrelation (Global Moran's I = 0.381) for the dependent variable. That is, tracts with similar percentages of

Table 6.1 Descriptive statistics

Variables	N	Mean	Std Dev	Minimum	Maximum	Moran's I	
Number of Respondents in each Tract	959	20.811	11.978	1.00	110.00	0.236	***
% People with Poor/Fair Health	959	19.963	14.357	0.00	100.00	0.381	***
Food Environment							
Difficulty in Finding Fruits and Vegetables (1–4 = Very easy to very difficult)	959	1.467	0.254	1.00	3.00	0.369	***
Quality of Grocery Stores (1–4 = Excellent to Poor)	959	1.748	0.387	1.00	4.00	0.581	***
Food-desert Tract	959	0.034	0.182	0.00	1.00	0.260	***
Neighborhood Social Capital							
Willingness to Help (1–5 = Always to Never)	959	2.321	0.397	1.00	5.00	0.221	***
% Collective Improvement	959	61.384	19.070	0.00	100.00	0.379	***
Sense of Belonging (1–4 = Strong agree to Strongly disagree)	958	1.824	0.222	1.00	4.00	0.118	***
Trust (1–4 = Strongly Agree to Strongly Disagree)	958	1.987	0.352	1.00	4.00	0.572	***
Health Behaviors							
# Servings of Fruit and Vegetables	958	2.838	0.632	0.00	6.80	0.260	***
Physical Activities (0–4 = None to more than 3 days per week)	959	2.647	0.424	0.00	4.00	0.104	***
Smoking Intensity (1–4 = Everyday to never smoked)	959	3.178	0.359	1.00	4.00	0.250	***
Socio-demographic Characteristics							
Average Age	959	52.755	5.409	28.00	79.00	0.177	***
% High School Graduate	959	31.622	17.432	0.00	100.00	0.394	***
% Some College	959	20.073	11.903	0.00	100.00	0.067	***
% College Graduate	959	24.338	14.647	0.00	100.00	0.349	***
% Post-College	959	16.540	14.565	0.00	100.00	0.442	***
% Married or Living with a Partner	959	55.101	19.979	0.00	100.00	0.523	***
% White (not Latino)	958	69.132	33.484	0.00	100.00	0.813	***
% Black (not Latino)	958	21.843	30.069	0.00	100.00	0.804	***
% Latino	958	5.279	10.517	0.00	100.00	0.481	***

(continued)

Table 6.1 (continued)

Variables	N	Mean	Std Dev	Minimum	Maximum	Moran's I	
% Asian	958	1.711	3.873	0.00	33.33	0.081	**
% Biracial or Multiracial	958	1.287	3.095	0.00	33.33	0.087	**
% Native American	958	0.411	1.476	0.00	12.50	0.005	
% Below 100% of Federal Poverty Level	959	8.698	12.076	0.00	100.00	0.504	***
% Male	958	33.243	13.070	0.00	85.71	0.053	***
Dummy for Philadelphia County	959	0.377	0.485	0.00	1.00	0.929	***

$p < 0.001$***, $p < 0.01$**

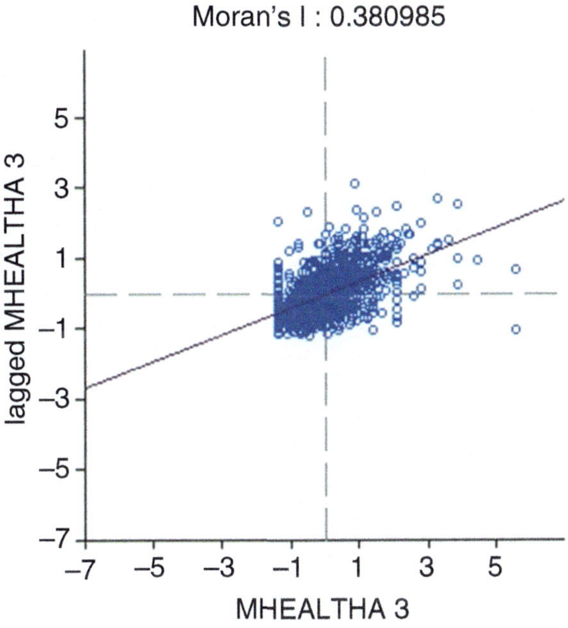

Fig. 6.2 Moran's I for % poor/fair health

the population reporting poor/fair health tended to be geographically nearby. The LISA map (Fig. 6.5) indicated that high percentages of poor/fair health tended to cluster in the county of Philadelphia, whereas low percentages were found to be clustering in Bucks and Chester. Therefore, we found supporting evidence for Hypothesis One.

The interactions between the food environment and SRH are visualized in three maps (see Figs. 6.5, 6.6, and 6.7). The green polygons in the three maps denote percentages of people who rated their health as poor or fair in each tract, and the red polygons in Figs. 6.5 and 6.6 represent the degree of difficulty in finding fruits and vegetables, and the quality of grocery stores in the neighborhood, respectively.

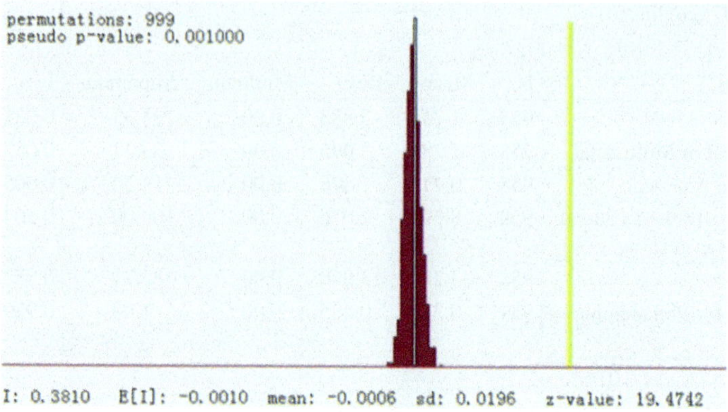

Fig. 6.3 Permutation empirical distribution for Moran's I

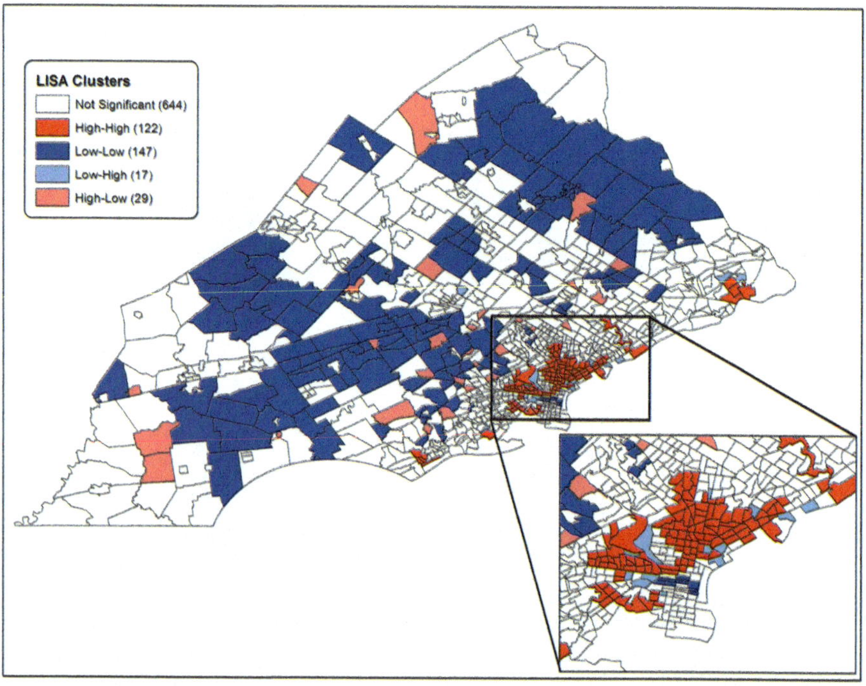

Fig. 6.4 LISA cluster map for % poor/fair health

As seen in Fig. 6.5, deeper green areas are associated with darker red areas, indicating that higher percentages of poor or fair health are found in areas where it is more difficult to find fruits and vegetables (Pearson correlation coefficient [ρ] = 0.498). Tracts with lower quality of grocery stores are associated with higher

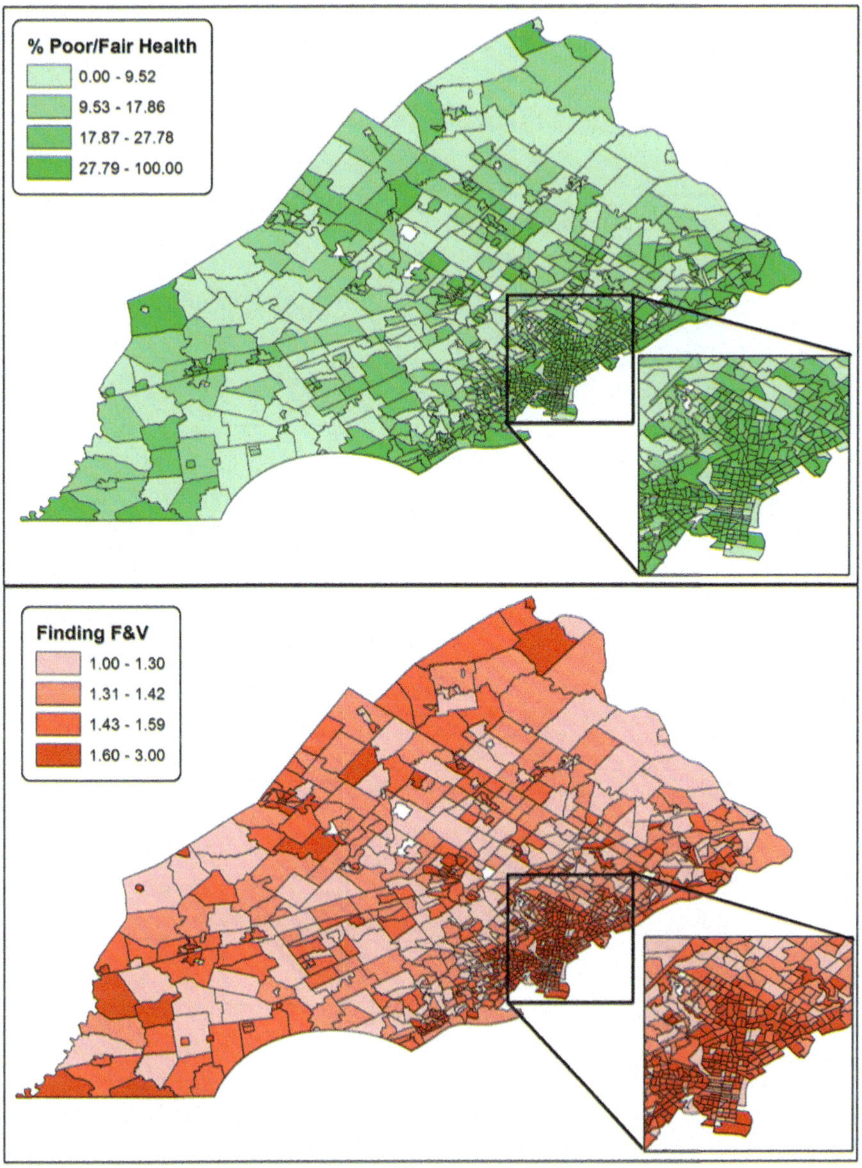

Fig. 6.5 Percentage poor/fair health vs. difficulty in finding fruits and vegetables

percentages of poor or fair health ($\rho = 0.528$). The red polygons in Fig. 6.7 simply denote food-desert tracts. Since there are only 33 food-desert tracts in the five counties, most of the tracts are blank. Still, it can be seen that the percentages of poor or fair health are very high in food deserts ($\rho = 0.168$).

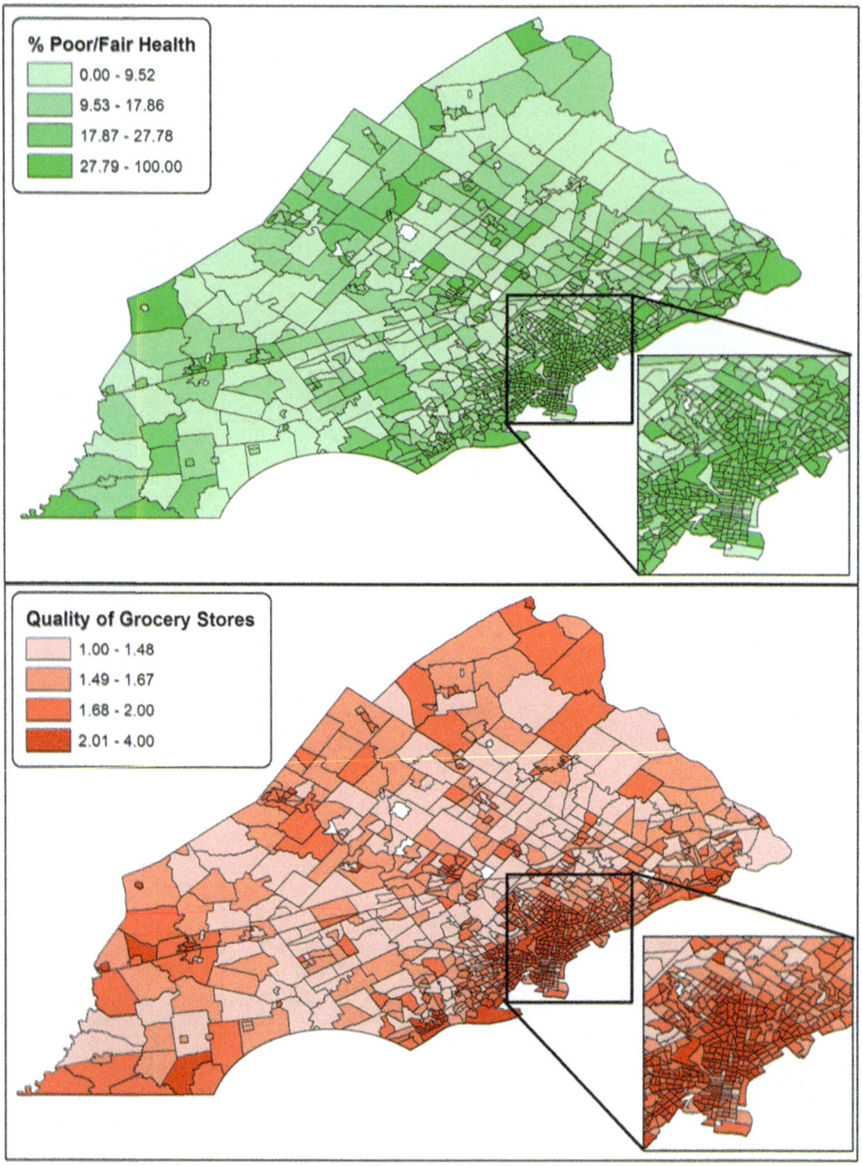

Fig. 6.6 Percentage poor/fair health vs. quality of grocery stores

Regression Results

Table 6.2 provides the results from the OLS, spatial lag, and SEM models with percentages of the population with poor or fair health as the dependent variable in all models. The Moran's I for the residuals of the OLS model was marginally significant

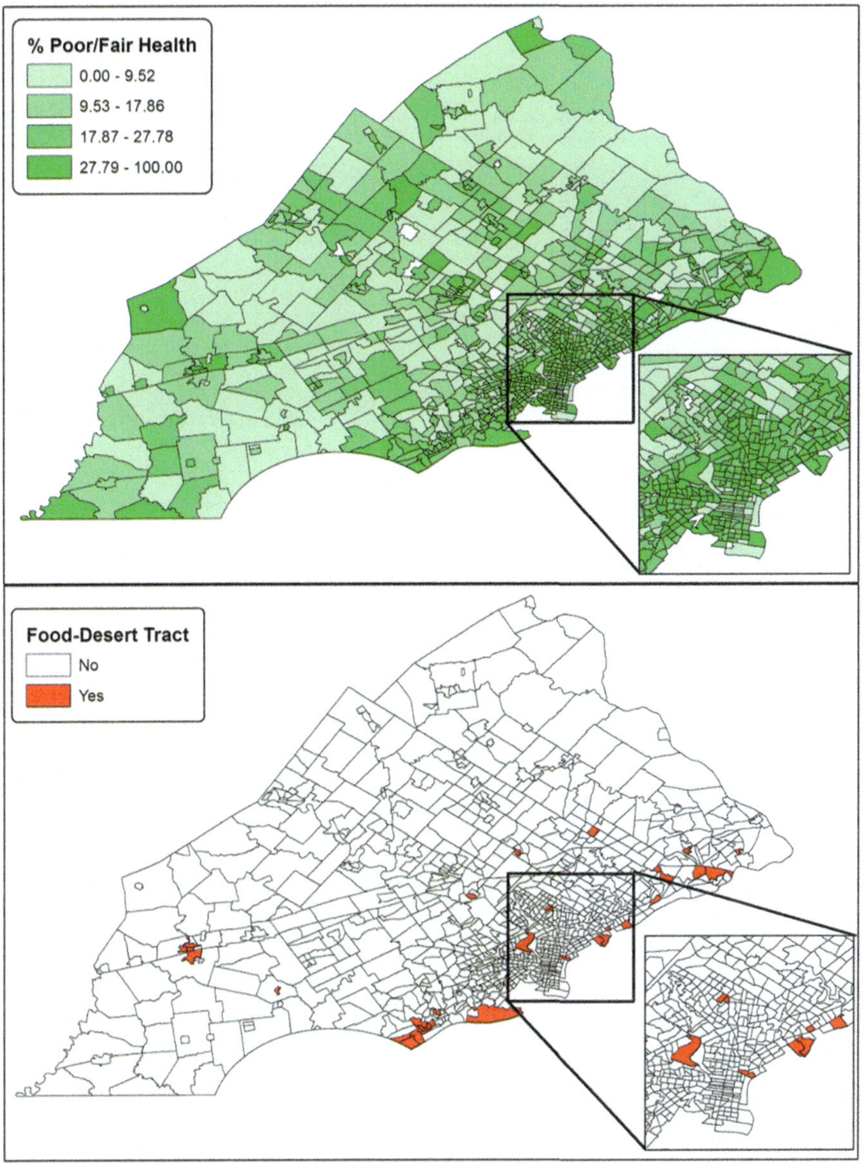

Fig. 6.7 Percentage poor/fair health vs. food-desert tracts

($p < 0.1$), implying the existence of spatial dependence in the errors after adjustment for all the independent variables in the regression. As such, the effects of the independent variables in the OLS model might have been underestimated or overestimated.

Table 6.2 Regression results (dependent variable: % poor/fair health)

	OLS				Spatial Lag Model				SEM		
	Coef.	T-stat			Coef.	T-stat			Coef.	T-stat	
Food Environment											
Difficulty in Finding Fruits and Vegetables(1–4 = Very easy to very difficult)	4.491	2.49	*		4.481	2.52	*		4.514	2.53	*
Quality of Grocery Stores (1–4 = Excellent to Poor)	1.259	0.88			0.943	0.67			1.216	0.86	
Food-desert Tract	6.287	3.44	***		6.130	3.40	***		6.341	3.46	***
Neighborhood Social Capital											
Willingness to Help (1–5 = Always to Never)	0.337	0.33			0.327	0.33			0.332	0.33	
% Collective Improvement	−0.002	−0.11			−0.005	−0.24			−0.002	−0.09	
Sense of Belonging (1–4 = Strong agree to Strongly disagree)	0.306	2.58	**		0.322	2.76	**		0.310	2.66	**
Trust (1–4 = Strongly Agree to Strongly Disagree)	5.499	3.32	***		5.259	3.22	**		5.415	3.30	***
Health Behaviors											
# Servings of Fruit and Vegetables	0.292	0.47			0.303	0.50			0.271	0.44	
Physical Activities(1–4 = Less than once a week to more than 3 days per week)	−3.672	−4.39	***		−3.622	−4.40	***		−3.553	−4.31	***
Smoking Intensity (1–4 = Everyday to never smoked)	−3.290	−3.07	**		−3.070	−2.91	**		−3.111	−2.94	**
Socio-demographic Characteristics											
Average Age	0.261	3.85	***		0.267	4.01	***		0.261	3.90	***
% High School Graduate	−0.151	−3.27	**		−0.143	−3.15	***		−0.150	−3.29	**
% Some College	−0.176	−3.59	***		−0.166	−3.41	***		−0.178	−3.67	***
% College Graduate	−0.256	−5.19	***		−0.242	−4.92	***		−0.258	−5.28	***
% Post-College	−0.243	−4.82	***		−0.223	−4.43	***		−0.241	−4.84	***
% Married or Living with a Partner	0.015	0.61			0.014	0.61			0.012	0.51	
% Black (not Latino)	0.017	0.92			0.014	0.75			0.018	0.93	
% Latino	0.070	1.82	†		0.063	1.65	†		0.070	1.83	†

% Asian	−0.157	−1.90	†		−0.159	−1.96	†		−0.157	−1.92	†
% Biracial or Multiracial	0.278	2.83	**		0.279	2.88	**		0.279	2.88	**
% Native American	0.496	3.50	***		0.518	3.72	***		0.504	3.61	***
% Below 100% of Federal Poverty Level	0.223	5.22	***		0.217	5.15	***		0.220	5.21	***
% Male	0.048	1.88	†		0.051	2.03	*		0.049	1.94	†
Dummy for Philadelphia County	1.403	1.43			1.009	1.03			1.398	1.37	
Constant	16.916	1.85	†		14.144	1.56			16.459	1.83	†
Spatially Lagged Dependent Variable (Rho)					**0.095**	**2.19**	*				
Lamda									0.078	1.48	
	MI/DF	Value	P		DF	Value	P		DF	Value	P
Moran's I (error)	0.028	1.72	0.085	LR Test	1	4.81	0.028	LR Test	1	2.09	0.149
Lagrange Multiplier (lag)	1	5.03	0.025								
Robust LM (lag)	1	3.46	0.063								
Lagrange Multiplier (error)	1	2.07	0.150								
Robust LM (error)	1	0.51	0.476								
Lagrange Multiplier (SARMA)	2	5.53	0.063								
R^2	0.5459				0.5489				0.5474		
Log likelihood	−3536.71				−3534.31				−3535.67		
AIC	7123.43				7120.62				7121.34		
BIC	7245.08				7247.13				7242.99		

$p < 0.1$†, $p < 0.05$*, $p < 0.01$**, $p < 0.001$***

A common remedial approach for spatially dependent errors is to expand the OLS model into spatial regression models. Following Anselin (2005), the Lagrange Multiplier (LM) tests suggested that a spatial lag model fitted our data better than an SEM. We implemented a spatial lag model and compared its results with those from an SEM to further confirm that a spatial lag model outperformed SEM for our data. Specifically, Lamda in the SEM was not significant, which confirmed the LM tests in the OLS model, so it was unlikely that the observations of the dependent variable were correlated across distances because of omitted variables in the error terms. Furthermore, the spatial lag model yielded the highest R^2 and log-likelihood value, and the lowest AIC, suggesting that the spatial lag model had the best fit of all the three models. Therefore, regression results from the spatial lag model should be adopted for our data.

The coefficient for the spatial lag of the dependent variable was small (0.095) but significant ($p < 0.05$), indicating a possible feedback process between health statuses in neighboring tracts. Therefore, our second hypothesis was confirmed.

The spatial lag model indicated that two of the three food environment variables, including difficulty in finding fruits and vegetables and food-desert tracts, were significant, but the effect of grocery stores was not significant. The signs of all the significant predictors followed expectations in the spatial lag model. A worse food environment was associated with a higher percentage of poor or fair health. Living in food-desert tracts was particularly unfavorable for SRH. With other variables held constant, the percentage of the population with poor or fair SRH in food-desert tracts, on average, was about 6% higher than that in a non-food-desert tracts. However, the average number of servings in fruits and vegetables per day was not a significant predictor for the percentage of poor or fair health. Lower levels in the sense of belonging and trust in the neighborhood were also related to higher percentages of poor or fair health. Participating in more physical activities and smoking less intensively tended to benefit the SRH. The percentage of poor or fair health was positively related to the average age of the population, percentages of minorities, percentages of the male population, and the percentages of people who were below the poverty line in a tract. However, it was negatively associated with educational attainment and the percentage of the Asian population.

Discussions

This study investigated the relationship between the food environment and SRH while taking into spatial autocorrelation in the percentage of people with poor or fair health in a census tract. The Moran's I and LISA map results indicated that SRH were spatially autocorrelated, meaning that tracts with a high concentration of poor/fair health respondents were geographically close. Our finding echoes previous research on the geographical disparities in health outcomes (Dwyer-Lindgren et al. 2017; Fowler et al. 2019). Therefore, we confirmed our first hypothesis, which led to our exploration of the mechanism behind this observation. Past studies have documented the correlation between geographical disparities in health outcomes and geographical differences in health risk behaviors (Dwyer-Lindgren et al. 2017).

First, we tested for spatial dependence of four groups of explanatory variables: the food environment, neighborhood social capital, health behaviors, and socio-demographic characteristics. Almost all the variables were spatially autocorrelated except for the percentage of Native Americans. In particular, percentages of White or Black population in a tract were strongly correlated with those of neighboring tracts. This finding supported past evidence that the Philadelphia metropolitan area is one of the most segregated metropolitan areas in the United States for Blacks, although indices for its residential segregation are declining over time (Bureau of Census 2002). In contrast to Bucks, Chester, Delaware, Montgomery, the County of Philadelphia had a much more significant presence of Black population, a higher proportion of people living below the poverty line, a more inferior food environment, less social capital, and a higher percentage of people with poor or fair health. The agglomeration of deprived neighborhoods with high percentages of people below the poverty line, poor quality of grocery stores, and high neighborhood distrust level might all contribute to the clustering of people who have poor or fair health in those neighborhoods.

Second, we evaluated the impact of the four groups of explanatory variables, especially the neighborhood food environment, on the percentage of the population reporting poor or fair health. Using a data-driven approach, we found that a spatial lag model fitted our data best. In particular, a higher proportion of poor or fair health was significantly correlated with that for the neighboring communities, even after adjusting for the covariates in the model. Therefore, we found supporting evidence that the spatial dependence of SRH might have been partly driven by a direct feedback or diffusion process of health outcomes (Hypothesis Two), which was not accounted for in the four groups of explanatory variables. Past studies have found strong evidence for this diffusion process among influenza-related health outcomes (Crighton et al. 2007; Greene et al. 2006). This process might be facilitated by the similarity in the food environment, social capital, health behaviors, and socio-demographic characteristics in the spatial clusters.

Past studies have documented the geographical inequalities in healthy food access and found spatial clusters of low food access (Kolak et al. 2018; Walker et al. 2010). Results from the spatial lag model agreed with past findings on the impact of the disparities in food access on health outcomes. In contrast to previous studies that primarily assessed the impact of one aspect of neighborhood food environment (e.g., food stores) on health outcomes (Brown et al. 2008; Ford and Dzewaltowski 2011; Morland et al. 2006; Powell et al. 2007), this study adopted multiple measures of the neighborhood food environment. Difficulty in finding fruits and vegetables, in particular, is a global measure, implicitly evaluating various aspects including food stores, restaurants, schools, and worksites of the neighborhood food environment. Both difficulty in finding fruits and vegetables and food desert status were found to be significantly associated with the percentage of poor or fair health in each tract.

The results for the socio-demographic variables were mostly consistent with the results from Yang and Matthews (2012), based on individual-level data of the elderly population in 2008 PHMC. The percentage of the population with poor or fair health was significantly lower in communities where people had a stronger sense of belonging and a higher level of trust and engaged in more physical activities but less smoking intensity. Surprisingly, the coefficients for the percentage of Blacks and the dummy variable for the tracts in Philadelphia County were not significant after controlling for other socio-demographic variables. One plausible explanation is that the geographical disparities in health could be largely attributed to other variables in the analysis, such as differences in neighborhood affluence, food environment, social capital, and health behaviors. Therefore, we also found supporting evidence for our third hypothesis: the spatial dependence of SRH might be partially explained by the spatial clustering of groups of people who shared similar food environments, social capital, health behaviors, and socio-demographic characteristics.

Though one contribution of this study is to provide ecological evidence for the relationship between food environment and SRH by aggregating individual-level data to census tracts, this data aggregation approach naturally constitutes one limitation due to the loss of individual variability. The modifiable area unit problem (MAUP) arise when micro-data in small units are aggregated into larger areas, since the constructed variables may possess distinctive statistical properties that affect the signs, magnitudes and significance levels of regression coefficients (Fotheringham and Wong 1991; Mobley and Andrews 2008; Openshaw 1984). There is no solution to MAUP, but future research should attempt to examine the findings in this study with more detailed individual-level data and well-defined neighborhood definitions. Since the food environment is related to dietary behaviors, which in turn affect health outcomes, the impacts of the food environment on health outcomes are likely to be mediated by dietary behaviors. In other words, the availability of healthy foods in the neighborhood does not guarantee healthy eating habits. It will still be interesting to test this pathway using a structural equation modeling approach in future research.

Some policy implications can be drawn from the findings. First, as the food environment in a neighborhood is associated with SRH, enhancing access to fruits and vegetables may help improve the overall population health. Second, given the strong relationship between food-desert and SRH, it is crucial to focus on the population health in these food-desert tracts and to improve their food environment (e.g., easy access to quality and inexpensive food). Finally, the importance of neighborhood social capital indicates that, beyond the food environment, developing an environment where residents can easily interact with one another will benefit overall population health.

Conclusion

While many studies have investigated the socioeconomic inequalities in health, the geographical disparities in SRH were understudied. This study filled this research gap by investigating the spatial dependence of SRH, and how it was related to the disparities in the neighborhood environment, especially the food environment.

We found strong evidence for the spatial clustering of SRH and neighborhood characteristics. Using a data-driven approach, we employed a spatial lag model to study the impact of the neighborhood food environment on SRH, while controlling for neighborhood social capital, health behaviors, and socio-demographic characteristics. The coefficient for the spatial lag of SRH in the model was statistically significant, indicating a diffusion process between people living nearby. Results from the spatial lag model also indicated that the geographical disparities in SRH could be partially attributed to the spatial disparities in neighborhood characteristics. Specifically, the percentage of poor or fair health was higher in a food desert and in places where it was challenging to obtain fruits and vegetables. Possessing a higher level of social capital and engaging in healthy behaviors were also associated with a lower percentage of poor or fair health in a tract. Therefore, it is imperative to enhance the built food environment and foster social interaction and healthy behaviors in a community.

References

Anselin, L. (1995). Local indicators of spatial association—LISA. *Geographical Analysis, 27*(2), 93–115.

Anselin, L. (1996). The Moran scatterplot as an ESDA tool to assess local instability in spatial association. In M. M. Fischer, H. J. Scholten, & D. J. Unwin (Eds.), *Spatial analytical perspectives on GIS* (pp. 111–125). Boca Raton: CRC Press.

Anselin, L. (2005). *Exploring spatial data with GeoDa: a workbook, revised version*. Urbana: University of Illinois, Urbana-Champaign.

Brown, A. F., Vargas, R. B., Ang, A., & Pebley, A. R. (2008). The neighborhood food resource environment and the health of residents with chronic conditions. *Journal of General Internal Medicine, 23*(8), 1137–1144.

Bureau of Census. (2002). Racial and ethnic residential segregation in the U.S.: 1980–2000. http://www.census.gov/hhes/www/housing/housing_patterns/pdftoc.html

Chi, G., & Zhu, J. (2008). Spatial regression models for demographic analysis. *Population Research and Policy Review, 27*(1), 17–42.

Clifton, K. J. (2004). Mobility strategies and food shopping for low-income families – A case study. *Journal of Planning Education and Research, 23*(4), 402–413.

Crighton, E., Elliott, S., Moineddin, R., Kanaroglou, P., & Upshur, R. (2007). An exploratory spatial analysis of pneumonia and influenza hospitalizations in Ontario by age and gender. *Epidemiology and Infection, 135*(2), 253–261.

Dwyer-Lindgren, L., Mackenbach, J. P., van Lenthe, F. J., & Mokdad, A. H. (2017). Self-reported general health, physical distress, mental distress, and activity limitation by US county, 1995-2012. *Population Health Metrics, 15*(1), 16.

Ford, P. B., & Dzewaltowski, D. A. (2011). Neighborhood deprivation, supermarket availability, and BMI in low-income women: a multilevel analysis. *Journal of Community Health, 36*(5), 785–796.

Ford, E. S., & Mokdad, A. H. (2001). Fruit and vegetable consumption and diabetes mellitus incidence among US adults. *Preventive Medicine, 32*(1), 33–39.

Fotheringham, A. S., & Wong, D. W. S. (1991). The modifiable areal unit problem in multivariate statistical analysis. *Environment and Planning A, 23*(7), 1025–1044.

Fowler, M. E., DiBlasio, C., Crowe, M., & Kennedy, R. E. (2019). Geographic disparities in county-level prevalence of Alzheimer's disease across the United States. *Innovation in Aging, 3* (Suppl 1), S782.

Glanz, K., Sallis, J. F., Saelens, B. E., & Frank, L. D. (2005). Healthy nutrition environments: concepts and measures. *American Journal of Health Promotion, 19*(5), 330–333.

Greene, S. K., Ionides, E. L., & Wilson, M. L. (2006). Patterns of influenza-associated mortality among US elderly by geographic region and virus subtype, 1968–1998. *American Journal of Epidemiology, 163*(4), 316–326.

Hanibuchi, T., Kondo, K., Nakaya, T., Nakade, M., Ojima, T., Hirai, H., & Kawachi, I. (2011). Neighborhood food environment and body mass index among Japanese older adults: Results from the Aichi Gerontological Evaluation Study (AGES). *International Journal of Health Geographics, 10*(1), 1–9.

He, K., Hu, F., Colditz, G., Manson, J., Willett, W., & Liu, S. (2004). Changes in intake of fruits and vegetables in relation to risk of obesity and weight gain among middle-aged women. *International Journal of Obesity, 28*(12), 1569–1574.

Inagami, S., Cohen, D. A., Finch, B. K., & Asch, S. M. (2006). You are where you shop. *American Journal of Preventive Medicine, 31*(1), 10–17.

Kolak, M., Bradley, M., Block, D. R., Pool, L., Garg, G., Toman, C. K., et al. (2018). Urban foodscape trends: disparities in healthy food access in Chicago, 2007–2014. *Health & Place, 52*, 231–239.

Laraia, B. A., Siega-Riz, A. M., Kaufman, J. S., & Jones, S. J. (2004). Proximity of supermarkets is positively associated with diet quality index for pregnancy. *Preventive Medicine, 39*(5), 869–875.

Larson, N. I., Story, M. T., & Nelson, M. C. (2009). Neighborhood environments: disparities in access to healthy foods in the US. *American Journal of Preventive Medicine, 36*(1), 74–81.

McKinnon, R. A., Reedy, J., Morrissette, M. A., Lytle, L. A., & Yaroch, A. L. (2009). Measures of the food environment: A compilation of the literature, 1990–2007. *American Journal of Preventive Medicine, 36*(4), S124–S133.

Mobley, L. R., & Andrews, L. (2008). How sensitive are multilevel regression findings to defined area of context? A case study of mammography use in California. *Medical Care Research and Review, 65*(3), 315–337.

Moore, L. V., & Diez Roux, A. V. (2006). Associations of neighborhood characteristics with the location and type of food stores. *American Journal of Public Health, 96*(2), 325–331.

Morland, K., & Evenson, K. R. (2009). Obesity prevalence and the local food environment. *Health & Place, 15*(2), 491–495.

Morland, K., Wing, S., & Roux, A. D. (2002). The contextual effect of the local food environment on residents' diets: the atherosclerosis risk in communities study. *American Journal of Public Health, 92*(11), 1761–1768.

Morland, K., Diez Roux, A. V., & Wing, S. (2006). Supermarkets, other food stores, and obesity: the atherosclerosis risk in communities study. *American Journal of Preventive Medicine, 30*(4), 333–339.

Openshaw, S. (1984). *The modifiable areal unit problem*. Norwich: Elsevier Science Geo Abstracts.

Powell, L. M., Auld, M. C., Chaloupka, F. J., O'Malley, P. M., & Johnston, L. D. (2007). Associations between access to food stores and adolescent body mass index. *American Journal of Preventive Medicine, 33*(4), S301–S307.

Reisig, V., & Hobbiss, A. (2000). Food deserts and how to tackle them: A study of one city's approach. *Health Education Journal, 59*(2), 137–149.

Schuurman, N., Peters, P. A., & Oliver, L. N. (2009). Are obesity and physical activity clustered? A spatial analysis linked to residential density. *Obesity, 17*(12), 2202–2209.

Steinmetz, K. A., & Potter, J. D. (1996). Vegetables, fruit, and cancer prevention: A review. *Journal of the American Dietetic Association, 96*(10), 1027.

Voss, P. R., Long, D. D., Hammer, R. B., & Friedman, S. (2006). County child poverty rates in the US: A spatial regression approach. *Population Research and Policy Review, 25*(4), 369–391.

Walker, R. E., Keane, C. R., & Burke, J. G. (2010). Disparities and access to healthy food in the United States: A review of food deserts literature. *Health & Place, 16*(5), 876–884.
Yang, T.-C., & Matthews, S. A. (2012). Understanding the non-stationary associations between distrust of the health care system, health conditions, and self-rated health in the elderly: A geographically weighted regression approach. *Health & Place, 18*(3), 576–585.
Zenk, S. N., Schulz, A. J., Hollis-Neely, T., Campbell, R. T., Holmes, N., Watkins, G., et al. (2005). Fruit and vegetable intake in African Americans income and store characteristics. *American Journal of Preventive Medicine, 29*(1), 1–9.

Chapter 7
Constructing Life Tables from the Kaiser Permanente Smoking Study and Applying the Results to the Population of the United States

David A. Swanson, Simeon Chow, and Tom Bryan

Introduction

Abelin (1965) used a set of studies on smoking and health that relied on mortality rates and ratios as a jumping off point to illustrate the advantages of using survivorship and mortality measures based on life tables.[1] Specifically, Abelin (1965) converted age-specific mortality rates (deaths divided by number of person-years of exposure by age groups and smoking categories) found in three studies, Dorn (1959), Hammond (1964), and Hammond and Horn (1958) into the probabilities of dying using a standard actuarial formula:

[1]To our knowledge, adequate life tables have not been constructed from any of the major cohort studies other than the work in this paper, which uses data from the KP Smoking Study. The other major cohort studies include Hammond and Horn (1958), CPS I, CPS II and CPS-3 (American Cancer Society n.d.), the 16 cohorts in the seven country study (https://www.sevencountriesstudy.com/home/). There have, however, been studies that attempt to construct approximate life tables from other data sources, including Cancer Research UK Cancer Survival Group (2009), Ederer Axtell and Cutler (1961) and Ellis Coleman, and Rachet (2014). Blakely and Wilson (2005) point out shortcomings in some of these approximations.

D. A. Swanson (✉)
Edward A. Dickson Emeritus Professor, Department of Sociology, University of California Riverside, Riverside, CA, USA
e-mail: dswanson@ucr.edu

S. Chow
Independent Management Consultant, Richmond, VA, USA

T. Bryan
BryanGeoDemo, Richmond, VA, USA

B. Jivetti, Md. N. Hoque (eds.), *Population Change and Public Policy*, Applied Demography Series 11, https://doi.org/10.1007/978-3-030-57069-9_7

$$_{n}q_{x} = [({}_{n}m_{x})(n)]/[1 + (0.5)(n)({}_{n}m_{x})] \quad (7.1)$$

where

x = the beginning of an age group
n = the width of the age group in question
m = deaths per person-year
and q = the probability of dying between age x and age x

If n = 1, then Eq. (7.1) becomes

$$q_{x} = (m_{x})/[1 + (0.5)(m_{x})] \quad (7.2)$$

As an example of using this formula to convert an age-specific mortality rate to the probability that those alive at the start of the age group (x) will die before reaching the end of it (n), suppose that deaths per-person year for those aged 60–64 is 0.015. Using Eq. (7.1), we find that the for those who reach their 60th birthday, the probability of dying before reaching the 65th birthday is 0.0723, where 0.0723 = (0.015*5)/[1 + (.5*5*0.015)]. Once the probability of dying is estimated for a set of age groups, a complete life table can be generated (Keyfitz 1970; Kintner 2004), which provides a range of mortality measures, notably: (1) the probability of surviving from birthday to birthday as well as from age group to age group; and (2) expected years of life remaining at the start of any given age group. Life tables can be constructed for a wide range of variables – gender, race, ethnicity, and smoking status, for example.

Following in Abelin's (1965) pioneering footsteps, we construct life tables from cohort mortality data widely employed in efforts to examine smoking and health, namely the Kaiser Permanente Smoking Study (Friedman et al. 1997). The mortality data in this study have been used in terms of relative mortality and risk rates as we shortly show. However, they have never been used to generate life tables.[2] Before moving to this topic, we first provide a brief description of the KP Smoking Study taken from Friedman et al. (1997).

Between 1979 and 1986, the Kaiser Permanente Medical Care Program cohort study obtained detailed information on smoking habits from more than 60,000 subjects, age 35 and older, following up through 1987 on mortality. Based in the San Francisco Bay area of California, the study population was racially, ethnically and socioeconomically diverse. All of the subjects were members of Northern California Kaiser-Permanente HMO, who number about 30% of the population in the areas served. The subjects, most of whom joined Kaiser-Permanente through

[2]Friedman et al. (1997) provided a data set that addresses the mortality of current cigarette smokers, former cigarette smokers, and never smokers at various ages. These data were obtained from a KP Medical Care Program Cohort (KP) study. The study obtained baseline information on more than 60,000 subjects between 1979 and 1986, age 35 years and older and followed the cohort for mortality through 1987. The data set contains information about how long a person in the study has been a cigarette smoker and whether the person had quit smoking and if so how long the person has been a former cigarette smoker. The study subjects logged over 372,000 person-years, with 54% categorized as never smokers, 19% as former smokers and 26% as current smokers.

employer-based health insurance plans, were reflective of the local population during the study period except they all had health insurance, a slightly higher average educational attainment and lacked low and high income extremes. Eighty-three percent of Kaiser Permanente members who had routine health appraisals were asked to complete a detailed questionnaire about their smoking habits for a follow-up study. About 2000 persons who received another form of preventive health examination in San Francisco in 1981 also completed these questionnaires and are included in the total cohort of more than 100,000 persons. Follow-up for mortality used the California Automated Mortality Linkage System program for linkage

We believe that it is worthwhile to convert the mortality rates found in the Kaiser-Permanente (KP) Smoking Study data into life table data because the study has been fairly widely cited. Here are illustrative examples of how these data have been employed: (1) to examine the decline in excess risk of chronic obstructive pulmonary disease following quitting smoking Lee and Forey, 2013); (2) in reviews of the effects of smoking on health (Ernster et al. 2000; Fry et al. 2013; Jacobs et al. 1999; Lee and Forey 2013; Lee et al. 2012; Thun et al. 2000); (3) to assess the potential population health benefit of modified risk tobacco products (Bachand and Sulsky 2013; Bachand et al. 2018, Muhammad-Kah et al. 2016); (4) as a basis for comparing study results (Bach et al. 2003; Bain et al. 2004; De Matteis et al. 2012; Blizzard and Dwyer 2003; Thun et al. 2006); (5) as a reference point in a given study (Hunt et al. 2005; Lee et al. 2015; Levy et al. 2007; Poland and Teischinger 2017; van Iersel et al. 2006).

Virtually all of these studies would have benefited from the use of mortality measures constructed from life tables, especially those aimed at assessing the population health benefit of modified risk tobacco products (Bachand and Sulsky 2013; Bachand et al. 2018; Muhammad-Kah et al. 2016). We make this claim because a life table is, in essence, one form of combining mortality rates of a population at different ages into a single statistical model. A life table is specifically designed to measure the level of mortality of a population of interest (Kintner 2004). In the paper that follows, we first develop life tables from the KP smoking study mortality rates and then extend the life tables by first adjusting them to reflect the US population and then using models to create life tables for a much more comprehensive set of age groups than are found in the KP smoking study.

The remaining paper is organized as follows. After briefly discussing life tables in general, we describe the KP smoking study data, then discuss the method used to generate a set of preliminary life tables from them. Following these descriptions and discussion, we show the preliminary life tables developed from the KP smoking study. We then assess the preliminary life tables and identify inconsistencies and then describe how we minimized these inconsistencies and go on to produce revised life tables that are displayed in terms of survivorship and life expectancy by age. We discuss the methods used to produce these revised life table and then discuss the fact that the KP population has higher life expectancy than its temporally equivalent US population. We then show how the KP life tables can be extended to their temporally equivalent and more recent US populations and include a discussion of the strengths and weaknesses of the KP life tables and the extension of them to US populations over time. The paper concludes with a discussion of the application of the paper's results to the assessment of the net population health.

Life Tables: General Comments and an Example of Research

Life tables represent an important component of demography. Not only do they serve as methodological and conceptual tools (Burch 2018; Swanson and Tedrow 2012; Yusuf et al. 2014), but they support a wide range of both applied work (Abelin 1965; Jiang et al. 2010; Kintner and Swanson 1994; Owen et al. 2018; Preston 1970a, b; Retherford 1972; Richards and Donaldson 2010; Siegel 2002; Smith et al. 2013; Thomas and Bao 2016); and academic work (Swanson and Sanford 2012; Trovato and Lalu 2001; Villavicencio and Riffe 2016; Wrycza 2014; Yashin et al. 2016).

It should not be surprising that life tables have been widely studied and that many of their defining characteristics, such as their internal mathematical equalities, have been described (Kintner 2004). As one of many possible examples, there is a line of research that has examined equalities and inequalities in the life table. In addition to the obvious equalities such as the crude birth rate being equal to the crude death rate, this research has revealed that: (1) mean years lived is equal to mean years remaining, which is known as Carey's Equality Theorem (Vaupel 2009); and (2) the distribution of age composition is equal to the distribution of remaining lifetimes (Carey et al. 2008; Rao and Carey 2014, Vaupel 2009). To these equalities, the following can be added: (1) mean age is equal to mean years lived (Rao and Carey 2014); and (2) mean age is equal to mean years remaining (Kim and Aron 1989), which implies that mean years lived is equal to mean years remaining (Swanson and Tedrow 2019). From this body of work, Swanson and Tedrow (2019) identified two hitherto unknown inequalities. The first is that at any given age, the sum of mean years lived and mean years remaining exceeds life expectancy at birth when age is greater than zero and less than the maximum lifespan. The second inequality applies to the entire population and shows that the sum of mean years lived and mean years remaining exceeds life expectancy at birth.

As an example of an application of this line of research, Vaupel (2009) used Carey's Equality Theorem (Carey et al. 2008; Rao and Carey 2014) and a 2005 US life table to estimate that nearly half of the US population in 2009 will still be alive in 2050. Using the same US 2005 life table, Swanson and Tedrow (2019) found that on average the population lived 40.60 years and will live another 40.60 years on average, or 81.3 years in total, which is 3.67 years more than their life expectancy at birth of 77.63 years.

The KP Smoking Study Data

The mortality data from the KP smoking study we use as input for the initial life tables are found in tables 3, 4, 5, 6, 7, and 8 of the study by Friedman et al. (1997). By gender, these tables provide mortality data for all causes by selected age groups and cigarette smoking status, which are categorized as:

Table 7.1 Female life expectancy by age: total, never and prior smokers by duration since quitting

Smoking status	Age 35[a]	Age 50[a]	Age 65[a]	Age 75[a]
Never smoker	47.92 (37, 45786)	33.41 (118, 49,744)	19.35 (171, 24,159)	10.41 (299, 12,285)
Duration since quitting (years)				
2–10	46.24 (1, 5,493)	31.35 (15, 3,750)	17.83 (15, 1,572)	9.12 (15, 394)
11–19	46.94 (4, 6,027)	32.33 (16, 5,467)	18.45 (21, 2,505)	9.63 (23, 723)
20+	45.60 (2, 1,279)	33.26 (7, 4,405)	18.88 (20, 2,641)	9.98 (27, 852)

[a]Life expectancy references expected years of life remaining at an exact age. The numbers in parentheses shown within each cell represent, respectively, all-cause deaths and person years

Table 7.2 Male life expectancy by age: total, never prior smokers by duration since quitting

Smoking status	Age 35[a]	Age 50[a]	Age 65[a]	Age 75[a]
Never smoker	43.70 (49, 29,916)	29.60 (97, 24,020)	15.98 (161, 11,466)	7.66 (203, 4,486)
Duration since quitting (years)				
2–10	40.65 (12, 5,571)	26.55 (26, 3,625)	13.73 (14, 977)	4.30 (16, 253)
11–19	42.81 (5, 6,210)	28.24 (29, 6,107)	14.77 (52, 2,548)	7.02 (40, 671)
20+	39.32 (3, 1,499)	29.74 (19, 4,670)	16.15 (43, 3,507)	7.62 (67, 1,442)

[a]Life expectancy references expected years of life remaining at an exact age. The numbers in parentheses shown within each cell represent, respectively, all-cause deaths and person years

1. Never
2. Current
   (a) Quantity (<20 cigarettes daily; 20+ daily
   (b) Duration smoking (<20 years; 20–39 Years; 40+ years)
3. Former
   (a) Duration since quitting smoking (2–10 years, 11–20 years, 21+ years)

The KP smoking study provides overall mortality rates for all causes by selected racial groups, gender, and age in Table 7.1, but age groups are not provided, which precludes the construction of life tables from these data. In Table 7.2, mortality rates for all causes are provided by race, gender, and age group, but the small numbers preclude the construction of life tables from these data.

Methods

In constructing the preliminary life tables from the KP smoking study, we employ a different method than that employed by Abelin (1965) to convert age-specific mortality rates into the probabilities of dying using the formula shown as

Eq. (7.1), which assumes that deaths occur evenly throughout a given age interval. Instead, we employ a conversion formula that assumes that deaths occur in increasing numbers within a given age interval, specifically in an exponential manner (Fergany 1971):

$$_{n}q_{x} = 1 - e^{(-n*_{n}m_{x})} \tag{7.3}$$

where, as in Eq. (7.1),

x = the beginning of an age group
n = the width of the age group in question
m = deaths per person-year
q = the probability of dying between age x and age x
and where $e \approx 2.71828$

As noted earlier, once the probability of dying is estimated for a set of age groups, a complete life table can be generated (Kintner 2004). We can start with the number of Person-years lived during a given year by (lx) people alive at the start of that year, which is denoted by (L_x). This can be estimated as follows.

Let (l_x) be the number of survivors aged x at the beginning of a given age group, x. Of these ($_{n}d_{x}$) will die before reaching age x + n, so that l_{x+n} would be alive, aged x + n at the beginning of the next age group x + n. Obviously, l_{x+n} lived for n whole years. To this, one must add the average period lived by the people who died aged $_{n}d_{x}$. To do this, we divide the number of expected deaths, $_{n}d_{x}$ generated by Eq. (7.2) by the age-specific mortality rate, $_{n}m_{x}$

As defined in Eq. (7.1) (Fergany 1971; Kintner 2004). Thus,

$$_{n}L_{x} = {_{n}d_{x}} / {_{n}m_{x}} \tag{7.4}$$

where ω is the highest attainable age. This can be done rather easily by cumulating (L_x) from age ω to age x.

Because (T_x) is the total person-years lived beyond age x, and (l_x) is the number of persons alive at age x, the average number of person-years lived beyond age x, also referred to as the life expectancy at age x (e_x), is calculated as follows:

$$e_{x} = T_{x}/l_{x}$$

Life tables not only reflect the mortality regimes of actual populations, they represent hypothetical populations that will evolve as a result of the following conditions affecting these same populations over long periods of time:

- the population will remain closed to migration
- every year it will be augmented by a constant number of births (l_0), which is equal to the number of deaths that deplete the population each year ($\sum d_x$)
- it will experience a constant schedule of mortality ($_{n}q_{x}$) every year
- deaths will generally occur uniformly within each year.

Under these conditions, the characteristics of this "hypothetical" population (also known as a "complete" life table stationary population since its age structure and size remain constant over time) are as follows:

- L_x is the number of persons at each age x
- T_0 is the total size of the population
- l_0 is the number of births (which is equal to the number of total deaths)
- d_x is the number of deaths at each age x
- l_0/T_0 = the crude birth rate = the crude death rate

where

$T_0 = \sum L_x,$
$d_x = l_x\, q_x$

A complete life table uses single years from birth to the terminal age (beyond which no one lives). If the terminal age is 115, then the life table uses 0, 1, 2, 3, 4, . . ., 115.

An "abridged" life table uses age groups, where the width of the age group is provided by n. In this case, the characteristics of the hypothetical population are

- ${}_nL_x$ is the number of persons in the age group that is n years wide starting at age x,
- T_0 is the total size of the population
- l_0 is the number of births (which is equal to the number of total deaths)
- ${}_nd_x$ is the number of deaths in age group that is n years wide starting at age x
- l_0/T_0 = the crude birth rate = the crude death rate

Because we use Fergany's (1971) method to construct the preliminary life tables, there are several points that need discussion that are specific to this method. Fergany's (1971) method is advantageous because only the age-specific death rates are needed to construct an abridged life table. "In addition to its simplicity, it is, in contrast to other methods, self-contained in the sense that beyond making only the assumption of approximating the force of mortality by a step function (which is all we observe anyway) no further assumptions, approximations, or parameter estimates are required to compute all the life table functions." (Fergany 1971: 334). One disadvantage of this method in terms of the KP mortality data is that for the terminal open ended age group, where the hazard rate (${}_{\omega}q_x$) is 1.00, an adjustment has to be made because the calculation of "Years lived (${}_{\omega}L_x$) requires an age specific death rate for the terminal, open ended age group, which is not available for the KP mortality data (Friedman et al. 1997).

With these conventions and characteristics in mind, we now turn to the construction of preliminary life tables from the KP Smoking Study. We describe the initial tables as preliminary because we expected – and encountered – some anomalies due to the small samples and conventions that characterize the KP Smoking Study data.

Constructing Preliminary Life Tables from the KP Smoking Study

Because of space constraints, we do not show any of the preliminary life table constructed from the "all-cause" mortality data reported in the KP Smoking study data by gender and smoking status (Friedman et al. 1997), which are available from the corresponding author.

In evaluating our preliminary life tables, we employed a set of a priori assumptions:

1. Those who are younger have a longer life expectancy than those who are older, all else equal;
2. Women will generally have longer life expectancies than men, all else equal;
3. Current smokers will have a shorter life expectancy than non-smokers, all else equal; and
4. Among prior smokers – at any given age, those who stopped smoking more recently will have a *lower* life expectancy than those who stopped smoking in the more distant past, all else equal;

In the course of constructing the preliminary life tables, results comport with these assumptions – with the exception of life expectancy among prior smokers, which can be seen in Tables 7.1 and 7.2

For females, an anomaly, contrary to our a priori assumption, is found at age 35, where, the life expectancy of those who quit smoking more than 20 years ago (45.60), is less than both those who quit 2–10 years ago (46.29) and those who quit 11–20 years ago (46.94). For males, anomalies contrary to our a priori assumptions are found at age 35, and at both age 50 and age 65, as follows:

1. At age 35, male life expectancy of those who quit smoking more than 20 years ago (39.32), is less than both those who quit 2–10 years ago (40.46) and those who quit 11–20 years ago (42.81);
2. At age 50, male life expectancy of those who quit more than 20 years ago (29.74) is, as expected, higher than both those who quit 2–10 years ago (26.55) and those who quit 11–20 years ago (28.24). However, it is slightly above the life expectancy of those males who never smoked (29.60); and
3. Similarly, at age 65, male life expectancy of those who quit more than 20 years ago (16.15) is, as expected, higher both those who quit 2–10 years ago (13.73) and those who quit 11–20 years ago (14.77). However, it is slightly above the life expectancy of those males who never smoked (15.90).

In the original KP smoking study publication (Friedman et al. 1997), there are acknowledgements to these anomalies. The first of these acknowledgments is for women, "In the youngest group, 35- to 49-year-olds, all-cause mortality was the highest among those who had quit smoking for more than twenty years, but this was based on only two deaths." (Friedman et al. 1997: 490). The second acknowledgment is in terms of men, "All-cause deaths among men showed decreasing risks with increased duration of quitting only in the 50- to 64 and 75+ year age groups..." (Friedman et al. 1997: 490). A third acknowledgement generalizes the anomalies, "An inverse relationship of risk with duration of quitting was often but not consistently seen." (Friedman et al. 1997: 493).

We agree with Friedman et al. (1997: 490) that small numbers play a role in these anomalies in a specific instance they note where an anomaly is "...based on only two deaths." We also believe that small numbers play a role in all of the anomalies they note, which are more explicitly stated by us. The issue of small numbers is important in the KP smoking data because they can reflect either the stochastic uncertainty associated with a small population, sampling uncertainty, or a combination of both.[2]

As an example of these uncertainties, we can use the coefficient of variation (CV), which is designed to make comparisons of relative uncertainty, whether stochastic or sample-based. First consider the case of males aged 35–49 who have never smoked. The mean is 49 (deaths), where 49 = 29,916*0.00164 and .00164 = 49/29,916. The variance is 48.98, where 48.98 = 29,916*0.00164*(1 − 0.00164) and the standard deviation is 6.99 = $(48.98)^{0.5}$. Thus, the coefficient of variation is 0.1427 = 6.99/49. Next, consider the case of males aged 35–49 who quit smoking 20 or more years ago. The mean is 3, where 3 = 1,149*0.00261 and 0.00261 = 3/1,149. The variance is 2.99, where 2.99 = 1,149*0.00261*(1 − 0.00261) and the standard deviation is 1.729 = $(2.99)^{0.5}$. Thus, the coefficient of variation is 0.576 = 1.729/3. The CV for males aged 35–49 who quit smoking 20 or more years ago is four times higher than the CV for males aged 35–49 who have never smoked (0.576/0.1427 = 4.04), which indicates that the uncertainty for the latter is four times greater than the former. A similar situation exists in terms of the CVs for females aged 35–49 who never smoked and females aged 35–49 who quit smoking 20 or more year ago.

Because of the widespread use of the KP Smoking Study data, we believe that it is worth the effort to resolve the anomalies identified here.[3] To this end, we first interpolate the hazard rates ($_nq_x$ values) found in the preliminary life tables so that we have a set of hazard rates for age groups of five –year widths, starting at age 35–40 and ending at age 80–85. We then use these interpolated hazard rates as input to Gompertz-type regression models, which are used to generate a "smoothed" set of

[3]We employ here the definition that stochastic uncertainty is the manifestation of a process representing numerical values of some system randomly changing over time (Doob 1953); and the definition of sampling uncertainty as an estimate attached to a test result that characterizes the range of values within which the true value is asserted to lie (Ramsey and Ellison 2007). Stochastic uncertainty is far more likely to yield noticeable changes in a small population than in a large population (Swanson, forthcoming).

We also believe that there is an unacknowledged confounding factor that may play a role in the inconsistencies by duration since quitting. This unacknowledged factor is length of time smoked (and or quantity smoked) before quitting. As an extreme example of this potentially confounding effect, consider those aged 35 who reported that they quit smoking more than 20 years ago, which implies that they would have started smoking at age 14 or younger. How long could somebody age 14 years or younger been smoking before they quit in order to state they had not smoked for 20 or more years? This potentially confounding factor may not play as large a role as small numbers, but it may contribute to the inconsistencies.

hazard rates specific to each group associated with the 12 preliminary life tables that encompass a wider range of five year age groups, where feasible (e.g., for never smokers, these estimated hazard rates start at age 20–24 and end at age 90–94; however, for current smokers who have smoked for more than 40 years, the estimated hazard rates start at age 55–59 and end at age 90–94). These estimated hazard rates were examined in terms of the a priori assumptions. At this point, an anomaly remained for males who reported that they were current smokers, but had smoked for less than 20 years. The adjustment consisted of replacing zero deaths with three deaths in each of two oldest age groups, recalculating the hazard rates and using these revised hazard rates as the input to the Gompertz-type model.

With the estimated hazard rates in hand, an adjustment was used to convert them so that they would apply to the US population in 1990. To this end, a 1990 US life table was used (details here) as a "standard table" (Kintner 2004; United Nations 1982) and a gender-specific ratio, 1/(US e35/KPe35), was formed for all smoking groups employed in the KP Smoking Study. These adjusted hazard rates were then graphed and examined for anomalies. Three anomalies were found. The first was that females who reported being former smokers who had quit more than 20 years ago generally had lower hazard rates than females who reported never smoking. The second was that males who reported being former smokers who had quit more than 20 years ago generally had lower hazard rates than males who reported never smoking. The third anomaly was that males who reported being former smokers who had quit between 2 and 10 years ago generally had lower hazard rates than those who quit 11–20 years ago. The first and second anomalies were resolved using simple averages at each group between the hazard rates for former smokers who had quit 20+ and 11–20 years ago, respectively. The third anomaly was resolved by using simple averages at each group between the hazard rates for former male smoker who had quit between 11–20 years ago and 2–10 years ago, respectively. There is a sound justification for using this approach to resolve the each of the three anomalies.[4] Recall that age-specific death rates (${}_nm_x$), life-table death rates, also known as hazard rates, (${}_nq_x$), and survival ratios (${}_nS_x$), though differently derived, are closely related to each other. If one of these functions is known, reference to a system for constructing life tables makes it possible to estimate immediately the approximate levels of the other two functions. Fergany's method (1971), for example, converts ${}_nm_x$ into ${}_nq_x$, and ${}_nS_x$ is simply found by subtracting ${}_nq_x$ from 1. Because ${}_nq_x$ directly generates l_x and ${}_nd_x$, and in combination with ${}_nm_x$ generates ${}_nL_x$, and, hence, T_x, it is considered to be the fundamental life table function.

The expectation of life at a given age, e_x, is in a different category than ${}_nq_x$. It is both the result of the cumulative addition of specific values (T_x) and a ratio because

[4]In addition to following the United Nations (1982) guidelines for constructing life tables from incomplete data, we also use principles developed by Green and Armstrong (2015) who compared the efficacy of complex forecasting models to simple ones, and found justification for using simple methods in that they suggest that clients who prefer accuracy should use forecasts based on simple evidence-based procedures. We believe that the adjustment methods we employ in this paper is consistent with their principles and advice.

$e_x = T_x/l_x$. It is powerful in that it represents the one synthetic measure by which the "general" level of mortality can be summarized in a single figure (United Nations 1982: 25). This is evident in tables 12 and 13, where the life expectancy values clearly show the inconsistencies we have noted. However, life expectancy (e_x) cannot be used to construct $_nq_x$ because, T_x, the numerator used to created e_x, is the result of the cumulative addition of $_nL_x$ while lx, the denominator used to create e_x, is the result of the cumulative subtraction of $_nd_x$ from l_x. What these relationships suggest is that inconsistencies in the KP life tables need to be dealt with by revising the underlying $_nq_x$ values (or equivalently, the underlying $_nS_x$ values). Thus, In terms of resolving inconsistency that females who quit smoking 20+ years ago have a lower life expectancy at age 35 than females who quit smoking 2–10 years ago and females who quit smoking 11–20 years ago, we can take the average $_nq_x$ at each age between females who never smoked and those who quit smoking 20+ years ago.

Constructing Revised Life Tables for the Kaiser Permanente Population

It should be clear from the preceding discussion that the complexities found in the life table lead us to a method that allows us to directly assess the $_nq_x$ values via their reciprocals, $_nS_x$ values. we can do this because $_nS_x = 1 - {_nq_x}$ and can do this using a method described by Swanson and Tedrow (2012). In this approach, note that when the radix of a life table is equal to 1 ($l_0 = 1.00$) life expectancy at birth can be computed directly from the expression:

$$e_0 = S_0 + (S_0 * S_1) + (S_0 * S_1 * S_2) + , \ldots, + (S_0 * S_1 * S_2, \ldots, *S_x) \quad (7.5)$$

where

e_0 = life expectancy at birth
S_0 = survivorship from t = 0 (e.g., birth) to t = 1 (e.g., age 1)
S_1 = survivorship from t = 1 (e.g., age 1) to t = 2 (e.g., age 2)
and so on through S_x
and $S_x = {_1L_x}/ {_1L_{(x-n)}}$

Equation (7.5) is set up for single year age groups. However, we can generalize it to other age groups: $_nS_x = {_nL_x}/ {_nL_{(x-n)}}$, so that

$$e_0 = {_nS_0} + ({_nS_0}*{_nS_1}) + ({_nS_0}*{_nS_1}*{_nS_2}) + , \ldots, + ({_nS_0}*{_nS_1}*{_nS_2}, \ldots, *{_nS_x}) \quad (7.6)$$

As Eq. (7.5) and Eq. (7.6) both imply, the fundamental life table function is inherent in this method. That is via the $_nS_x$ values, we have $_nq_x$ values. Swanson and Tedrow (2012) show a derivation of the relationship between survivorship rates and life expectancy as shown in Eq. (7.5) and generalized to Eq. (7.6). Using this

Table 7.3 Revised life table for Kaiser Permanente females, never smokers: survivorship and life expectancy by age

Smoking status'	Age	$_5S_x$	Female e_x
Never	**20**	0.99989	66.25
Never	**25**	0.99968	61.25
Never	**30**	0.99921	56.27
Never	**35**	0.99830	51.32
Never	**40**	0.99671	46.41
Never	**45**	0.99411	41.56
Never	**50**	0.99008	36.81
Never	**55**	0.98411	32.17
Never	**60**	0.97557	27.69
Never	**65**	0.96372	23.39
Never	**70**	0.94767	19.27
Never	**75**	0.92640	15.33
Never	**80**	0.89875	11.55
Never	**85**	0.86338	7.85
Never	**90**	0.81877	4.09

Table 7.4 Revised life table for Kaiser Permanente males, never smokers: survivorship and life expectancy by age

Smoking status'	Age	$_5S_x$	Male e_x
Never	**20**	0.99972	62.05
Never	**25**	0.99922	57.07
Never	**30**	0.99822	52.12
Never	**35**	0.99642	47.21
Never	**40**	0.99344	42.38
Never	**45**	0.98880	37.66
Never	**50**	0.98194	33.08
Never	**55**	0.97215	28.69
Never	**60**	0.95866	24.52
Never	**65**	0.94055	20.57
Never	**70**	0.91677	16.87
Never	**75**	0.88616	13.40
Never	**80**	0.84740	10.13
Never	**85**	0.79905	6.95
Never	**90**	0.73951	3.70

approach, we have generated "revised" life tables from the preliminary life tables (to include the processing we did to resolve the anomalies noted earlier).

Again because of space constraints we do not show all the revised life tables corresponding to each of the 12 preliminary life tables, which as was the case with the preliminary tables, are available from the corresponding author. We show instead the results for two smoking categories by sex: (1) Never Smokers, and (2) Current smokers who smoke more than 20 cigarettes daily (Tables 7.3, 7.4, 7.5, and 7.6).

Table 7.5 Revised life table for Kaiser Permanente females, current smokers, 20 or more cigarettes daily: survivorship and life expectancy by age

Smoking status'	Age	$_5S_x$	Female e_x
Current 20+ cigarettes daily	**20**	0.99962	59.39
Current 20+ cigarettes daily	**25**	0.99897	54.41
Current 20+ cigarettes daily	**30**	0.99766	49.47
Current 20+ cigarettes daily	**35**	0.99532	44.58
Current 20+ cigarettes daily	**40**	0.99145	39.79
Current 20+ cigarettes daily	**45**	0.98545	35.13
Current 20+ cigarettes daily	**50**	0.97660	30.65
Current 20+ cigarettes daily	**55**	0.96404	26.39
Current 20+ cigarettes daily	**60**	0.94676	22.37
Current 20+ cigarettes daily	**65**	0.92363	18.63
Current 20+ cigarettes daily	**70**	0.89332	15.17
Current 20+ cigarettes daily	**75**	0.85439	11.98
Current 20+ cigarettes daily	**80**	0.80521	9.02
Current 20+ cigarettes daily	**85**	0.74397	6.21
Current 20+ cigarettes daily	**90**	0.66870	3.34

Table 7.6 Revised life table for Kaiser Permanente males, current smokers, 20 or more cigarettes daily: survivorship and life expectancy by age

Smoking status'	Age	$_5S_x$	Male e_x
Current 20+ cigarettes daily	**20**	0.99889	55.37
Current 20+ cigarettes daily	**25**	0.99738	50.43
Current 20+ cigarettes daily	**30**	0.99471	45.57
Current 20+ cigarettes daily	**35**	0.99041	40.81
Current 20+ cigarettes daily	**40**	0.98396	36.20
Current 20+ cigarettes daily	**45**	0.97475	31.79
Current 20+ cigarettes daily	**50**	0.96211	27.62
Current 20+ cigarettes daily	**55**	0.94530	23.70
Current 20+ cigarettes daily	**60**	0.92352	20.08
Current 20+ cigarettes daily	**65**	0.89589	16.74
Current 20+ cigarettes daily	**70**	0.86149	13.68
Current 20+ cigarettes daily	**75**	0.81931	10.88
Current 20+ cigarettes daily	**80**	0.76830	8.28
Current 20+ cigarettes daily	**85**	0.70734	5.78
Current 20+ cigarettes daily	**90**	0.63524	3.18

Discussion

Table 7.7a provides a summary of the e_{35} results for never and former smokers by gender, including for those revised life tables not shown here. It shows that the results both within and across gender by smoking status are consistent. Life expectancy at age 35 is higher for females at each smoking status and within gender is highest for never smokers, lowest for those who quit smoking 2–10 years ago and increases monotonically for former smokers as the years increase since quitting.

Table 7.7a Revised life table summary results by smoking status for Kaiser Permanente members

	Life expectancy at age 35	
Smoking status	Female	Male
Never	51.32	47.21
Former, 2–10 years since quitting	47.32	44.32
Former, 11–20 years since quitting	49.75	44.54
Former, 20+ years since quitting	50.48	45.83

Table 7.7b Revised life table summary results by smoking status for Kaiser Permanente members

	Life expectancy at age 35	
Smoking status	Female	Male
Never	51.32	47.21
Current, <20 cigarettes daily	48.34	42.97
Current, ≥20 cigarettes daily	44.58	40.81

Table 7.7c Revised life table summary results by smoking status for Kaiser Permanente members

	Life expectancy at age 55	
Smoking status	Female	Male
Never	32.17	28.69
Former, 20+ years since quitting	31.44	27.28
Former, 11–20 years since quitting	30.54	25.99
Former, 2–10 years since quitting	27.70	26.26
Current, <20 years duration	30.77	28.87
Current, 20–39 years duration	29.05	23.75
Current, 40+ years duration	26.27	22.06

Table 7.7b provides a summary of the e_{35} results for never and current smokers by gender and number of cigarettes smoked daily, including for those revised life tables not shown here. It shows that the results both within and across gender by smoking status are consistent. Life expectancy at age 35 is highest for female never smokers and lowest for males who smoke 20 or more cigarettes daily.

Table 7.7c provides a summary of the e_{55} results by gender for never smokers, duration since quitting for former smokers, and duration smoked for current smokers, including for the revised life tables not shown here. It shows that the results both within and across gender by smoking status are consistent in terms of: (1) never smokers v. former and current smokers; and (2) never smokers v. current smokers. It also shows that females who quit smoking 20 or more years ago have higher e_{55} values than either those who quit more recently or current smokers. In addition, life expectancy at age 55 is highest for female never smokers and lowest for males who have smoked for 40 or more years. However, for males, there are two inconsistencies: (1) the highest e_{55} value among former and current smokers is found for males who are current smokers but have smoked less than 20 years; and (2) e_{55} for males 2–10 years since quitting is higher than e_{55} for males who quit smoking 11–20 years ago.

Discussion of Revised Life Tables for Kaiser Permanente Population

Given the two remaining anomalies for males, we nonetheless find the results encouraging in that the life tables by smoking status are otherwise consistent, especially considering the small sample size as represented by the KP study population and other limitations, namely, that we do not know: (1) how many cigarettes were smoked daily by duration for current smokers; and (2) how long former smokers smoked and how many cigarettes they smoked daily. These factors would clearly cause differences in mortality and are likely to be underlying this specific anomaly and others that ae not apparent to us (See also Footnote 3). This situation is known as "hidden heterogeneity" among demographers (Vaupel and Missov 2014). Unfortunately, what is hidden to us in the KP study is likely to remain hidden.

Discussion of Revised KP Life Tables Applied to the US Population

The development of adequate life tables and survivorship information from the KP Smoking Study is of interest historically but to be relevant, the information needs to be applied to a more current and broader range of people, such as found in the work of Bachand and Sulsky (2013) and Bachand et al. (2018). To effect this application, the starting point is the recognition that the Kaiser Permanente population as represented in the smoking study (Friedman et al. 1997) has higher life expectancy values than does the comparable US population from the same time period.

We came to this working conclusion after weighting the revised life expectancy values by age and smoking statuses to produce a mean life expectancy by selected age and smoking status and comparing the results to US life tables produced by the Human Mortality Database project (Wilmoth et al. 2017) for the period 1985–89, which corresponds to the time period of the KP smoking study (Friedman et al. 1997).[5] The weighted results take into account proportions of the KP study

[5]The Human Mortality Database (HMD) is a collaborative project sponsored by the University of California at Berkeley and the Max Planck Institute for Demographic Research in Rostock, Germany. The purpose of the database is to provide researchers around the world with easy access to detailed and comparable national mortality data via the Internet. The database contains original life tables for almost 40 countries or areas, as well as all raw data used in constructing the tables. By

population in the smoking status categories (never current by time, former by years since quitting, and current by years smoked).[6] For example, e_{55} for US females 1985–89 is shown by the Human Mortality Database (HMD) is 26.69 while e_{55} for the weighted KP population (never, former by years since quitting, and current by years smoked) is 30.85; for US males 1985–89, the HMD life expectancy at age 55 is $e_{55} = 21.75$ while the weighted KP population (never, former by years since quitting, and current by years smoked) is 27.05. These differences suggest that the KP life tables need to be adjusted in order to better approximate the US population. This can be done by applying scalar values to the ${}_{n}S_{x}$ values found in the revised KP life tables (tables 14–31), a process we describe later.

The reason that the KP subject population has a higher life expectancy than the comparable US population is due to differences in healthcare access, income, race, ethnicity, and smoking status. The authors of the KP smoking study did not report ethnic and income groups (likely because the study population's sub-sets would have become too small to allow meaningful analysis). Similarly, they will become too small if we attempted to weight it not only by age and gender, but also race. Thus, we weight the e_x values by age and gender by smoking status proportions. Among females, CDC found that US females age 17 and over, 54.7% were never smokers, 20.9% were former smokers, and 24.4% were current smokers. Among US males age 17 and over, CDC found that 27% were never smokers, 35.1% were former smokers, and 37.9% were current smokers. Using the KP "person years," we find that the estimated proportions are very different: Among females 70.71% were never smokers and the remaining 29.29% were either current or former smokers; among males, 61.7% were never smokers and the remaining 38.3% were either current or former smokers (Friedman et al. 1997). Thus, the weighting of the KP population is done by applying information of the percent of the US population in each of three smoking status categories, never, former, and current by the CDC (US Center for Disease Control 1990) as described in endnote 6. Because the "current" and "former" smoking status categories were not broken down by the CDC in terms of years smoked or years since quitting (or by the number of cigarettes smoked by daily by those who were listed as current smokers), we distributed the

design, populations in the HMD are restricted to those with data (both vital statistics and census information) that cover the entire population and that are very nearly complete. As such, the HMD covers almost all of Europe, plus Australia, Canada, Japan, New Zealand, Chile, Israel, Taiwan and the United States. Outside this set, there are very few countries that possess the kind of data required for the HMD. Details are available at either of the following addresses: www.mortality.org and www.humanmortality.de (Wilmoth et al. 2017).

[6]The US Centers for Disease Control (1990) estimated that at that time among: (1) US females age 17 and over, 54.7% were never smokers, 20.9% were former smokers, and 24.4% were current smokers; and (2) US males age 17 and over, 27% were never smokers, 35.1% were former smokers, and 37.9% were current smokers. Using the KP "person years," we find that the estimated proportions are very different: Among females 70.71% were never smokers and the remaining 29.29% were either current or former smokers; among males, 61.7% were never smokers and the remaining 38.3% were either current or former smokers (Friedman et al. 1997).

proportions in the current and former categories to the same categories by age as found in the KP smoking study. For example, among females, CDC found that US females age 17 and over, 54.7% were never smokers, 20.9% were former smokers, and 24.4% were current smokers. Consequently, we weighted the KP e_x values for female never smokers by .547 and the female former smokers e_x values of those who quit 2–10 years ago by 0.209/3 = 0.0697, the female former smoker e_x values of those who quit 11–19 years ago by 0.209/3 = 0.0697, and the female former smoker e_x values of those who quit 20+ years ago by 0.209/3 = 0.0697. We weighted the e_x values of current KP female smokers who smoked for less than 20 years by 0.244/3 = 0.081333, the e_x values of current KP female smokers who smoked between 20 and 39 years by 0.244/3 = 0.081333, and the e_x values of current KP female smokers who smoked for 40 or more years by 0.244/3 – 0.081333 in each smoking status by age and sex (Friedman et al. 1997).

One of the reasons we selected the approach described by Swanson and Tedrow (2012) to develop the revised tables is that it allows us to use a simple adjustment such that the adjustment will modify them and their two functions (survivorship and life expectancy) in order to represent other populations, namely US populations. We use the term "scalar" adjustment for this process because we need apply only a single number to effect the desired changes so that the life tables approximate the Human Mortality database life tables for the US population for the any of the following relevant periods, 1985–1989, 1990–1994, 1995–1999, 2000–2004, 2005–2009, 2010–2014, 2015–2016. These periods are relevant because the first corresponds with the time period over which the KP smoking study data were gathered and the latter (especially 2000–2004, which corresponds to the 2000 US decennial census correspond to potential US populations that can be used in the population health modeling process, which is the final outcome desired for this project.

As an example of the scalar process, consider the adjustment of the life table functions (survivorship and life expectancy) for KP females by smoking status to US life table for females, 2000–2004. The approximation was accomplished by an iteration process that started with a scalar equal to 1.000 (no adjustment) and used successive scalar values to bracket the target life expectancy in the HMD life table for US females for 2000–2004 by applying the scalar to the KP ${}_nS_x$ values for females. The iterative process yielded a final scalar value of 0.9775, which was found by applying the successive scalar values to the ${}_nS_x$ values by smoking status and calculating the resulting mean life expectancy for all smoking status groups by weighting the e_x values using the proportion in each smoking status shown in Exhibits 7.1 and 7.2.

Exhibit 7.1: Example of an Adjustment of KP Life Tables for Females by Smoking Status to Serve as a Life Table for US Females, 2000–2004

	Original female person years	Adjusted female
II. Adjustment using proportion smoking in US 1986 weighted KP study e55	30.85	27.55
Smoking status	Proportion	Proportion
Never	0.547	0.547
Former, 20+ years since quitting	0.069666667	0.069666667
Former, 11–20 years since quitting	0.069666667	0.069666667
Former, 2–10 years since quitting	0.069666667	0.069666667
Current, <20 years duration	0.081333333	0.081333333
Current, 20–39 years duration	0.081333333	0.081333333
Current, 40+ years duration	0.081333333	0.081333333
	1	1
Weighted KP study e55 (1985–89)	30.85	27.55
HMD US 1985–89 e55	26.69	26.69
HMD US 1990–94 e55	27.17	27.17
HMD US 1995–99 e55	27.55	27.55
HMD US 2000–04 e55	27.55	27.55
HMD US 2005–09 e55	28.42	28.42
HMD US 2010–14 e55	29.04	29.04
HMD US 2015–16 e55	29.2	29.2

References

Centers for Disease Control (1990). Smoking and Health: A National Status Report, 2nd Edition: A Report to Congress. Public Health Service. USDHHS Publication no. 87-8369. Rockville, MD, US Department of Health and Human Service

Human Mortality Database. US Female Life Tables 5 × 5, 1933–2015

Human Mortality Database. US Male Life Tables 5 × 5, 1933–2015

Exhibit 7.2: Example of an Adjustment of KP Life Tables for Males by Smoking Status to Serve as a Life Table for US Males, 2000–2004

	Original male person years	Adjusted male
II. Adjustment using proportion smoking in US 1986 weighted KP study e55	27.05	24.01
Smoking status	Proportion	Proportion
Never	0.27	0.27
Former, 20+ years since quitting	0.117	0.117
Former, 11–20 years since quitting	0.117	0.117
Former, 2–10 years since quitting	0.117	0.117
Current, <20 years duration	0.126333333	0.126333333
Current, 20–39 years duration	0.126333333	0.126333333
Current, 40+ years duration	0.126333333	0.126333333
	1	1
Weighted KP Study e55 (1985–89)	27.05	24.01
HMD US 1985–89 e55	21.75	21.75
HMD US 1990–94 e55	22.57	22.57
HMD US 1995–99 e55	23.19	23.19
HMD US 2000–04 e55	24.02	24.02
HMD US 2005–09 e55	25.06	25.06
HMD US 2010–14 e55	25.71	25.71
HMD US 2015–16 e55	25.85	25.85

References

Centers for Disease Control (1990). Smoking and Health: A National Status Report, 2nd Edition: A Report to Congress. Public Health Service. USDHHS Publication no. 87-8369. Rockville, MD, US Department of Health and Human Service

Human Mortality Database. US Female Life Tables 5 × 5, 1933–2015

Human Mortality Database. US Male Life Tables 5 × 5, 1933–2015

Application of Results to Assess Population Health Impacts of Tobacco Products

Once the revised KP life table functions are adjusted to reflect the desired US population life table (i.e., the mortality of the 2000 US general population as found in the HMDB life US life tables for 2000–2004). The next step is to develop a model so that the needed survivorship rates can be fed into a population health model for the US (e.g., an agent based model). The process is grounded in the Gompertz model, which views the human mortality rate increasing exponentially as a function of age (Greenwood 1928). In its general form, the model is expressed as:

$$h(\text{age}) = B * c^{\text{age}} \tag{7.7}$$

where h(age) is the mortality hazard rate for the given age, and B and c are constants.

If we let B = e^α and c = e^β then the model translates into:

$$h(\text{age}) = e^{(\alpha+\beta*\text{age})} \tag{7.8}$$

where α = intercept and β = the age coefficient

While mortality rates and doubling periods differ across various populations, the Gompertz model has been empirically validated over two centuries across a wide range of countries (Lee et al. 2014; Olshansky and Carnes 1997). The model is widely used in various fields, including insurance and actuarial science (Bowers et al. 1997; Dickson et al. 2009) demography (Impagliazo 1985; Smith and Keyfitz 1977) and biology (Greenwood 1928).

Previously, the Gompertz model was thought to be appropriate only thru around age 80, as mortality appeared to decelerate at higher ages. However recent research suggest otherwise, showing the Gompertz model applies up to low to mid 100's. Gavrilov et al. (2014) find that the mortality trajectory at advanced ages follows the Gompertz model up to the ages 102–105 years without a noticeable deceleration. The Gompertz model may be viewed as a sub-model under a broader range of risk models known as proportional hazards model, first proposed by David Cox (1972). The general model is given by:

$$h\_i(t) = \lambda\_0(t) * e^{\left(\beta\_1*x\_i1+\cdots+\beta\_k*x\_ik\right)} \tag{7.9}$$

For individual i at time t, the hazard is the product of: (1) the function λ 0(t), which is left unspecified with the only condition that its value cannot be negative, and (2) the linear function β1 * xi1 + . . . + βk * xik comprised of k covariates that are then exponentiated. If for example one wanted to model US general population mortality starting in the year 2000, the revised KP life tables are not appropriate "as is" because they are not representative of the US year 2000 population. This is due to issues discussed earlier: (1) the KP study observed mortality was collected during the

1980s (2) the KP study population has different characteristics, particularly access to health care in that KP is both the insurer and the provider. Thus, the KP revised life tables reflect lower mortality rates than found in the general US population and would not be representative of that of ALCS ABM's target US population. The Human Mortality Database mortality tables provide the data for the required 2000 US population mortality experience. Using this approach, we created models applicable to US females and males in 2000. The input data for females are shown in Exhibit 7.1a and the model in Exhibit 7.1b. For the model applicable to US males, the input data are shown in Exhibit 7.2a and the model in Exhibit 7.2b.

Exhibit 7.1a: Data Input for the Female Comprehensive Gompertz Model for $_nq_x$ Using 4 Covariates, Age, Smoking Status and Age, US Female Population, 2000 (2000–04)

Females						
Adjusted KP $_nS_x$	$_nq_x$ ($_nq_x = 1 - _nS_x$)	LN ($_nq_x$)	Age	Smoking status	Years smoked (or since last smoked)	Age × years smoked
0.9773953	0.0226047	−3.78960	20	0	0	0
0.9771847	0.0228153	−3.78032	25	0	0	0
0.9767236	0.0232764	−3.76031	30	0	0	0
0.9758366	0.0241634	−3.72292	35	0	0	0
0.9742816	0.0257184	−3.66055	40	0	0	0
0.9717395	0.0282605	−3.56629	45	0	0	0
0.9678033	0.0321967	−3.43589	50	0	0	0
0.9619685	0.0380315	−3.26934	55	0	0	0
0.9536226	0.0463774	−3.07094	60	0	0	0
0.9420347	0.0579653	−2.84791	65	0	0	0
0.9263459	0.0736541	−2.60838	70	0	0	0
0.9055589	0.0944411	−2.35978	75	0	0	0
0.8785280	0.1214720	−2.10807	80	0	0	0
0.9774940	0.0225060	−3.79397	20	−1	6	−6
0.9774676	0.0225324	−3.79280	25	−1	6	−6
0.9773710	0.0226290	−3.78852	30	−1	6	−6
0.9770853	0.0229147	−3.77598	35	−1	6	−6
0.9763597	0.0236403	−3.74480	40	−1	6	−6
0.9747171	0.0252829	−3.67763	45	−1	6	−6
0.9713181	0.0286819	−3.55149	50	−1	6	−6
0.9647745	0.0352255	−3.34599	55	−1	6	−6
0.9529005	0.0470995	−3.05549	60	−1	6	−6
0.9323921	0.0676079	−2.69403	65	−1	6	−6

0.8984221	0.1015779	−2.28693	70	−1	6	−6
0.8441387	0.1558613	−1.85879	75	−1	6	−6
0.7600545	0.2399455	−1.42734	80	−1	6	−6
0.9774367	0.0225633	−3.79143	20	−1	15.5	−15.5
0.9772835	0.0227165	−3.78466	25	−1	15.5	−15.5
0.9769089	0.0230911	−3.76831	30	−1	15.5	−15.5
0.9761183	0.0238817	−3.73464	35	−1	15.5	−15.5
0.9746167	0.0253833	−3.67366	40	−1	15.5	−15.5
0.9719833	0.0280167	−3.57496	45	−1	15.5	−15.5
0.9676430	0.0323570	−3.43092	50	−1	15.5	−15.5
0.9608364	0.0391636	−3.24001	55	−1	15.5	−15.5
0.9505886	0.0494114	−3.00757	60	−1	15.5	−15.5
0.9356753	0.0643247	−2.74381	65	−1	15.5	−15.5
0.9145884	0.0854116	−2.46027	70	−1	15.5	−15.5
0.8854993	0.1145007	−2.16717	75	−1	15.5	−15.5
0.8462212	0.1537788	−1.87224	80	−1	15.5	−15.5
0.9773489	0.0226511	−3.78755	20	−1	20	−20
0.9770820	0.0229180	−3.77583	25	−1	20	−20
0.9765296	0.0234704	−3.75202	30	−1	20	−20
0.9755055	0.0244945	−3.70931	35	−1	20	−20
0.9737524	0.0262476	−3.64018	40	−1	20	−20
0.9709282	0.0290718	−3.53799	45	−1	20	−20
0.9665907	0.0334093	−3.39892	50	−1	20	−20
0.9601824	0.0398176	−3.22345	55	−1	20	−20
0.9510139	0.0489861	−3.01622	60	−1	20	−20
0.9382462	0.0617538	−2.78460	65	−1	20	−20
0.9208733	0.0791267	−2.53670	70	−1	20	−20

(continued)

Females						
Adjusted KP $_nS_x$	$_nq_x$ ($_nq_x = 1 - _nS_x$)	LN ($_nq_x$)	Age	Smoking status	Years smoked (or since last smoked)	Age × years smoked
0.8977027	0.1022973	−2.27987	75	−1	20	−20
0.8673366	0.1326634	−2.01994	80	−1	20	−20
0.9774137	0.0225863	−3.79041	20	1	10.5	10.5
0.9772219	0.0227781	−3.78195	25	1	10.5	10.5
0.9767761	0.0232239	−3.76258	30	1	10.5	10.5
0.9758749	0.0241251	−3.72450	35	1	10.5	10.5
0.9742257	0.0257743	−3.65838	40	1	10.5	10.5
0.9714258	0.0285742	−3.55525	45	1	10.5	10.5
0.9669428	0.0330572	−3.40952	50	1	10.5	10.5
0.9600936	0.0399064	−3.22122	55	1	10.5	10.5
0.9500235	0.0499765	−2.99620	60	1	10.5	10.5
0.9356846	0.0643154	−2.74396	65	1	10.5	10.5
0.9158138	0.0841862	−2.47472	70	1	10.5	10.5
0.8889098	0.1110902	−2.19741	75	1	10.5	10.5
0.8532106	0.1467894	−1.91876	80	1	10.5	10.5
0.9771221	0.0228779	−3.77758	20	1	29.5	29.5
0.9765223	0.0234777	−3.75170	25	1	29.5	29.5
0.9753745	0.0246255	−3.70397	30	1	29.5	29.5
0.9734014	0.0265986	−3.62690	35	1	29.5	29.5
0.9702615	0.0297385	−3.51531	40	1	29.5	29.5
0.9655455	0.0344545	−3.36811	45	1	29.5	29.5
0.9587743	0.0412257	−3.18869	50	1	29.5	29.5
0.9493972	0.0506028	−2.98375	55	1	29.5	29.5
0.9367891	0.0632109	−2.76128	60	1	29.5	29.5
0.9202494	0.0797506	−2.52885	65	1	29.5	29.5

0.8989998	0.1010002	−2.29263	70	1	29.5	29.5
0.8721828	0.1278172	−2.05715	75	1	29.5	29.5
0.8388603	0.1611397	−1.82548	80	1	29.5	29.5
0.9483469	0.0516531	−2.96320	50	1	40	40
0.9354641	0.0645359	−2.74053	55	1	40	40
0.9187893	0.0812107	−2.51071	60	1	40	40
0.8976650	0.1023350	−2.27950	65	1	40	40
0.8713859	0.1286141	−2.05094	70	1	40	40
0.8391997	0.1608003	−1.82759	75	1	40	40
0.8003076	0.1996924	−1.61098	80	1	40	40

1 Current smoker
−1 Former smoker
0 Never smoker

Exhibit 7.1b: Female Comprehensive Gompertz Model for $_nq_x$ Using 4 Covariates, Age, Smoking Status and Age, US Female Population, 2000 (2000–04)

Summary output	
Regression statistics	
Multiple R	0.948948392
R square	0.900503051
Adjusted R square	0.895528204
Standard error	0.220639935
Observations	85

ANOVA

	df	SS	MS	F	Significance F
Regression	A	35.24793331	8.811983327	181.0111893	3.02572E-39
Residual	80	3.894558502	0.048681981		
Total	84	39.14249181			

	Coefficients	Standard Error	t Stat	P-value	Lower 95%	Upper 95%	Lower 95.0%	Upper 95.0%
Intercept	−4.875850538	0.077135785	−63.21126466	4.66553E-70	−5.029355643	−4.722345433	−5.029355643	−4.722345433
X Variable 1	0.033065319	0.001302638	25.3833487	5.01548E-40	0.030472986	0.035657652	0.030472986	0.035657652
X Variable 2	−0.099276455	0.063382377	−1.566310065	0.121223304	−0.225411404	0.026858495	−0.225411404	0.026858495
X Variable 3	0.004328614	0.002668883	1.62188189	0.108763444	−0.000982634	0 009639861	−0.000982634	0.009639861
X Variable 4	0.007554462	0.003378203	2.236236939	0.028121348	0.000831624	0.014277301	0.000831624	0.014277301

Variable
1 = age
2 = smoking status
3 = years smoked (+)/years since last smoked (−)
4 = Age*Years smoked (+)/years since last smoked (−)
The estimated equation is $\ln(_nq_x) = -4.87585015861879 + 0.0330653163522129 * \text{age} - 0.0992771811245758 * \text{smokingstatus} + 0.00432862862681633 * \text{years} + 0.00755450916062156 * \text{age} \times \text{years}$

Exhibit 7.2a: Data Input for the Male Comprehensive Gompertz Model for $_nq_x$ Using 4 Covariates, Age, Smoking Status and Age, US Male Population, 2000 (2000–04)

Males						
Adjusted KP $_nS_x$	$_nq_x$ ($_nq_x = 1 - {_nS_x}$)	LN ($_nq_x$)	Age	Smoking status	Years smoked (or since last smoked)	Age × years smoked
0.9822230	0.0177770	−4.02985	20	0	0	0
0.9817371	0.0182629	−4.00288	25	0	0	0
0.9807544	0.0192456	−3.95047	30	0	0	0
0.9789852	0.0210148	−3.86253	35	0	0	0
0.9760556	0.0239444	−3.73202	40	0	0	0
0.9714993	0.0285007	−3.55783	45	0	0	0
0.9647515	0.0352485	−3.34533	50	0	0	0
0.9551418	0.0448582	−3.10425	55	0	0	0
0.9418883	0.0581117	−2.84539	60	0	0	0
0.9240918	0.0759082	−2.57823	65	0	0	0
0.9007298	0.0992702	−2.30991	70	0	0	0
0.8706514	0.1293486	−2.04524	75	0	0	0
0.8325719	0.1674281	−1.78720	80	0	0	0
0.9770568	0.0229432	−3.77473	20	−1	6	−6
0.9763246	0.0236754	−3.74332	25	−1	6	−6
0.9748921	0.0251079	−3.68457	30	−1	6	−6
0.9723843	0.0276157	−3.58937	35	−1	6	−6
0.9683297	0.0316703	−3.45237	40	−1	6	−6
0.9621549	0.0378451	−3.27425	45	−1	6	−6
0.9531794	0.0468206	−3.06143	50	−1	6	−6
0.9406108	0.0593892	−2.82364	55	−1	6	−6
0.9235400	0.0764600	−2.57099	60	−1	6	−6
0.9009376	0.0990624	−2.31201	65	−1	6	−6

(continued)

Males

Adjusted KP ${}_nS_x$	${}_nq_x$ (${}_nq_x = 1 - {}_nS_x$)	LN (${}_nq_x$)	Age	Smoking status	Years smoked (or since last smoked)	Age × years smoked
0.8716495	0.1283505	−2.05299	70	−1	6	−6
0.8343930	0.1656070	−1.79814	75	−1	6	−6
0.7877538	0.2122462	−1.55001	80	−1	6	−6
0.9822467	0.0177533	−4.03119	20	−1	15.5	−15.5
0.9817937	0.0182063	−4.00599	25	−1	15.5	−15.5
0.9808462	0.0191538	−3.95525	30	−1	15.5	−15.5
0.9790681	0.0209319	−3.86648	35	−1	15.5	−15.5
0.9759814	0.0240186	−3.72893	40	−1	15.5	−15.5
0.9709313	0.0290687	−3.53809	45	−1	15.5	−15.5
0.9630455	0.0369545	−3.29807	50	−1	15.5	−15.5
0.9511897	0.0488103	−3.01981	55	−1	15.5	−15.5
0.9339181	0.0660819	−2.71686	60	−1	15.5	−15.5
0.9094191	0.0905809	−2.40151	65	−1	15.5	−15.5
0.8754569	0.1245431	−2.08310	70	−1	15.5	−15.5
0.8293077	0.1706923	−1.76789	75	−1	15.5	−15.5
0.7676920	0.2323080	−1.45969	80	−1	15.5	−15.5
0.9821484	0.0178516	−4.02566	20	−1	20	−20
0.9816281	0.0183719	−3.99693	25	−1	20	−20
0.9806338	0.0193662	−3.94423	30	−1	20	−20
0.9788865	0.0211135	−3.85784	35	−1	20	−20
0.9759942	0.0240058	−3.72946	40	−1	20	−20
0.9714195	0.0285805	−3.55503	45	−1	20	−20
0.9644429	0.0355571	−3.33662	50	−1	20	−20
0.9541206	0.0458794	−3.08174	55	−1	20	−20
0.9392374	0.0607626	−2.80078	60	−1	20	−20

0.9182552	0.0817448	−2.50415	65	−1	20	−20
0.8892564	0.1107436	−2.20054	70	−1	20	−20
0.8498828	0.1501172	−1.89634	75	−1	20	−20
0.7972689	0.2027311	−1.59587	80	−1	20	−20
0.9815774	0.0184226	−3.99418	20	1	10.5	10.5
0.9804523	0.0195477	−3.93490	25	1	10.5	10.5
0.9785721	0.0214279	−3.84306	30	1	10.5	10.5
0.9756868	0.0243132	−3.71673	35	1	10.5	10.5
0.9715214	0.0284786	−3.55860	40	1	10.5	10.5
0.9657772	0.0342228	−3.37486	45	1	10.5	10.5
0.9581336	0.0418664	−3.17327	50	1	10.5	10.5
0.9482483	0.0517517	−2.96130	55	1	10.5	10.5
0.9357594	0.0642406	−2.74512	60	1	10.5	10.5
0.9202853	0.0797147	−2.52930	65	1	10.5	10.5
0.9014256	0.0985744	−2.31694	70	1	10.5	10.5
0.8787620	0.1212380	−2.11000	75	1	10.5	10.5
0.8518589	0.1481411	−1.90959	80	1	10.5	10.5
0.9817980	0.0182020	−4.00622	20	1	29.5	29.5
0.9807071	0.0192929	−3.94802	25	1	29.5	29.5
0.9786430	0.0213570	−3.84637	30	1	29.5	29.5
0.9751286	0.0248714	−3.69404	35	1	29.5	29.5
0.9695814	0.0304186	−3.49270	40	1	29.5	29.5
0.9613094	0.0386906	−3.25216	45	1	29.5	29.5
0.9495083	0.0504917	−2.98595	50	1	29.5	29.5
0.9332591	0.0667409	−2.70694	55	1	29.5	29.5
0.9115253	0.0884747	−2.42504	60	1	29.5	29.5
0.8831507	0.1168493	−2.14687	65	1	29.5	29.5

(continued)

Males						
Adjusted KP $_nS_x$	$_nq_x$ ($_nq_x = 1 - _nS_x$)	LN ($_nq_x$)	Age	Smoking status	Years smoked (or since last smoked)	Age × years smoked
0.8468573	0.1531427	−1.87639	70	1	29.5	29.5
0.8012434	0.1987566	−1.61567	75	1	29.5	29.5
0.7447814	0.2552186	−1.36563	80	1	29.5	29.5
0.9369493	0.0630507	−2.76382	50	1	40	40
0.9177752	0.0822248	−2.49830	55	1	40	40
0.8933006	0.1066994	−2.23774	60	1	40	40
0.8626890	0.1373110	−1.98551	65	1	40	40
0.8250541	0.1749459	−1.74328	70	1	40	40
0.7794619	0.2205381	−1.51168	75	1	40	40
0.7249314	0.2750686	−1.29073	80	1	40	40

1 Current smoker
−1 Former smoker
0 Never smoker

Exhibit 7.2b: Male Comprehensive Gompertz Model for $_nq_x$ Using 4 Covariates, Age, Smoking Status, Years, and Age by Years, US Male Population, 2000 (2000–04)

Summary output	
Regression statistics	
Multiple R	0.979717457
R Square	0.959846296
Adjusted R Square	0.95783861
Standard Error	0.171856908
Observations	85

ANOVA

	df	SS	MS	F	Significance F
Regression	4	56.48069777	14.12017444	478.0860504	5.55205E-55
Residual	80	2.362783759	0.029534797		
Total	84	58.84348153			

	Coefficients	Standard error	t Stat	P-value	Lower 95%	Upper 95%	Lower 95.0%	Upper 95.0%
Intercept	−5.194625011	0.060081225	−86.46003847	8.80513E-81	−5.314190459	−5.075059563	−5.314190459	−5.075059563
X Variable 1	0.041879635	0.001014628	41.27586765	1.07031E-55	0.039860462	0.043898808	0.039860462	0.043898808
X Variable 2	−0.221450075	0.049368666	−4.485640231	2.40472E-05	−0.319696851	−0.123203298	−0.319696851	−0.123203298
X Variable 3	0.002517455	0.002078799	1.211014189	0.229455356	−0.001619487	0.006654397	−0.001619487	0.006654397
X Variable 4	0.014044315	0.002631289	5.337426912	8.58834E-07	0.008807882	0.019280748	0.008807882	0.019280748

Variable
1 = age
2 = smoking status
3 = years smoked (+)/years since last smoked (−)
4 = Age*years smoked (+)/years since last smoked (−)
The estimated equation is $\ln(_nq_x) = -5.19462495486153 + 0.0418796505958745 * \text{age} - 0.221450058632551 * \text{smokingstatus} + 0.00251742357642365 * \text{years} + 0.0140443131891837 * \text{age} \times \text{years}$

The model for females appears to be adequate, with the exception that multi-collinearity is present and affects the significance tests. It has a high coefficient of variation ($R^2 = 0.9005$) and the adjusted value is $R^2 = 0.8955$, which suggests that there is no distortion of the goodness of fit due to a small sample size ($n = 85$) relative to the number of independent variables ($k = 4$). Two of the four coefficients are statistically significant at the .05 level, with years smoked ($p = 0.1088$) and smoking status ($p = 0.1212$) not statistically significant. One would expect both of these variable to be statistically significant and it is likely the case that they are not because each of them is highly correlated with age x years smoked: $r = 0.90$ for smoking status and age x years smoked; and $r = 0.57$ for years smoked and age x years. High correlations like these tends to inflate the standard errors estimated for their coefficients, which can lead to a determination that any or all are not statistically significant.

Given that the input data are not from a random sample, the fact that one or more of the coefficients is not statistically significant is not a huge concern. However, this correlation may have led to some level of inaccuracy in the estimation of the values of the coefficients for either or both of these variables, and, as such, degrade the predictability of the model. This suggests that it may be wise to omit the variable, age x years. In total, the diagnostic evaluation suggests that with the exception of multicollinearity, the model does not substantially violate the underlying assumptions of OLS regression models and is adequately specified. When, however, the variable age x years is removed, the indications of multicollinearity disappear without a noticeable decline in the coefficient of variation ($R^2 = 0.8953$), which supports the use of this revised model:

$$\begin{aligned} \mathrm{Ln}({}_{n}q_{x}) = & -4.9068227324992 + 0.0332947010597161 * \mathrm{age} \\ & +0.0275251439138526 * \mathrm{smokingstatus} \\ & +0.0075854124754397 * \mathrm{years} \end{aligned}$$

As was the case for the female model, the model for males appears to be adequate except for the presence of multi-collinearity, which affects the significance tests. It has a high coefficient of variation ($R^2 = 0.9598$) and the adjusted value is $R^2 = 0.9578$, which suggests that there is no distortion of the goodness of fit due to a small sample size relative to the number of independent variables. Three of the four coefficients are statistically significant at the .05 level, with years smoked being the only one that is not ($p = 0.2295$). One would expect this variable to be statistically significant and it is likely not because this variable is highly correlated with age x years smoked. ($r = 0.89$). High correlation like this between any two of the independent variables tends to inflate the standard errors estimated for their coefficients, which can lead to a determination that either or both are not statistically significant.

Given that the input data are not from a random sample, the fact that one or more of the coefficients is not statistically significant is not a huge concern. However, the presence of multicollinearity may also have led to some level of inaccuracy in the estimation of the values of the coefficients for either or both of these variables, and, as such, degrade the predictability of the model. This suggests that it may be wise to omit the variable, age x years. In sum, the diagnostic evaluation suggests that with the exception of multicollinearity, the model does not substantially violate the underlying assumptions of OLS regression models and is adequately specified. As was the case with the female Gompertz model when the variable age x years is removed, the indications of multicollinearity disappear without a noticeable decline in the coefficient of variation ($R^2 = 0.9455$), which supports the use of this revised model for males:

$$\begin{aligned}\mathrm{Ln}({}_{n}q_{x}) = &-5.2522049403514 + 0.0423060913618448 * \mathrm{age} \\ &+0.0142835314779996 * \mathrm{smokingstatus} + 0.00857199196558667 * \mathrm{years}\end{aligned}$$

Conclusion

Following the path laid out in Abelin's seminal 1965 article, we illustrate how we developed life tables from cohort mortality data widely employed in efforts to examine smoking and health, which in this case is the Kaiser Permanente Smoking Study. The mortality data in this study have been used in terms of relative mortality and risk rates in regard to smoking behaviors. However, they have never been used to generate life tables. After describing life tables in general, we described the KP smoking study data, then discussed the methods and procedures used to generate the life tables from them. We showed the life tables developed from the KP smoking study and then discussed the methods used to extend these life tables to the US population and created hazard rate and survivorship data that can be used as input to models designed to assess the population health impact of tobacco products.

Acknowledgements The authors thank Lai Wei, Thad Hannel, and Raheema Muhammad-Kah for their comments and assistance.

References

Abelin, T. (1965). Application of life table methods to results of epidemiologic follow-up studies on smoking and mortality. *American Journal of Epidemiology, 81*(2), 254–269.

American Cancer Society. (n.d.). *History of the cancer preventions studies.* Available online at: https://www.cancer.org/research/we-conduct-cancer-research/epidemiology/history-cancer-prevention-study.html

Bach, P., Kattan, M., Thornquist, M., Kris, M., Tate, R., Barnett, M., Hsieh, L., & Begg, C. (2003). Variations in lung cancer risk among smokers. *Journal of the National Cancer Institute, 95*(6), 470–478.

Bachand, A., & Sulsky, S. (2013). A dynamic population model for estimating all-cause mortality due to lifetime exposure history. *Regulatory Toxicology and Pharmacology, 67*, 246–251.

Bachand, A., Sulsky, S., & Curtin, G. (2018). Assessing the likelihood and magnitude of a population health benefit following the market introduction of a modified-risk tobacco product: Enhancements to the dynamic population modeler, DPM (+1). *Risk Analysis, 38*(1), 151–162.

Bain, C., Feskanich, D., Speizer, F., Thun, M., Hertzmark, E., Rosner, B., & Colditz, G. (2004). Lung cancer rates in men and women with comparable histories of smoking. *Journal of the American Cancer Institute, 96*(11), 826–834.

Blakely, T., & Wilson, N. (2005). The contribution of smoking to inequalities in mortality by education varies over time and by sex: Two national cohort studies, 1981–84 and 1996–99. *International Journal of Epidemiology, 34*, 1054–1062.

Blizzard, L., & Dwyer, T. (2003). Case-control study of lung cancer during 1994–1997 in the birth cohort in Tasmania, Australia, with an excess of female cases during 1983–1992. *Cancer Causes and Control, 14*, 123–129.

Bowers, N., Gerbe, H., Hickman, J., Jones, D., & Nesbitt, C. (1997). *Actuarial mathematics: Society of actuaries.*

Burch, T. (2018). *Model-based demography: Essays on integrating data, technique, and theory.* Dordrecht: Springer.

Cancer Research, UK Cancer Survival Group. (2009). *Life tables for England and Wales by sex, calendar period, region and deprivation.* London School of Hygiene & Tropical Medicine. http://www.lshtm.ac.uk/eph/ncde/cancersurvival/tools/index.html

Carey, J. R., Papadopoulos, T., Müller, H.-G., Katsoyannos, B., Kouloussis, N., Wang, J.-L., Wachter, K., Yu, W., & Liedo, P. (2008). Age structure and extraordinary life span in wild medfly populations. *Aging Cell, 7*, 426–437.

Cox, D. R. (1972). Regression models and life tables (with discussion). *Journal of the Royal Statistical Society. Series B: Statistics in Society, 34*(2), 33.

De Matteis, S., Consonni, D., Pesatori, A., Beren, A., Bertazzi, P., Caporaso, N., Lubin, J., Wacholder, S., & Landi, M. (2012). Are women who smoke at higher risk for lung cancer than men who smoke? *American Journal of Epidemiology, 177*(7), 601–612.

Dickson, D., Hardy, M., & Waters, H. (2009). *Actuarial mathematics for life contingent risks.* Cambridge: Cambridge University Press.

Doob, J. (1953). *Stochastic processes.* New York: Wiley.

Dorn, H. (1959). Tobacco consumption and mortality from cancer and other diseases. *Public Health Reports, 74*, 581–593.

Ederer, F., Axtell, L., & Cutler, S. (1961). The relative survival rate: A statistical methodology. *National Cancer Institute Monograph, 6*, 101–121.

Ellis, L., Coleman, M., & Rachet, B. (2014). The impact of life tables adjusted for smoking on socio-economic difference in net survival for laryngeal and lung cancer. *British Journal of Cancer, 111*(1), 195–202.

Ernster, V., Kaufman, N., Nichter, M., Samet, J., & Yoon, S. (2000). Women and tobacco: Moving from policy to action. *Bulletin of the World Health Organization, 78*(7), 891–901.

Fergany, N. (1971). On the human survivorship function and life table construction. *Demography, 8* (3), 331–334.

Friedman, G., Tekawa, I., Sadler, M., & Sidney, S. (1997). Smoking and mortality: The Kaiser Permanente experience. In D. Shopl, D. Burns, L. Garfinkel, & J. Samet (Eds.), *Changes in cigarette-related disease risks and their implication for prevention and control* (pp. 477–499). Rockville: Department of Health and Human Services, Public Health Service, National Institutes of Health, National Cancer Institute.

Fry, j., Lee, P., Forey, B., & Coombs, K. (2013). Dose-response relationship of lung cancer to amount smoked, duration, and age starting. *World Journal of Meta-Analysis, 1*(2), 57–77.

Gavrilov, L., Gavrilova, N., Stone, C., & Zissu, A. (2014). New findings on older people's life expectancies confirm Gompertz law: The impact on the value of securitized life settlements. *Journal of Structured Finance, 20*(2), 66–73.

Green, K., & Armstrong, J. S. (2015). Simple versus complex forecasting: The evidence. *Journal of Business Research, 68*, 1678–1685.

Greenwood, M. (1928). "Laws" of mortality from the biological point of view. *Journal of Hygiene, 28*(3), 267–294.

Hammond, E. (1964). Smoking in relation to mortality and morbidity. Findings in the first 34 months of follow-up in a prospective study started in 1959. *Journal of the National Cancer Institute, 32*, 1161–1188.

Hammond, E., & Horn, D. (1958). Smoking and death rates—report on forty-four months of follow-up of 187,783 men. I. Total mortality. II. Death rates by cause. *Journal of the American Medical Association, 166*, 1159–1172. 1294–1308.

Hunt, D., Blakely, T., Woodward, A., & Wilson, N. (2005). The smoking-mortality association varies over time and by ethnicity in New Zealand. *International Journal of Epidemiology, 24*, 1020–1028.

Impagliazo, J. (1985). *Deterministic aspects of mathematical demography: An investigation of the stable theory of population including an analysis of the population statistics of Denmark.* Berlin: Springer.

Jacobs, D., Adachi, H., Mulder, I., Kromhout, D., Menotti, A., Nissinen, A., & Blackburn, H. (1999). Cigarette smoking and mortality risk: Twenty-five-year follow-up of the seven countries study. *Archives of Internal Medicine, 159*, 733–740.

Jiang, J., Liu, B., Sitas, F., Li, J., Zeng, X., Han, W., Zou, X., Wu, Y., & Zhao, P. (2010). Smoking-attributable deaths and potential years of life lost from a large, representative study in China. *Tobacco Control, 19*, 7–12. https://doi.org/10.1136/tc.2009.031245.

Keyfitz, N. (1970). Finding probabilities from observed rates or how to make a life table. *The American Statistician, 24*(1), 28–33.

Kim, Y., & Aron, J. (1989). On the equality of average age and average expectation of remaining life in a stationary population. *SIAM Review, 31*(1), 110–113.

Kintner, H., & Swanson, D. A. (1994). Estimating vital rates from corporate databases: How long will general motors' salaried retirees live? In H. Kintner, T. Merrick, P. Morrison, & P. Voss (Eds.), *Demographics: A casebook for business and government* (pp. 265–297). Boulder: Westview Press.

Kintner, H. (2004). The life table. In J. Siegel & D. Swanson (Eds.), *The methods and materials of demography* (2nd ed., pp. 301–340). San Diego: Elsevier Academic Press.

Lee, P., & Forey, B. (2013). Indirectly estimated absolute lung cancer mortality rates by smoking status and histology type based on a systematic review. *BMC Cancer, 13*, 1–90.

Lee, P., Forey, B., & Coombs, K. (2012). Systematic review with meta-analysis of the epidemiological evidence in the 1900s relating smoking to lung cancer. *BMC Cancer, 12*, 385.

Lee, S. J., Boscardin, W., Kirby, K., & Covinsky, K. (2014). Individualizing life expectancy estimates for older adults using the Gompertz law of human mortality. *PLoS One, 9*(9), e108540. https://doi.org/10.1371/journal.pone.0108540.

Lee, P., Hamling, J., Fry, J., & Forey, B. (2015). Using the negative exponential model to describe changes in risk of smoking-related diseases following changes in exposure to tobacco. *Advances in Epidemiology*, ID 487876.

Levy, D., Mumford, E., & Gerlowski, D. (2007). Examining trends in quantity smoked. *Nicotine & Tobacco Research, 9*(11), 1287–1296.

Muhammad-Kah, R., Pithawalla, Y., Gogova, M., Wei, L., & Boone, E. (2016). *An agent-based modeling approach for tobacco product risk assessments.* Paper presented at the Joint statistical meeting, Chicago.

Olshansky, S. J., & Carnes, B. (1997). Ever since Gompertz. *Demography, 34*(1), 1–15.

Owen, A., Maulida, S., Zomer, E., & Liew, D. (2018). Productivity burden of smoking in Australia: A life table modelling study. *Tobacco Control*, 1–8. https://doi.org/10.1136/tobaccocontrol-2018-054263.

Poland, B., & Teischinger, F. (2017). Population modeling of modified risk tobacco products accounting for smoking reduction and gradual transitions of relative risk. *Nicotine & Tobacco Research, 19*(11), 1277–1283.

Preston, S. (1970a). An international comparison of excessive adult mortality. *Population Studies, 24*, 5–20.
Preston, S. (1970b). *Older male mortality and cigarette smoking*. Berkeley: University of California Press.
Ramsey, M., & Ellison, S. (Eds.). (2007). *Eurachem/EUROLAB/CITAC/Nordest/AMC guide: Measurement uncertainty arising from sampling, a guide to methods and approaches*. Tornio: Eurachem Secretariat.
Rao, A., & Carey, J. (2014). Generalization of Carey's equality and a Theorem on stationary population. *Journal of Mathematical Biology*. https://doi.org/10.1007/s00285-014-0831-6. http://entomology.ucdavis.edu/files/203430.pdf.
Retherford, R. (1972). Tobacco smoking and the sex mortality differential. *Demography, 9*(2), 203–216.
Richards, H., & Donaldson, M. (2010). *Life and worklife expectancies*. Tucson: Lawyers and Judges Publishing.
Siegel, J. (2002). *Applied demography*. San Diego: Academic.
Smith, D., & Keyfitz, N. (1977). *Mathematical demography: Selected papers*. Berlin: Springer.
Smith, S., Tayman, J., & Swanson, D. (2013). *A practitioner's guide to state and local population projections*. Dordrecht: Springer.
Swanson, D. A. (forthcoming). Estimating the underlying infant mortality rates for small populations, even those reporting zero infant deaths: A case study of 66 local health areas in British Columbia. *Canadian Studies in Population*.
Swanson, D., & Sanford, A. (2012). Socio-economic status and life expectancy in the United States, 1990–2010: Are we reaching the limits of human longevity? *Population Review, 51*(2), 16–41.
Swanson, D., & Tedrow, L. (2012). Using cohort change ratios to estimate life expectancy in populations with negligible migration: A new approach. *Canadian Studies in Population, 39*, 83–90.
Swanson, D., & Tedrow, L. (2019). On mathematical equalities and inequalities in the life table: Something old and something new. *Canadian Studies in Population*. (forthcoming).
Thomas, J., & Bao, L. (2016). Modeling the dynamics of an HIV epidemic. In R. Schoen (Ed.), *Dynamic demographic analysis* (pp. 91–114). Dordrecht: Springer.
Thun, M., Calle, E., Rodriguez, C., & Wingo, P. (2000). Epidemiological research at the American Cancer Society. *Cancer Epidemiology, Biomarkers and Prevention, 9*, 861–868.
Thun, M., Henley, S., Burns, D., Jemal, A., Shanks, T., & Calle, E. (2006). Lung cancer rates in lifetime nonsmokers. *Journal of the National Cancer Institute, 98*(10), 691–699.
Trovato, F., & Lalu, N. M. (2001). Narrowing sex differences in life expectancy: Regional variations, 1971–1991. *Canadian Studies in Population, 28*(1), 89–110.
United Nations. (1982). *Model life tables for developing countries* (Department of International Economic and Social Affairs, Population Studies No. 77). New York: United Nations.
United States Centers for Disease Control. (1990). *Smoking and health: A national status report, 2nd Edition: A report to Congress* (Public Health Service. USDHHS Publication No. 87–8369). Rockville: Department of Health and Human Service.
Van Iersel, C., de Koning, H., Draisma, G., Mali, W., Scholtem, E., Nackaerts, K., Prokop, M., Habbema, J., Oudkerk, M., & van Klaveren, R. (2006). Risk-based selection from the general population in a screening trial: Selection criteria, recruitment and power for the Dutch-Belgian randomized lung cancer multi-slice CT screening trial (NELSON). *International Journal of Cancer, 120*, 868–874.
Vaupel, J. (2009). Life lived and left: Carey's equality. *Demographic Research, 20*, 7–10.
Vaupel, J., & Missov, T. (2014). Unobserved population heterogeneity: A review of formal relationships. *Demographic Research, 31*(22), 659–686.
Villavicencio, F., & Riffe, T. (2016). Symmetries between life lived and left in finite stationary populations. *Demographic Research, 35*, 381–398.
Wilmoth, J., Andreev, K., Jdanov, D., Glei, D. A., & Riffe, T. with the assistance of Boe, C., Bubenheim, M., Philipov, D., Shkolnikov, V., Vachon, P., Winant, C., Barbieri, M. (2017). *Methods protocol for the human mortality database* (Version 6). Available online at: www.mortality.org

Wrycza, T. (2014). Variance in age at death equals average squared remaining life expectancy at death. *Demographic Research, 30*(50), 1405–1412.

Yashin, A., Stallard, E., & Land, K. (2016). *Biodemography of aging: Determinants of healthy life span and longevity*. Dordrecht: Springer.

Yusuf, F., Martins, J., & Swanson, D. (2014). *Methods of demographic analysis*. Dordrecht: Springer.

Chapter 8
Prevalence of Tobacco Smoking and Alcohol Consumption in Australia

Farhat Yusuf and Julian de Meyrick

Introduction

In Australia, as in most countries, the prevalence of tobacco smoking and alcohol consumption continue to be regularly measured in general surveys such as the Australian Bureau of Statistics' four-yearly National Health Survey (Australian Bureau of Statistics 2016) and more focused surveys such as the National Drug Strategy Household Survey (Australian Institute of Health and Welfare 2017), also run at approximately four-yearly intervals. Various patterns in tobacco, alcohol and other substance usage across different groups in populations are regularly the subject of research. For example, Baker et al. (2006) looked at patterns of smoking within sub-populations. Patterns of tobacco and alcohol usage among a sample of socio-economically disadvantaged people were the focus of Twyman et al. (2016), and across larger groups in Bonevski et al. (2014). Studies such as Morley and Hall (2007) examined different changes within different segments in the population. They found that males had reduced their progression to current smoker status while females had increased their progression. As a result, there has been a convergence in prevalence across the sexes. Morley and Hall make the interesting point that

---

Disclaimer: This is to certify that this article is an original research by the above authors and is being submitted exclusively to the International Association of Applied Demography for publication consideration into an edited book on Applied Demography by Springer Publishing Company.

F. Yusuf (✉)
Menzies Centre for Health Policy, School of Public Health, Faculty of Medicine & Health, The University of Sydney, Sydney, NSW, Australia

Macquarie Business School, Macquarie University, Macquarie Park, NSW, Australia
e-mail: farhat.yusuf@sydney.edu.au

J. de Meyrick
Macquarie Business School, Macquarie University, Macquarie Park, NSW, Australia

B. Jivetti, Md. N. Hoque (eds.), *Population Change and Public Policy*, Applied Demography Series 11, https://doi.org/10.1007/978-3-030-57069-9_8

young people are taking up smoking despite being aware of the health risks associated with smoking and at a time when smoking prevalence is very low and continuing to decline.

Reis et al. (2017) found that tobacco smoking and hazardous drinking are not related in terms of addiction but they did identify segments which are more likely to engage in both these behaviours. Many other studies examine the relationship between tobacco and alcohol consumption and other conditions such as mental health issues (See for example: Bowden et al. (2011) and Falk et al. (2008)).

Both tobacco smoking and risky alcohol consumption have been identified as causing serious, avoidable morbidity and mortality in the populations of most countries for many years. The WHO Global Status Report on Alcohol and Health 2014 (WHO 2014) estimates that "Globally, alcohol consumption results in approximately 3.3 million deaths each year and this number has already been adjusted to take into account the beneficial impact of lowrisk (sic) patterns of alcohol consumption. Of all deaths worldwide, 5.9% are attributable to alcohol consumption."

The US Surgeon General, Luther Terry, identified tobacco smoking as the leading cause of avoidable morbidity and mortality in the US in his 1964 report (Smoking and Health 1964). Collins and Lapsley (1996) estimate that tobacco smoking in Australia is responsible for 23,000 avoidable deaths and $18 billion per year in lost productivity. This huge cost has been confirmed many times since then (see for example: Banks et al. (2015)). Another approach is to explore the benefits of reducing tobacco smoking prevalence. Magnus et al. (2012) found that reducing smoking prevalence from 23% to 15% saved AU$491 million in health costs.

While smoking prevalence in Australia has declined (Australian Institute of Health and Welfare 2017), Yusuf and Leeder (2015) found that alcohol consumption has increased in Australia in recent years.

This paper uses data from the large-scale National Health Survey to identify segments within the Australian population which might be at increased risk of concurrently smoking tobacco and consuming alcohol at risky levels. The findings can be used to "fine tune" tobacco and alcohol control initiatives for these segments which should be much more effective than undifferentiated initiatives directed at the population as a whole.

Data and Methods

The Australian National Health Survey (NHS) was the main source of data for this study. It was conducted by the Australian Bureau of Statics (ABS) in all states and territories and across urban, rural and remote areas of Australia (other than very remote areas).

The sample was selected using a multistage, stratified probability sampling procedure (Australian Bureau of Statistics 2016). The survey was conducted from 6 July 2014 to 4 July 2015.

People living in very remote areas were excluded from both surveys. This limitation of the scope had only a minor impact on aggregate estimates – overall such exclusions accounted for about 3% of the population.

The survey focussed on the health status of Australians and health-related aspects of their lifestyle. Information was collected about respondents' long-term medical conditions, consultation with health professionals, and other actions recently taken in regard to their health (e.g., used medication, taken days off from work). Information was also collected on lifestyle factors which may affect health, for example smoking and alcohol consumption.

The authors were provided with a confidentialised unit record file through an arrangement between the ABS and the University of Sydney. The file contained records of 19,257 persons. Each record also contained a weighting factor to be used in calculating the weighted estimates and 60 replicate weights to facilitate the calculation of standard errors for each weighted estimate. The relative standard error (RSE) for an estimate was calculated as the ratio of its standard error divided by the weighted estimate.

Since the questions on smoking and alcohol were only asked from persons aged 15 and over, the 3,863 persons under 15 were excluded from our analysis. Another 122 persons were also excluded because they refused to answer the questions on smoking. Thus, the final sample consisted of 15,272 persons who were aged 15 and over and who had answered the questions on alcohol consumption and smoking behaviour.

The following eight independent variables have been used in this study:

- Demographic variables – age and sex of respondents.
- Socio-economic variables – education and occupation of respondents and their household income (available in quintiles).
- SEIFA Index of disadvantage – a composite index calculated by the ABS (Australian Bureau of Statistics 2013) for each area based on various aspects of disadvantage. This index represented the level of socio-economic disadvantage (expressed in terms of quintiles) prevalent in an area rather than the socio-economic characteristics of the individual respondents.
- Distress levels were determined by using the scores on Kessler's K10 scale (Australian Bureau of Statistics 2012) grouped into four categories – low (scores 10–15), moderate (scores 16–21), high (scores 22–29) and very high (30–50). These scores were only calculated for person 18 years of age or older.
- Respondents were asked if they were suffering from any mental illness. Those who reported in the affirmative were asked if they were taking or not taking medications for the particular condition(s) affecting them. Since the most common mental health issue reported was depression we decided to use only this condition in our analysis. As with all other health conditions, the data on mental health were as reported by the respondents and did not involve use of clinical (psychiatric or psychological) tools of diagnosis.

Table 8.1 Cross-classification of the 15 year and older sample by smoking status and alcohol consumption status: Australia, National Health Survey, 2014–15

	Smoking status			
Alcohol consumption status	Current smokers	Ex-smokers	Non-smokers	Total
Risky drinkers	1529 (A)	2249 (E)	2420 (I)	6198
Light drinkers	577 (B)	1900 (F)	3380 (J)	5857
Ex-drinkers	209 (C)	409 (G)	648 (K)	1266
Non-drinkers	137 (D)	212 (H)	1602 (L)	1951
Total	2452	4770	8050	15,272

The dependent variable was based on the following two components:

- Various questions were asked about the smoking habits of respondents. These included questions such as age at which smoking was started, frequency and type of tobacco products used. The smoking status for each respondent was classified into *current smokers*, including all those who smoke daily, weekly or less than weekly but regard themselves as regular smokers. *Ex-smokers* had smoked in the past but not in the last 12 months, and *non-smokers* included those who claimed never to have smoked.
- Respondents were asked about their alcohol consumption during the past one year and detailed questions about their consumption during the week prior to the survey. These data were used to determine whether or not respondents had exceeded the short-term alcohol consumption guidelines of the National Health and Medical Research Council (NHMRC) (National Health and Medical Research Council 2009). Respondents who had not consumed alcohol during the past one year but had consumed before that were classified as *ex-drinkers*. Those consuming alcohol during the past one year were classified as *current drinkers*. Of these, respondents who reported having 5 or more standard drinks on any occasion were considered having exceeded the NHMRC guidelines (*risky drinkers*), and the remainder were classified as *non-risky drinkers*.

The sample was divided into 15 groups by cross-classifying the smoking and alcohol consumption status as shown in Table 8.1. For each group the dependent variable was the combination of their smoking and alcohol consumption status. Thus, for example, in group A the dependent variable was coded as 1 for all persons who were current smokers and risky drinkers while the remainder were coded as 0 (zero), and for group F the dependent variable was coded as 1 for all respondents who were ex-smokers and non-risky drinkers while the remainder were coded as 0 (zero).

We ranked groups A through L on sample size and selected the top six groups (A, E, F, I, J and L) as they accounted for 86.4% of the total sample. We also included an additional group consisting of all current drinkers (risky and non-risky) because it was of particular interest.

Binary logistic regression (SPSS version 24) was used as the main method for modelling purposes. We computed the binary logistic regression models for each of the six groups. On the basis of the very small values of the Nagelkerke R^2 we decided to exclude groups E and J from further analysis.

Findings

Table 8.2 gives the distribution of selected characteristics of the sample as well as weighted estimates (in thousands) for the Australian population using the ABS weights and the RSEs for each estimate. It may be noted that all the subsequent analyses presented in this paper are based on weighted estimates representing the 15 years and older population of Australia.

Overall, 15% of the total sample were current smokers, 30% were ex-smokers and 54% had never smoked, which is broadly in line with surveys reported elsewhere (Australian Bureau of Statistics 2016). Alcohol consumption is much more prevalent in the sample with the largest group (43%) reporting that they consumed alcohol at a risky level, 36% consumed alcohol but not to a risky level, 8% were ex-drinkers and 13% had never drunk alcohol.

Among current smokers, 65% drank to a risky level and 22% consumed alcohol but not to a risky level, leaving only 13% for those who were either ex-drinkers or had never drunk alcohol. Among those who had never smoked, 28% either never drank alcohol or had given it up. Eighty-three percent of those who had never drunk alcohol had also never smoked tobacco.

Table 8.3 gives the results of binary logistic regression modelling for each of the groups identified in the previous section.

In Australia alcohol and tobacco are only legally available to people aged 18 years and over, so the odds ratio of a person in the lowest age group (15–19) being both a current smoker and a consumer of alcohol at levels in excess of NHMRC guidelines (*risky drinker*) were low but significantly above those for the oldest respondents. The highest OR was in the 20–29 group and then steadily declined with increase in age. When non-risky drinkers were added to the analysis, the model's Nagelkerke R^2 decreased to 0.144 and the age gradient was flattened (see Fig. 8.1). The ORs for the other independent variables were not affected substantially.

Both tobacco smoking and risky drinking have been associated with young males rather than young females although this pattern has been steadily changing. The odds of a male being in this group were nearly double those of a female.

The model indicates that, as income increases, so too do the odds of simultaneously smoking and consuming alcohol at risky levels. This is contrary to other strong indicators of membership of this group; being unemployed, being in a blue-collar occupation compared with a white-collar occupation and having a lower level of post-school qualification. These factors tend to be associated with lower, not higher, income and greater rather than less socio-economic disadvantage. Moreover, there was no clear relationship between membership of this group and living in an area of distinct socio-economic disadvantage.

People who reported a mental health issue were more likely than those who did not report such an issue to belong to this group. Among those who did report a mental health issue, those who were not taking medication for depression were even

Table 8.2 Marginal distributions in the sample and weighted estimates for various characteristics used: Australia, National Health Survey, 2014–15

	Sample		Weighted estimates		
Characteristic	No.	%	No. in 000's	RSE	%
Alcohol status					
Risky drinkers	6198	40.6	7902	0.0115	42.7
Non-risky drinkers	5857	38.4	6630	0.0152	35.9
Ex-drinkers	1266	8.3	1487	0.0343	8.0
Non-drinkers	1951	12.8	2468	0.0313	13.4
Smoking status					
Current smoker	2452	16.1	2859	0.0283	15.5
Ex-smoker	4770	31.2	5547	0.0165	30.0
Non-smoker	8050	52.7	10,081	0.0093	54.5
Age					
15–19	1134	7.4	1488	0.0286	8.0
20–29	1958	12.8	3225	0.0132	17.4
30–44	4091	26.8	4841	0.0008	26.2
45–64	4887	32.0	5698	0.0010	30.8
65+	3202	21.0	3236	0.0029	17.5
Gender					
Males	7043	46.1	9126	0.0014	49.4
Females	8229	53.9	9361	0.0012	50.6
Education					
Bachelor and above	4002	26.2	4824	0.0185	26.1
Diploma	1592	10.4	1996	0.0288	10.8
Certificate	3591	23.5	4499	0.0204	24.3
No post-school education	6087	39.9	7169	0.0145	38.8
Occupation					
White collar occupations	6751	44.2	8289	0.0133	44.8
Blue collar occupations	2640	17.3	3615	0.0212	19.6
Unemployed	517	3.4	704	0.0625	3.8
Not in labour force	5364	35.1	5880	0.0151	31.8
Income					
Q1 (lowest)	2484	16.3	3260	0.0247	17.6
Q2	2927	19.2	3358	0.0208	18.2
Q3	2634	17.2	3196	0.0273	17.3
Q4	2675	17.5	3267	0.0220	17.7
Q5 (highest)	2802	18.3	3244	0.0241	17.5
Income not stated	1750	11.5	2162	0.0377	11.7
SEIFA index of disadvantage quintile					
Q1 (least disadvantaged)	2824	18.5	3553	0.0522	19.2
Q2	3012	19.7	3611	0.0485	19.5
Q3	3169	20.8	3753	0.0511	20.3
Q4	3084	20.2	3819	0.0484	20.7
Q5 (most disadvantaged)	3183	20.8	3751	0.0431	20.3

(continued)

Table 8.2 (continued)

Characteristic	Sample		Weighted estimates		
	No.	%	No. in 000's	RSE	%
Kessler's stress scale					
Low distress level	9826	64.3	12,011	0.0073	65.0
Moderate distress level	2759	18.1	3437	0.0230	18.6
High distress level	1189	7.8	1405	0.0332	7.6
Very high distress level	578	3.8	657	0.0549	3.6
Could not be determined	90	0.6	131	0.1326	0.7
Persons 15– < 18	830	5.4	846	0.0089	4.6
Reported mental illness					
None	13,625	89.2	16,694	0.0037	90.3
Used anti-depressants	1178	7.7	1251	0.0337	6.8
Not used anti-depressants	469	3.1	543	0.0780	2.9
Total	**15,272**	**100.0**	**18,488**	**0.0007**	**100.0**

more likely than those who did take medication to be currently smoking and consuming alcohol at risky levels.

Group F: Ex-smokers and Non-risky Drinkers (Nagelkerke R^2 *= 0.137)*

This group is of interest as, at least superficially, they seem to be heeding advice from health educators and policy-makers. Unfortunately, across all variables there were no clear gradients, to enable identification of groups of people who have given up smoking and who consume alcohol within NHMRC guidelines.

The typical respondent could have come from any of the age groups, was equally as likely to be male or female, could live in any area on the SEIFA disadvantage scale, did not report any particular level of distress and there was no real impact of reporting a mental health issue, either being treated with anti-depressants or not. While the typical respondent was not located in any particular income quintile, those in the highest income quintile were the least likely to be in this group. Lower levels of post-school education was also associated with higher odds of belonging to this group. Both blue-collar workers and those who were unemployed were slightly less likely than white-collar workers to be in this group.

Table 8.3 Odds ratios (OR) and 95% confidence intervals (CI) based on binary logistic regression models for selected groups of population: Australia, National Health Survey, 2014–15

	Group A: current smokers and risky drinkers		Group F: ex-smokers and non-risky drinkers		Group I: non-smokers and risky drinkers		Group L: non-smokers and non-drinkers	
Characteristic	OR	95% CI	OR	95% CI	OR	95% CI	OR	95% CI
Age								
65+	1.000		1.000		1.000		1.000	
45–64	5.594	5.543–5.646	0.517	0.515–0.519	1.741	1.730–1.753	0.842	0.837–0.846
30–44	8.959	8.875–9.044	0.251	0.250–0.253	2.692	2.674–2.709	1.194	1.187–1.201
20–29	10.233	10.136–10.331	0.086	0.085–0.087	6.785	6.741–6.829	0.870	0.864–0.875
15–19	4.386	4.331–4.442	0.040	0.040–0.041	11.186	11.094–11.279	1.649	1.634–1.664
Gender								
Females	1.000		1.000		1.000		1.000	
Males	1.823	1.816–1.829	0.948	0.944–0.951	1.674	1.670–1.679	0.505	0.504–0.507
Education								
Bachelor and above	1.000		1.000		1.000		1.000	
Diplomas	2.184	2.170–2.199	1.143	1.137–1.149	0.953	0.949–0.957	0.637	0.633–0.641
Certificates	3.297	3.279–3.316	1.287	1.281–1.293	0.925	0.921–0.928	0.316	0.314–0.318
No post-school education	3.711	3.691–3.732	1.204	1.199–1.210	0.855	0.852–0.859	0.610	0.607–0.613
Occupation								
White collar occupations	1.000		1.000		1.000		1.000	
Blue collar occupations	1.311	1.306–1.317	0.788	0.784–0.792	0.691	0.689–0.694	1.415	1.407–1.423
Unemployed	2.464	2.444–2.484	0.920	0.910–0.929	0.598	0.593–0.602	1.377	1.365–1.388
Not in labour force	1.002	0.997–1.008	0.995	0.991–1.000	0.415	0.413–0.417	1.985	1.975–1.994
Income quintile								
Q1 (lowest 20%)	1.000		1.000		1.000		1.000	
Q2	1.418	1.409–1.427	0.819	0.814–0.823	1.102	1.096–1.108	0.765	0.761–0.769
Q3	1.733	1.722–1.745	0.893	0.888–0.899	0.927	0.922–0.932	0.583	0.580–0.587

Q4	1.845	1.832–1.858	0.994	0.987–1.000	1.196	1.190–1.203	0.519	0.515–0.522
Q5 (highest 20%)	1.924	1.909–1.938	0.628	0.624–0.632	1.637	1.628–1.647	0.359	0.356–0.362
SEIFA index of disadvantage quintile								
Q5 (Most disadvantaged)	1.000		1.000		1.000		1.000	
Q4	1.156	1.150–1.162	0.867	0.863–0.871	0.944	0.940–0.947	1.366	1.358–1.373
Q3	1.077	1.071–1.083	0.917	0.912–0.921	0.853	0.850–0.856	1.600	1.591–1.609
Q2	1.154	1.148–1.161	0.932	0.927–0.937	0.729	0.726–0.733	1.779	1.769–1.789
Q1 (least disadvantaged)	1.261	1.254–1.268	0.910	0.905–0.914	0.512	0.510–0.514	2.455	2.441–2.469
Kessler's stress scale								
Low distress level	1.000		1.000		1.000		1.000	
Moderate distress level	1.387	1.381–1.392	0.939	0.935–0.943	0.945	0.942–0.949	0.961	0.957–0.965
High distress level	1.779	1.769–1.789	0.936	0.930–0.941	0.791	0.787–0.795	0.838	0.832–0.843
Very high distress level	1.921	1.906–1.935	1.020	1.012–1.029	0.470	0.465–0.474	1.004	0.995–1.013
Reported mental illness								
None	1.000		1.000		1.000		1.000	
Used anti-depressants	1.212	1.205–1.220	0.999	0.993–1.005	0.953	0.947–0.958	0.555	0.550–0.559
Not used anti-depressants	1.580	1.567–1.594	1.079	1.070–1.088	0.830	0.822–0.837	0.810	0.802–0.818

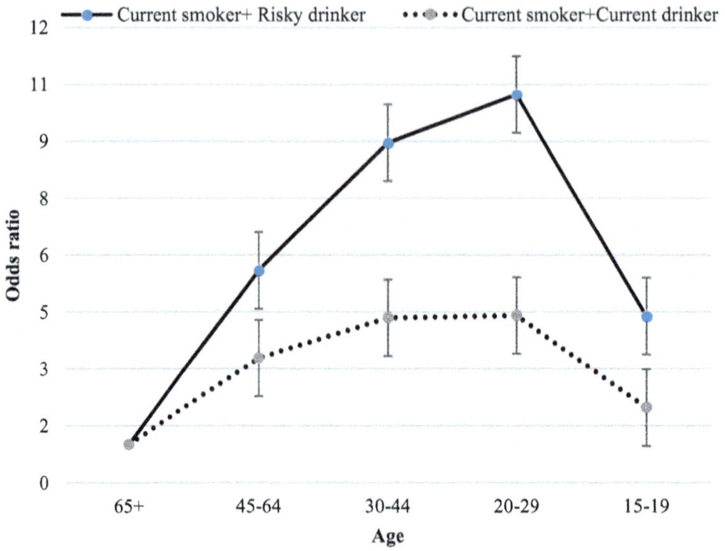

Fig. 8.1 Age-specific odds ratios for Current smokers & Risky drinkers compared with Current smokers and Current drinkers (*i.e., those drinking at* risky or non-risky levels): Australia, National Health Survey, 2014–15

Group I: Non-smokers and Risky Drinkers (Nagelkerke R^2 *= 0.189)*

This was a large group, containing the most frequently occurring behaviours, never having taken up smoking and consuming alcohol at a risky level.

There was a strong age gradient within this group with the odds of belonging increasing markedly with decreasing age. The OR of the youngest group (15–19), reporting this pattern of behaviour was 11 times that of the reference group. The odds steadily declined from there but the ratio for the second oldest group (45–60), was still well above that for the reference group.

The OR for males belonging in this group was again significantly higher compared with females. The odds of belonging to this group declined steadily with decreases in disadvantage. The OR for the least disadvantaged group was only half that of the most disadvantaged. This was consistent with OR for the highest income quintile also being the highest of those for all quintiles. There was no clear pattern in ORs for second, third and fourth quintiles. It was also consistent with a lower level of post-school education and with being in a blue-collar occupation or unemployed – all being associated with reduced ORs of belonging to this group.

An increase in distress level was associated with reduced odds of belonging to this group but there was no association between reporting a mental health issue (either taking medication or not) and membership of this group.

Group L: Non-smokers and Non-drinkers (Nagelkerke $R^2 = 0.259$*)*

This model achieved the highest Nagelkerke R^2 of the models considered and is of particular interest as it represents the ultimate objective of health educators and policy makers working to reduce the impact of avoidable morbidity and mortality associated with tobacco smoking and excessive alcohol consumption.

As is to be expected with such a large sample, all ORs were significantly different from the reference group (65+) but there was no clear association between membership of this group and any particular age group. In contrast with previous models, this was the only one where the odds ratio for males was about half that for females.

Once again, the relationship between disadvantage and being a non-drinker and a non-smoker was complex. There was a strong negative association between disadvantage and membership of this group. Members of the least disadvantaged group were more than twice as likely as members of the most disadvantaged group to be non-smokers and non-drinkers. A higher level of post-school qualification was also associated with being a non-smoker and non-drinker. On the other hand, increasing income was associated with reduced odds of belonging to this group. Persons in the highest income quintile were only one-third as likely as those in the lowest income quintile, the reference group, to be members of this group. Similarly, the ORs were higher for those in blue-collar occupations and those who were unemployed compared with the reference group.

Level of distress shows no clear relationship between changes in level and membership of this group. Respondents who reported that they were taking anti-depressants were only half as likely compared to those reporting no mental health issues to belong to this group.

Discussion

Our analysis has shown that in Australia, most people do not smoke but they do drink alcohol. Two groups of particular interest, each representing approximately one tenth of the population, 15 and over are at the extremes of the behavioural patterns of interest: those who both currently smoke and consume alcohol at risky levels and those who have never smoked or consumed alcohol.

Tobacco smoking has been the target of sustained, comprehensive initiatives throughout Australia for more than 50 years, including reducing the availability of tobacco to minors; anti-smoking advertising in print, broadcasting and outdoor media; bans on pro-tobacco advertising and promotion; increasing restrictions on spaces where smoking is allowed; "plain" (*i.e.*, unadorned with promotional material) packages featuring large, graphic warnings about the damage that smoking causes; significant price increases generated by continual increases in government taxes on the sale of tobacco; and education campaigns in schools especially in primary schools. Alcohol has not been the target of similar initiatives and alcohol consumption remains pervasive across the Australian population. It has been suggested that alcohol control initiatives are unlikely to gain support in Australia due to alcohol's entrenched status in Australia, the power and resources of the alcohol industry, and the fact that community is reluctant to admit the damage that alcohol is doing to people, the community and the economy. These factors were also applicable to tobacco smoking when tobacco control campaigns began. It took many years but tobacco control campaigns are steadily paying off. Now is the time for alcohol control activists to learn from the tobacco activists and begin developing effective interventions. Policy-makers are charged with the responsibility to ensure that effective pressure is kept on tobacco control while alcohol control pressure is ramped up.

The next step is to identify the segments of the population most at risk and to tailor appropriate strategies to help them, as simple, undifferentiated strategies are unlikely to be equally effective across different groups.

Our findings indicate that young males are the prime target. These are the people who are most likely to smoke tobacco and drink at risky levels and least likely to belong to the never smoked, never drank group. Young males are also the most likely to be non-smokers but to consume alcohol to risky levels, suggesting that they and their circle have heeded the anti-smoking messages but there have been no similar, comprehensive alcohol initiatives. This reinforces the need for increased pressure on alcohol consumption without reducing the pressure on cigarette smoking.

Other groups in the community who are smoking and drinking at risky levels are those who might be "self-medicating" to deal with stress although the picture is complex. Because smoking is now an expensive habit, those with least disadvantage and in the highest income quintiles are the most likely to be in the current smoker, risky drinker group and those with the greatest disadvantage and lowest income being the least likely to be in the non-smoker, non-drinker group. On the other hand, those coping with stress are most likely to be current smokers and risky drinkers. Those who are dealing with mental health issues but are not taking medication, seem to be using tobacco and alcohol to help deal with the issues. Segments in the community might be dealing with the stress of being unemployed or feeling locked out of the employment market due to having a low level of education or being employed in a low-status occupation when low-status occupations are being taken over by machines or exported to low labour-cost countries, by turning to tobacco and alcohol for relief.

Initiatives directed at these groups should not focus principally on the damage that smoking and alcohol consumption are doing – thereby adding to the stress – but, rather on providing and promoting better solutions to the problems.

References

Australian Bureau of Statistics. (2012). *Information paper: Use of the Kessler Psychological Distress Scale in ABS Health Surveys*, Australia. http://www.abs.gov.au/ausstats/abs@.nsf/Lookup/4817.0.55.001Chapter192007-08

Australian Bureau of Statistics. (2013). *Technical paper: Socio-Economic Indexes for Areas (SEIFA)*. http://www.abs.gov.au/ausstats/abs@.nsf/mf/2033.0.55.001

Australian Bureau of Statistics. (2016). *National Health Survey. Users' Guide, 2014–15*. http://www.abs.gov.au/AUSSTATS/abs@.nsf/DetailsPage/4363.02014-15?OpenDocument

Australian Institute of Health and Welfare. (2017). *National drug strategy household survey 2016: Detailed findings* (Drug Statistics Series No. 31. Cat. No. PHE 214). Canberra.

Baker, A., Ivers, R. G., Bowman, J., Butler, T., Kay-Lambkin, F. J., Wye, P., Walsh, R. A., Pulver, L. J., Richmond, R., Belcher, J., Wilhelm, K., & Wodak, A. (2006). Where there's smoke, there's fire: High prevalence of smoking among some sub-populations and recommendations for intervention. *Drug and Alcohol Review, 25*, 85–96. https://doi.org/10.1080/09595230500459552.

Banks, E., Joshy, G., Weber, M., Liu, B., Grenfell, R., Egger, S., Paige, E., Lopez, A. D., Sitas, F., & Beral, V. (2015). Tobacco smoking and all-cause mortality in a large Australian cohort study: Findings from a mature epidemic with current low smoking prevalence. *BMC Medicine, 13*, 38–42. https://doi.org/10.1186/s12916-015-0281-z.

Bonevski, B., Regan, T., Paul, C., Baker, A., & Bisquera, A. (2014). Associations between alcohol, smoking, socioeconomic status and co-morbidities: Evidence from the 45 and up study. *Drug and Alcohol Review, 33*(2), 169–176. https://doi.org/10.1111/dar.12104.

Bowden, J. A., Miller, C. L., & Hiller, J. E. (2011). Smoking and mental illness: A population study in South Australia. *Australia and New Zealand Journal of Psychiatry, 45*, 325–331. https://doi.org/10.3109/00048674.2010.536904.

Collins, D. J., & Lapsley, H. M. (1996). *The social costs of drug abuse in Australia in 1988 and 1992* (Vol. 30). Canberra: Australian Government Publishing Service.

Falk, D. E., Yi, H., & Hiller-Sturmhafel, S. (2008). An epidemiologic analysis of co-occurring alcohol and tobacco use and disorders. *Alcohol Research and Health, 31*(2), 100–110. ISSN:1535-7414.

Magnus, A., Cadilhac, D., Sheppard, L., Cumming, T., Pearce, D., & Carter, R. (2012). Economic benefits of achieving realistic smoking cessation targets in Australia. *American Journal of Public Health, 102*, 1313–1319. https://doi.org/10.2105/AJPH.2009.191056.

Morley, K. I., & Hall, W. D. (2007). Explaining the convergence of male and female smoking prevalence in Australia. *Addiction, 103*, 487–495. https://doi.org/10.1111/j.1360-0443.2007.02087.x.

National Health and Medical Research Council. (2009). *Australian guidelines to reduce health risks from drinking alcohol*. Canberra. https://www.nhmrc.gov.au/guidelines-publications/ds10

Reis, A. M., Quintal, C., & Lourenco, O. (2017). Killing two birds with one stone? Association between tobacco and alcohol consumption. *Public Health, 154*, 136–143. https://doi.org/10.1016/j.puhe.2017.10.019.

Smoking and Health. (1964). *A report of the advisory committee to the Surgeon General of the Public Health Service* (Public Health Service Publication No. 1103). Washington, DC.

Twyman, L., Bonevski, B., Paul, C., Bryant, J., West, R., Siahpush, M., D'Este, C., Oldmeadow, C., & Palazzi, K. (2016). Factors associated with concurrent tobacco smoking and heavy alcohol consumption within a socioeconomically disadvantaged Australian sample. *Substance Use & Misuse, 51*(4), 459–470. https://doi.org/10.3109/10826084.2015.1122065.

WHO. (2014). *Global status report on alcohol and health*. http://www.who.int/substance_abuse/publications/global_alcohol_report/msb_gsr_2014_1.pdf?ua=1

Yusuf, F., & Leeder, S. (2015). Making sense of alcohol consumption data in Australia: Has alcohol consumption in Australia increased in recent years? *Medical Journal of Australia, 203*(3), 128–130. https://doi.org/10.5694/mja15.00151.

Chapter 9
A Cross-National Analysis of Premature Non-communicable Diseases (NCD) Mortality Differentials Among 183 Countries

Jinyuan Qi

Abbreviations

CVD	Cardiovascular Diseases
GBD	Global Burden of Diseases studies
HI	High-income country according to World Bank classification
IHME	Institute for Health Metrics and Evaluation
LI	Low-income country according to World Bank classification
LMI	Lower-middle-income country according to World Bank classification
LMIC	Low- and middle-income country according to World Bank classification
NCD	Non-communicable Diseases
SDG	Sustainable Development Goal
SES	Socioeconomic status
UMI	Upper-middle-income country according to World Bank classification
UN	United Nations
UNDP	United Nations Development Programme
WHO	World Health Organization

Introduction

Non-communicable diseases (NCD), which include cardiovascular diseases (CVD) and cancer (the full list of NCDs is in Appendix A), constitute the major burdens of illness and disability throughout the world. NCDs account for around 60% of all deaths and 44% of premature deaths (before age 70) worldwide, and 80% of

J. Qi (✉)
Princeton University, Princeton, NJ, USA
e-mail: jinyuanq@princeton.edu

B. Jivetti, Md. N. Hoque (eds.), *Population Change and Public Policy*, Applied Demography Series 11, https://doi.org/10.1007/978-3-030-57069-9_9

chronic-disease deaths occur in low- and middle-income countries (Daar et al. 2007). The annual decline rates of NCD mortality between age 50 and 70 from 1990 to 2015 showed a widening gap among low- and middle-income countries and high-income countries (Verguet et al. 2017). Further, population aging has increased the prevalence and mortality of NCDs for all ages globally, while age-specific NCD mortality rates had different patterns at different income levels (Verguet et al. 2017). The recent WHO global status report (2014b) on NCD gave encouraging evidence that premature NCD deaths could be significantly reduced worldwide since most premature NCD deaths are preventable. To enhance public health policies and address current health disparities, it is urgent to study premature NCD mortality differentials across countries and develop population metrics for policy and program surveillance. As many macro-level studies focus on the global impact of NCDs on macro-economic productivity (Chaker et al. 2015), less work has examined the macro-level demographic and social factors associated with premature NCD mortality differentials across countries. Since one of the Sustainable Development Goals (SDGs) is reducing by one third premature deaths from NCDs by 2030 (SDSN Target 3.4, 2016), a cross-national analysis is necessary to inform the demographic and epidemiological challenges to achieve the goal.

This study aims to quantify the impact of demographic changes including population growth and population aging on premature mortality of NCDs and explore associated social factors, such as GDP and education, to explain the premature NCD mortality differentials across countries. In the other words, this study analyzed country-level data from 183 WHO member states (the full list is in Appendix B) to investigate two main questions – how demographic change and epidemiological change affected premature NCD deaths and what the macro-level factors were associated with premature NCD mortality differentials across countries in last two decades. Measuring premature mortality in this study, rather than overall mortality, is to focus attention on deaths that could have been prevented. By examining premature mortality rates across countries and investigating the underlying determinants of premature mortality, resources can be targeted toward strategies that will efficiently extend the life expectancy of populations.

Background and Literature Review

Decomposition

To analyze the demographic and epidemiological drivers of premature NCD deaths, decomposition of the changes of premature NCD deaths are needed. An analysis on global cardiovascular (CVD) mortality (Roth et al. 2015) developed three counterfactual scenarios to represent the principal drivers of change in CVD deaths (population growth alone, population growth and aging, and epidemiologic changes in disease) from 1990 to 2013. To my knowledge, this method was originally developed by Mathers and Loncar (2006) in their analyses of the relative impact of demographic and epidemiological change on the projected numbers of deaths by cause by calculating three hypothetical alternatives from 2002 to 2030. They first calculated the expected number of deaths in 2030 given the 2030 projected age-specific rates under the baseline scenario and the 2002 population. The difference between this and the 2002 mortality estimate is a measure of the change in mortality expected solely on the basis of changing age-specific mortality rates, and is labelled "epidemiological change". Second, they calculated the expected number of deaths in 2030 by taking the 2002 age-specific death rates and applying them to the 2030 projected population. The difference between this and the 2002 mortality estimate is a measure of the change in mortality expected solely on the basis of changing demography (including size and age composition of the population). They then repeated this calculation, but applied the 2002 age-specific death rates to projected population numbers for 2030 that matched the total projected male and female population numbers for each country but retained the 2002 age distribution of the population. The difference between this and the 2002 mortality estimates is a measure of the change in mortality expected solely on the basis of population growth excluding changes in age composition. The difference between the total change in mortality due to demography and this latter estimate gives a measure of the effect of the change in age composition of the population alone. These two components of demographic change are labelled "population growth" and 'population ageing'. The total projected change in numbers of deaths between 2002 and 2030 is the sum of the population growth, population ageing, and epidemiological change components.

Since this study focuses on the premature NCD mortality as WHO (2018) defined premature NCD deaths are those NCD deaths occurring under age 70, the three-scenario method for all age groups between two years/periods needs to be modified. In the methods section, I formalized their method for a broad age group between age 50 and 69, which is the most representative age group for premature NCD mortality comparison across countries because most of premature NCD deaths occur between 50 and 69 in the most countries (Appendix C). Further, the proportion of the age group between 50 and 69 among the whole population is a better indicator of population aging for most countries than the proportion of the age group 30–69 including middle-aged population between 30 and 49. The effect of the change of age composition for all ages on NCD deaths increase could be called "population aging" in previous research because the death rates of NCD are usually higher in older age groups than the younger age group before age 30. However, this may not be true in many sub-Saharan African countries when we only look at premature NCD deaths as the newborns could have higher NCD death rates than those aged 30–49 because of the high mortality of congenital birth defects (IHME 2015). The change of age-composition with high fertility rate would not be necessarily called population aging in these countries. A detailed analysis to support this argument is in the Appendix C by showing the percentage of premature NCD deaths among all NCD deaths, and the percentage of NCD deaths occurred within three different age groups (30–69, 30–49, and 50–69) among all premature NCD deaths.

Determinants

To explore the macro-level factors associated with premature NCD mortality differentials, frameworks have been developed to explain population health differentials. One of the most influential conceptual frameworks to understand social determinants of health was originally developed by Solar and Irwin (2007) and amended in WHO's report (2008) of the Commission on Social Determinants of Health (CSDH) (Fig. 9.1). This framework has three domains (A-Health Inequity; B-Social Stratification and Intermediate Determinants; C-Structural Determinants) and divides social determinants of health and health inequalities into three interrelating categories – socioeconomic and political contexts including governance, policy, cultural and societal norms and values are structural determinants; social positions including education, income, gender and race as a result of social stratification are influenced by the context and will affect the distribution of health and well-being through intermediate determinants; health care system will target those intermediate determinants including material circumstances, social cohesion, psychological factors, behaviors and biological factors. For this macro-analysis of premature NCD mortality differentials across countries, macro-level determinants of premature NCD mortality are particularly needed but not specified in CSDH original framework. United Nations Development Programme (UNPD) applied this framework to NCD outcomes and risk factors in a discussion paper (2013) by using CVD and diabetes as illustrative examples given the wide scope of the NCD epidemic (and the reality of multiple, overlapping NCD epidemics). UNPD's framework clearly highlights root causes of the level and distribution of NCDs. As this macro-analysis focuses on the variability of premature NCD mortality across countries, the literature review will follow UNPD's framework of various major factors and focus on macro-level studies.

Domain A – Health Inequity

Sex

The literature has suggested that sex differences are evident at country and regional levels. UNPD (2013) specifically analyzed CVD and diabetes as the leading NCDs – sex-difference patterns are complex. While the finding that men tended to have higher CVD mortality rates in all but the oldest age groups remained consistent historically across countries and regions (Kelly and Fuster 2010), the findings varied for diabetes as women generally had higher premature diabetes deaths than men with a small difference (GBD 2015; WHO 2014b). The sex differences of NCD mortality could be due to the different levels of exposure and vulnerability to NCD risk factors. For example, social customs related to physical mobility may reduce women's opportunities for activity, reflected in disparities between men and women in

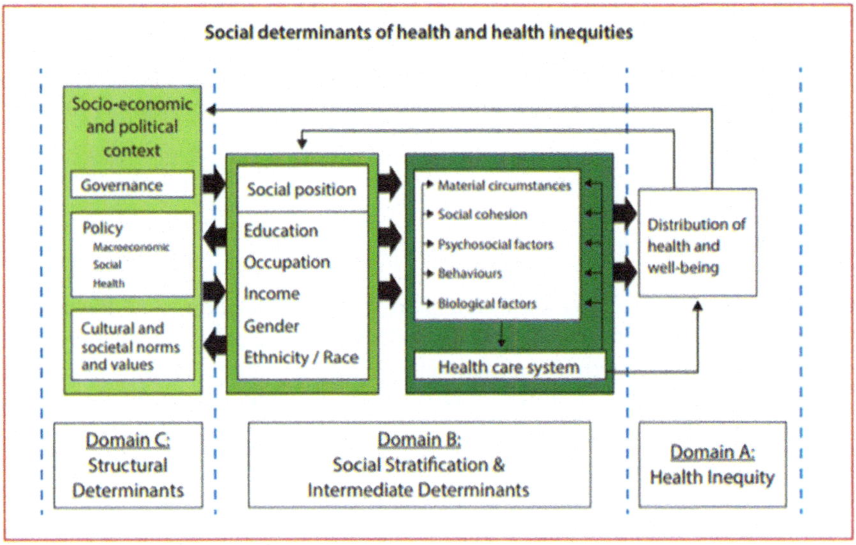

Fig. 9.1 The social determinants of health by Solar and Irwin (2007) as cited in WHO-CSDH 2008

physical activity levels (WHO 2009). Additionally, smoking was generally more prevalent among men, but the prevalence of smoking among women has been changing with gender empowerment (Hitchman and Fong 2011). Thus, NCD mortality differences between sexes were hypothesized to be significant but could have different patterns in different countries or regions within different periods.

Income

Many studies investigated the association between socioeconomic status (SES) and NCD mortality. SES is usually measured by income and education. A synthesis (Sommer et al. 2015) of 22 systematic reviews regarding this topic has shown that having low SES and/or living in low- and middle-income countries (LMICs) increased the risk of developing CVD, lung and gastric cancer, type 2 diabetes, and chronic obstructive pulmonary disease (COPD), and low SES increased the risk of mortality from lung cancer, COPD, and reduced breast cancer survival in high-income countries. UNDP (2013) suggested that CVD and diabetes disproportionately affected lower SES group within countries, but limited evidence has been found in low-income countries. In general, higher income was hypothesized to be associated with lower NCD mortality across countries but low-income countries do not have enough evidence.

SES is also associated with NCD risk factors. A recent systematic review (Allen et al. 2017) summarized the correlation that low SES groups were found to have a significantly higher prevalence of tobacco and alcohol use than high SES groups, while high SES groups were found to be less physically active and consume more fats, salt, and processed food than low SES groups. The patterns for tobacco use and physical activity were clear in the presented studies included in the review (Allen et al. 2017), while the heterogeneity between dietary outcome measures and a paucity of evidence around harmful alcohol use limited the certainty of these findings.

Education Attainment

As education attainment is a measurement of SES, the literature about the association between SES and NCD mortality and risk factors suggested that generally lower education attainment was associated with higher mortality. However, a cross-national study (Goyal et al. 2010) of 44 countries found that more education might have no protective effect on CVD in LMICs, particularly for women, while more education was associated with lower diabetes prevalence globally. UNPD (2013)'s analyses within countries showed that low education was associated with CVD in high-income countries, but the evidence was less clear in lower-income settings. Thus, we could hypothesize that higher education was generally associated with lower NCD mortality.

Domain B –Social Stratification and Intermediate Determinants

Social stratification could influence intermediate determinants as stated in the Domain A – sex and SES are associated with different risk factors. The health system influences the intermediate determinants, especially biological risk factors, such as raised blood pressure, raised cholesterol, and obesity. Many biological risk factors for NCD are driven by behavioral risk factors. Tobacco use, harmful use of alcohol, physical inactivity and unhealthy diet are four main behavior risk factors shaped by social stratification (UNDP 2013). All those biological/behavioral risk factors are hypothesized to be positively associated with higher NCD mortality. Therefore, the effectiveness of most biomedical approaches will be enhanced when combined with multispectral action on the social determinants that influence risk behaviors in the first place.

Obesity

Obesity is a major biological risk factor for NCD with increasing global prevalence (Lim et al. 2013). The obesity epidemic began in high-income countries in the 1970s and 1980s and has subsequently increased in LMICs (Swinburn et al. 2011). There are some hypotheses that explain the increasing trend of obesity – reductions in physical activity, increases in sedentary behaviors, increased reliance upon processed foods, and increased use of edible oils and sugar-sweetened beverages (Popkin et al. 2012). A cross-national study found that obesity was more prevalent among more educated women in low-income countries but lower among more educated women in upper middle-income countries (Monteiro et al. 2004). Another major study among 54 LMICs showed a positive association between income and overweight/obesity among women (Subramanian et al. 2011). Thus, women tended to have higher obesity prevalence than men. Obesity tended to affect those with higher incomes or education (SES) in lower income countries, while obesity affected those with lower incomes or education (SES) in higher income countries.

Tobacco Use

Tobacco use is the only modifiable risk factor common to all four major NCDs – CVD, cancer, chronic respiratory diseases and diabetes (Eriksen et al. 2013). A study in 2000 found that tobacco use and related mortality was increasingly concentrated in LMICs, and 70% of the projected tobacco-related deaths would be in LMICs (Gajalakshmi et al. 2000). Generally, men have higher smoking prevalence than women, but the tobacco use among women are increasing at alarming rates (Hitchman and Fong 2011; Lim et al. 2013). While many studies investigated tobacco use gradient within a country or region and found higher smoking prevalence among lower SES groups (Eriksen et al. 2013; Hiscock et al. 2012), a few studies showed that cross-national pattern was more complicated – tobacco-use levels did not vary simply by region, and historical economic conditions related to the timing of adoption and current national differences in smoking (Pampel 2010). Educational inequalities in smoking were found to follow markedly different patterns in different regions – among both men and women, the differences between the educational groups were quite small in Sub-Saharan Africa and Latin America; among women, in Latin America and the Eastern Mediterranean a positive gradient was observed, while in South-East Asia and East Asia a strong negative gradient was visible; among men, a slight positive gradient was observed in Eastern Europe and negative gradients in the other regions (Bosdriesz et al. 2014). Thus, for tobacco use, we could hypothesize men had higher prevalence than women but could not hypothesize a certain pattern across-countries based on the literature.

Harmful Use of Alcohol

Harmful use of alcohol is not only associated with NCD but also other health risks, such as sexually transmitted infections, domestic violence, and road traffic accidents (Lim et al. 2013; WHO 2014b). Harmful use of alcohol was the leading risk factor for premature deaths in males, yet there was evidence that women may be more vulnerable to alcohol-related harm (WHO 2014a). For health outcomes such as cancers, gastrointestinal diseases or CVDs, the same level of consumption led to more pronounced outcomes for women (Rehm et al. 2010). The gap between alcohol use levels among men and women might have been narrowed in line with economic development and changing gender roles (Wilsnack et al. 2014). Sex differences for health outcomes of harmful alcohol use were not clear when different health consequences were considered for different levels of alcohol use or drinking patterns. Most evidence of harmful use of alcohol was in high-income countries, while the evidence was quite limited in low-income countries. Research in high-income countries has shown harmful alcohol use was more prevalent among lower SES groups and more prevalent among men than among women (Schmidt et al. 2010). At macro-level, greater economic wealth was broadly associated with higher levels of alcohol consumption and lower abstention rates (WHO 2014a). However, for a given level or pattern of drinking, the alcohol-attributable mortality and burden of disease and injury would generally be greater in societies with lower economic development than in more affluent societies (WHO 2014a).

Physical Inactivity

Physical activity can prevent NCDs at population level as regular physical activity has been found to lower the risk of cancer, cardiorespiratory, metabolic, musculoskeletal and functional diseases (WHO 2010). An imbalance between insufficient energy expenditure through physical activity and too much energy input from dietary intake was associated with increased risk of overweight and obesity (UNDP 2013). The sex difference was evident – women were less likely than men to be physically active across most regions and countries (Guthold et al. 2008; WHO 2014b). Gender norms or concerns about safety might discourage women from engaging in physical activities outdoors (Addy et al. 2004). Higher-income countries tended to have lower overall levels of physical activity (WHO 2014b), but within higher-income countries higher SES groups tended to be more physically active (Giles-Corti and Donovan 2002).

Unhealthy Diet

Unhealthy diet, such as high levels of trans-fats, saturated fats, processed foods, sugar, salt and sugary drinks, was associated with increased risk of NCDs (WHO 2014b), while healthy diets with adequate intake of fruits and vegetables was associated with reduced risk of coronary heart disease and stroke (Dauchet et al. 2006; He et al. 2006). Unhealthy diet and insufficient physical activity were associated with higher risk of obesity (Cecchini et al. 2010). Sex and SES would impact diet. Low fruit intake was found to be a greater share of disease burden for men than women (Lim et al. 2013). Unhealthy diets were found to be associated with low SES groups with limited economic means and lower levels of education in high-income countries (Groth et al. 2001; Singh et al. 2010). Limited evidence was available in low-income countries.

Domain C – Structural Determinants

Structural determinants of health refer to macro-level socio-economic, political and cultural factors, policy choices and contexts. National and local contexts influence and are influenced by global factors, including global and regional governance arrangements, trade regimes and treaties (WHO 2008). Structural factors generate social stratifications of individuals and groups, including by income and education. As expressed in the CSDH framework, social stratifications (Domain B) ultimately result in health inequities (Domain A). Thus, structural determinants matter greatly for NCD prevention and development (UNDP 2013). Because structural determinants of NCDs have great variations at macro-level, their quantitative impacts and cross-national differentials on NCD outcomes are often difficult to determine. UNDP (2013) used case studies within countries to show that public health interventions did have certain or probable impacts on NCD dynamics and risk factors. To make cross-country comparisons, public health expenditure could be useful to indicate the investments to implement public health interventions, which are structural determinants. The performance of health system was found to increase with higher health expenditure per capita (Evans et al. 2001). Previous studies have shown the importance of GDP as a determinant of health spending in both developing and developed countries (Hitiris and Posnett 1992; Lu et al. 2010). Thus, we could hypothesize that NCD mortality was negatively associated with high public health expenditure per capita.

As mentioned above, for some major determinants, cross-national comparison is difficult due to data unavailability, especially in low-income countries. Some hypotheses cannot be clearly determined from previous literature with contradictory findings in different regions at different times. Data and methods section introduce available data for examining most of the listed determinants of NCD mortality.

Data and Methods

Decomposition

The first part of this study tried to answer how much of NCD mortality changes over past two decades were attributed to demographic changes by decomposing the NCD death counts. NCD death counts D occurred at age 50 to 69 in a certain year can be written as

$$D = K \cdot P \cdot M$$

Where K represents the total population in that year, P represents the proportion of those who aged 50–69 in the total population and M represents the age-specific (50–69) NCD death rate. If we have continuous data, we could get derivatives of above equation to decompose the factor of population growth, population aging and epidemiological change as

$$\frac{\dot{D}}{D} = \frac{\dot{K}}{K} + \frac{\dot{P}}{P} + \frac{\dot{M}}{M}$$

However, the available data are far from being continuous considering a lot of missing data. Thus, we need to decompose the three factors in discrete terms between any two time points.

$$\mathrm{D} = \mathrm{K} \cdot \mathrm{P} \cdot \mathrm{M} \text{ and } \mathrm{D} + \Delta\mathrm{D} = (\mathrm{K} + \Delta\mathrm{K}) \cdot (\mathrm{P} + \Delta\mathrm{P}) \cdot (\mathrm{M} + \Delta\mathrm{M})$$

$$\mathrm{D} + \Delta\mathrm{D} = \mathrm{K} \cdot \mathrm{P} \cdot \mathrm{M} + \Delta\mathrm{K} \cdot \mathrm{P} \cdot \mathrm{M} + \mathrm{K} \cdot \Delta\mathrm{P} \cdot \mathrm{M} + \mathrm{K} \cdot \mathrm{P} \cdot \Delta\mathrm{M}$$

$$+\Delta\mathrm{K} \cdot \Delta\mathrm{P} \cdot \mathrm{M} + \Delta\mathrm{K} \cdot \mathrm{P} \cdot \Delta\mathrm{M} + \mathrm{K} \cdot \Delta\mathrm{P} \cdot \Delta\mathrm{M} + \Delta\mathrm{K} \cdot \Delta\mathrm{P} \cdot \Delta\mathrm{M}$$

$$\Delta\mathrm{D} = \Delta\mathrm{K} \cdot \mathrm{P} \cdot \mathrm{M}\ (\textit{pop growth}) + \mathrm{K} \cdot \Delta\mathrm{P} \cdot \mathrm{M}(\textit{pop aging}) + \mathrm{K} \cdot \mathrm{P} \cdot \Delta\mathrm{M}(\textit{epi change})$$

$$+\Delta\mathrm{K} \cdot \Delta\mathrm{P} \cdot \mathrm{M} + \Delta\mathrm{K} \cdot \mathrm{P} \cdot \Delta\mathrm{M} + \mathrm{K} \cdot \Delta\mathrm{P} \cdot \Delta\mathrm{M}$$
$$+ \Delta\mathrm{K} \cdot \Delta\mathrm{P} \cdot \Delta\mathrm{M}\ (\textit{four interaction terms.})$$

Population growth refers to the change of NCD death counts (occurred at age 50–69) attributed to the change of the whole population size if age composition (the proportion of those aged 50–69 here) and age-specific (50–69) NCD death rates remained same. Similarly, population aging refers to the change of NCD death counts (occurred at age 50–69) attributed to the change of the proportion of people aged 50–69 if the whole population size and NCD death rates remained same. The epidemiological change refers to the change of NCD death counts (occurred at age 50–69) driven by the change of age-specific (50–69) NCD death rates if the population size and age composition remained same. There are four interactions the equation, and the sign of each interaction would depend on the positive or negative changes of three dominant factors over time. The total of the interaction effects would influence total NCD death counts, but the influence is expected to be not big. As the scale of change over one period is rather small and the changes could happen in opposite directions, all interaction terms having two or three small changing values with different signs would add up to a moderate number.

For decomposition analysis of premature NCD deaths, this study used population data of 183 countries including general population and death counts by age and sex from United Nations Population Division (1990–2015) (UN 2015), and the cause-of-death data for corresponding countries used the percentage of deaths caused by NCDs from the IHME Global Burden of Disease Database (1995–2015) (GBD 2015). Since the UN only provides death count estimates in 5-year period, this study used 5-year period as analysis unit and aggregated estimated mid-year population data (the sum of five continues years within the period) and NCD death percentage (the mean of five continues years within the period) into 5-year groups (1990–95, 1995–2000, 2000–05, 2005–10, and 2010–15). The decomposition of

NCD mortality change was performed between every two continuous periods (e.g., from 1990–95 to 1995–2000) using the equation above to see how much change of NCD deaths was driven up by demographic changes and how much was due to epidemiological change.

Determinants

The second part of the study attempted to explore the macro-level determinants of premature NCD mortality differentials across 183 countries over past two decades. Fixed-effects model is used for investigating the association between major determinants and premature NCD mortality to control for time-invariant and country-invariant effects. The dependent variable is age-specific (50–69) NCD death rate by sex (both, male, and female respectively) from the decomposition part. Appendix D shows a comparison of three NCD mortality measurements – age-specific (50–69) NCD death rate used here, the probability of dying because of NCD between age 50 and 69, and the probability of dying because of NCD between age 30 and 69. They are all significantly correlated and any of them can indicate the premature NCD mortality differentials across countries.

To examine the macro-level determinants of premature NCD mortality, the country-level data were used from various data sources in line with the major determinants listed in UNDP's framework. In the Domain A of UNDP's framework, sex differences are presented through premature NCD death rates by sex (both, male and female), and independent variables by sex. However, two independent variables including GDP per capita, and public health expenditure per capita do not have sex-specific data. The income in Domain A is represented as GDP per capita (2011 $International dollars) from 1990 to 2015, income-group classification in 2014 from the World Bank (2016) database. The education attainment in Domain A is represented as estimated years of schooling among those aged 50–69 by sex from 1990 to 2015 from the Wittgenstein Centre for Demography and Global Human Capital (2015).

Biological and behavior risk factors (obesity, tobacco use, alcohol use, insufficient physical activity) in Domain B are all from the WHO Global Health Observatory data repository (2017b). The obesity is measured by the age-standardized percentage of body mass index more than or equal to 30 by sex across countries from 1990 to 2014. The tobacco use is measured by the age-standardized prevalence estimates for current smoking of any tobacco product by sex for adults aged 15 years and over from 2000–2015. "Tobacco smoking" includes cigarettes, cigars, pipes or any other smoked tobacco products. "Current smoking" includes both daily and non-daily or occasional smoking. Since whether alcohol use is harmful depends on consumption levels and patterns and the threshold is different for different populations, this study used the recorded alcohol consumption per capita (in liters of pure alcohol) for both sexes for adults aged 15 years and over to provide a more comparable estimate of the level of alcohol use in a country from 1990 to 2014. For physical activity, WHO only has data for one year in 2010 by estimating the age-standardized and sex-specific percentage of defined population (adults) attaining less than 150 min of moderate-intensity physical activity per week, or less than 75 min of vigorous-intensity physical activity per week, or equivalent. Limited by the data availability, physical activity is not included in the model but examined in the results section too. The nutrition/diet data are not available across countries either. However, as stated from the previous literature, insufficient physical activity combined with unhealthy diet were associated with higher risk of obesity. Obesity, as the biological risk included in the model may reflect higher behavioral risk in either/both two determinants.

For Domain C, this study used public health expenditure per capita (in 2010 $International dollars) data from the WHO Global Health Expenditure Database (2017a) from 1990 to 2015. The public health expenditure per capita was calculated by the difference between total health expenditure per capita and private public health expenditure per capita in a country. Thus, it includes both governmental health expenditure and investment from non-government organizations. The analyses were performed at country level but are mainly presented by World Bank income levels (the classifications are shown in Appendix B).

Fixed-effects model was used to not only examine the determinants for premature NCD mortality, but also to investigate the association between SES and risk factors. Different models were tested to explain the cross-national differentials considering data availability and possibly non-linear association. Since different variables have very different scales and distributions, all variables were log-transformed except time to reduce skewness and make the relationship clearer. An example of the two-way fixed-effects model (Model 1in the results) for both sexes with all variables included is shown as below.

$$\begin{aligned}\ln M_{it} = {} & \beta_0 + \beta_1 \ln Y_{it} + \beta_2 \ln E_{it} + \beta_3 \ln OI_{it} + \beta_4 \ln TI_{it} + \beta_5 \ln AI_{it} \\ & + \beta_6 \ln PHE_{it} + \Sigma\mu_{l-1} T_{tl-1} + \Sigma\delta_{n-1} I_{in-1} + \varepsilon_{it}\end{aligned}$$

where M_{it} is the age-specific (50–69) NCD death rate for both sexes of country i ($i = 1, 2, 3, \ldots$, n) at time t ($t = 1, 2, \ldots$, l); $\ln Y_{it}$ is the log-transformed average income per capita, measured as gross domestic product (GDP) per capita (in 2011 international dollars); and $\ln E_{it}$ is the log-transformed average number of years of schooling among adults aged 50–69 for both sexes, referred to as "human capital"; T denotes time (period 1990–95, 1995–2000, 2000–05, 2005–10, 2010–15). In addition, $\ln OI_{it}$, $\ln TI_{it}$ and $\ln AI_{it}$ denote log-transformed tobacco use, alcohol use and obesity level for both sexes respectively. TI is measured by the age-standardized prevalence of smoking (%) among adults, and OI is measured by the age-standardized percentage of obesity (BMI $\geq$ 30) among adults. AI is measured by the liters of pure alcohol consumption (recorded) per capita per year. $\ln PHE_{it}$ is the log-transformed public health expenditure per capita; $\Sigma\mu_{l-1} T_{tl-1}$ and $\Sigma\delta_{n-1} I_{in-1}$ denote fixed effects for the l−1 time points and n−1 countries in the dataset, respectively; and ε_{it} is the error term. Additional analyses were performed for risk factors and public health expenditure to investigate the association between the prevalence of risk and income and education.

Results

Decomposition

The decomposition of premature NCD deaths (at age 50–69 for both sexes) change across 183 countries from period 1990–1995 to period 2010–2015 clearly shows that premature NCD deaths were driven up by population growth and population aging. Population aging became a dominant factor contributing to total death counts since 1995–2000, while epidemiological change of premature NCD death rate was declining in every period since 1990–95. The total interaction factor of four interaction terms was negative, however, the scale of the interaction was rather small compared to other three factors as expected. Globally (as shown in Figs. 9.2a and 9.2b), the annual rate of premature NCD mortality decline was staggering based on the density figure and the death counts driven by population aging has been steadily increasing over last two decades.

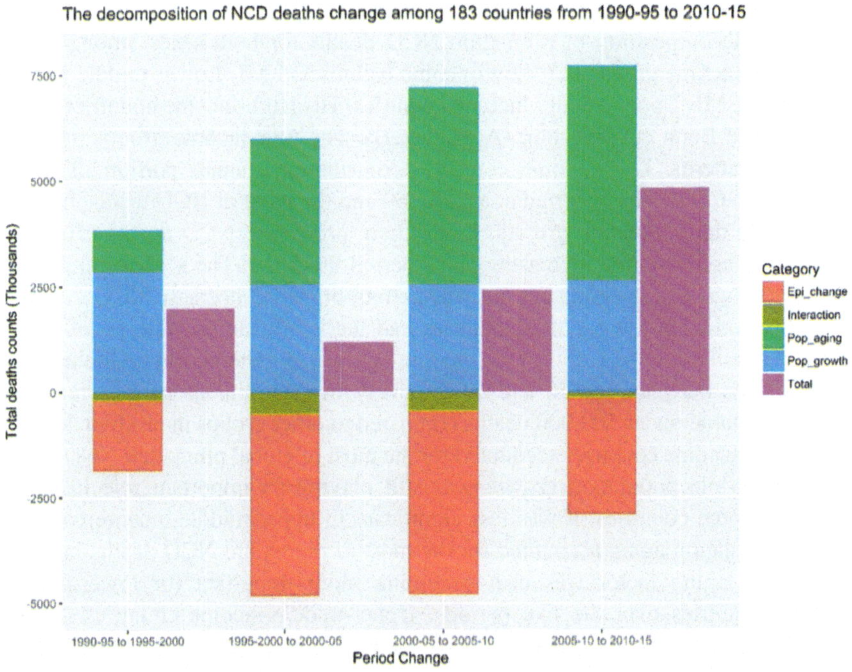

Fig. 9.2a The decomposition of NCD deaths change (age 50–69) for both sexes – global

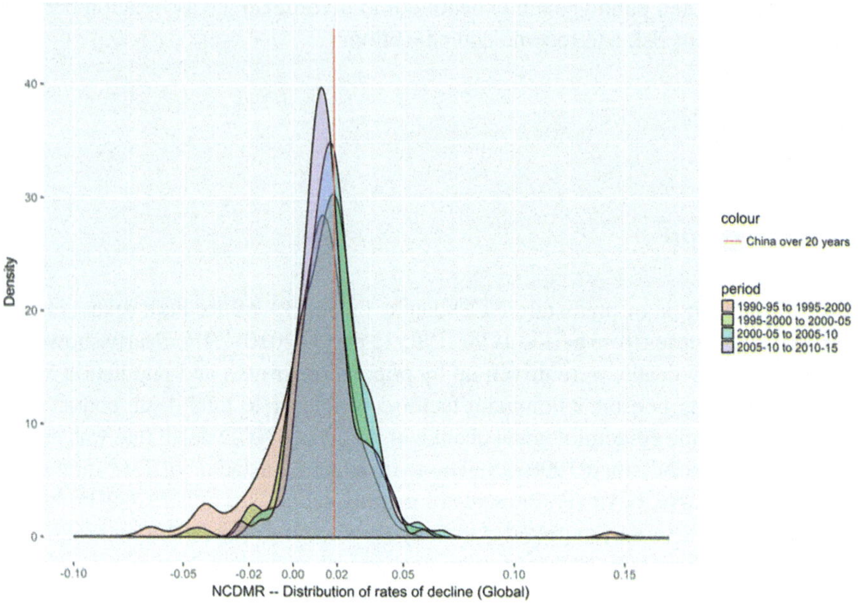

Fig. 9.2b The distribution of the annual rate of decline of premature NCD death rate – global

The following figures (Figs. 9.3a, 9.3b, 9.4a, 9.4b, 9.5a, 9.5b, 9.6a and 9.6b) show the decomposition of premature NCD deaths for both sexes among World Bank four income groups (LI- Low-income countries; LMI- Lower-middle-income countries; UMI- Upper-middle-income countries; HI- High-income countries) using 2014 World Bank classification (Appendix B). The four income groups had very different patterns. Low-income countries contributed a small portion of global change of NCD deaths occurred at age 50–69 and the most of NCD deaths increase was due to the population growth. In addition, premature NCD deaths caused by population aging started increasing since period 2005–10. The scale of epidemiological change was insignificant but was getting bigger to decrease NCD deaths in recent periods, but low-income countries had the lowest annual rate of premature NCD mortality decline among four income groups. In general, the total change of NCD deaths between age 50 and 69 for every two continuous periods remained small and stable under 500,000 deaths compared to other groups in last two decades.

Middle-income countries accounted for the most of global premature NCD deaths increase. While population growth was still playing an important role in lower-middle-income countries, it was less significant in upper-middle-income countries where population aging accounted for the most of premature NCD deaths increase. Population aging factor was also becoming more important for lower-middle-income countries over the five periods. Upper-middle-income countries (around 1.5% per year) generally performed better than lower-middle-income (around 1% per year) countries in terms of annual rate of premature NCD mortality decline from

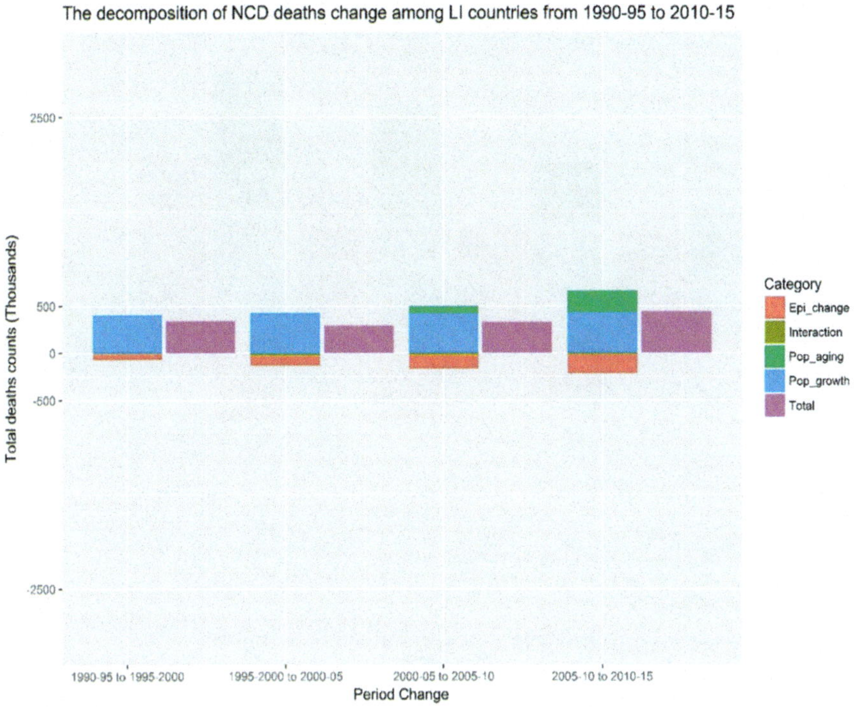

Fig. 9.3a The decomposition of NCD deaths change (age 50–69) for both sexes – LI

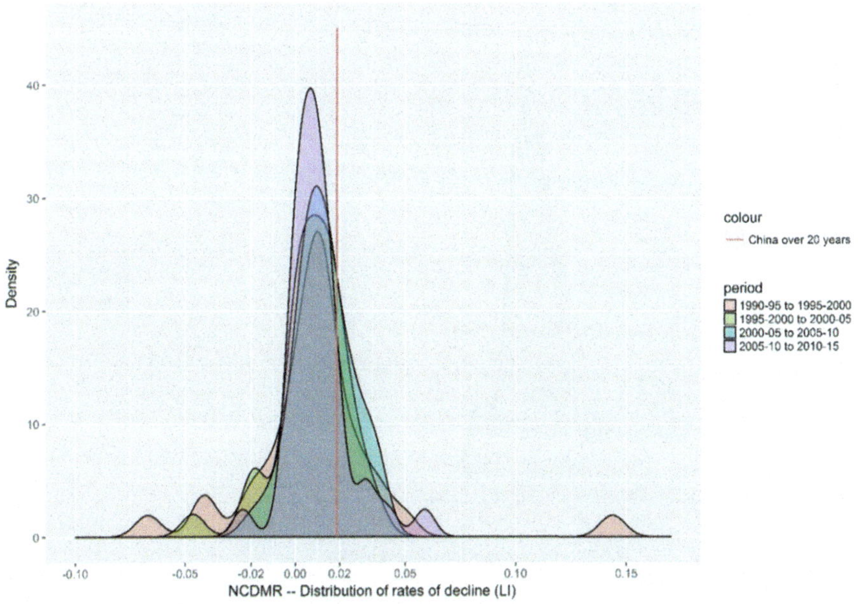

Fig. 9.3b The distribution of the annual rate of decline of premature NCD death rate – LI

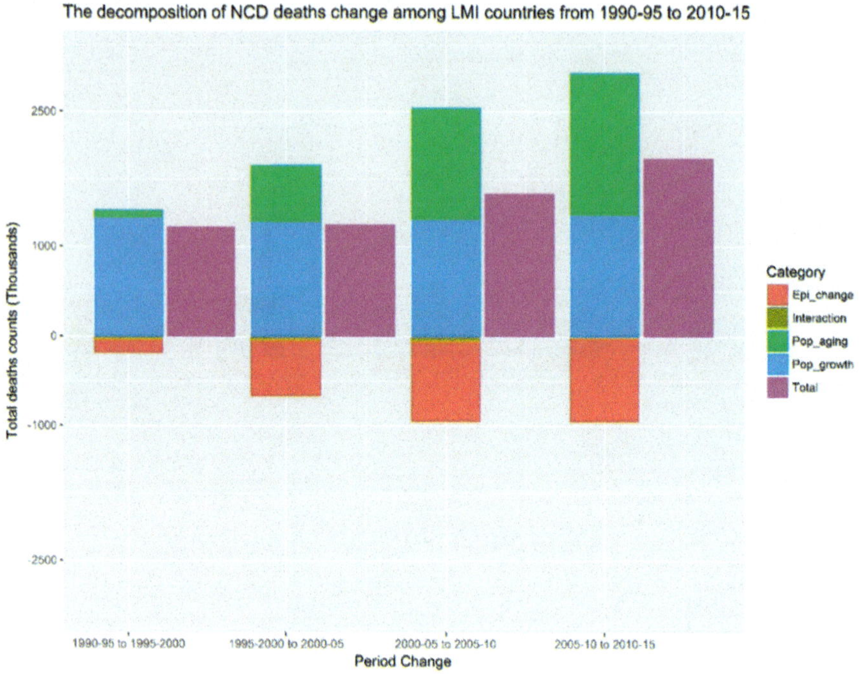

Fig. 9.4a The decomposition of NCD deaths change (age 50–69) for both sexes – LMI

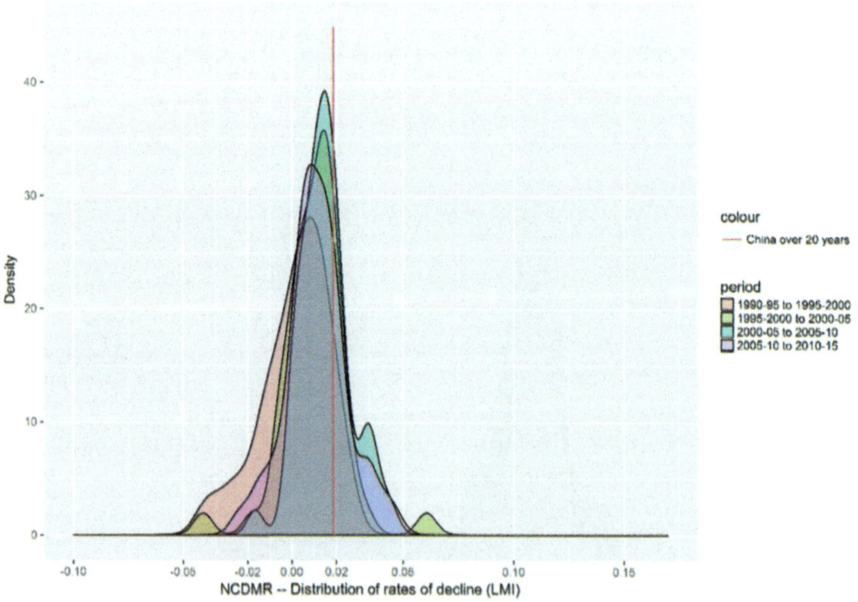

Fig. 9.4b The distribution of the annual rate of decline of premature NCD death rate – LMI

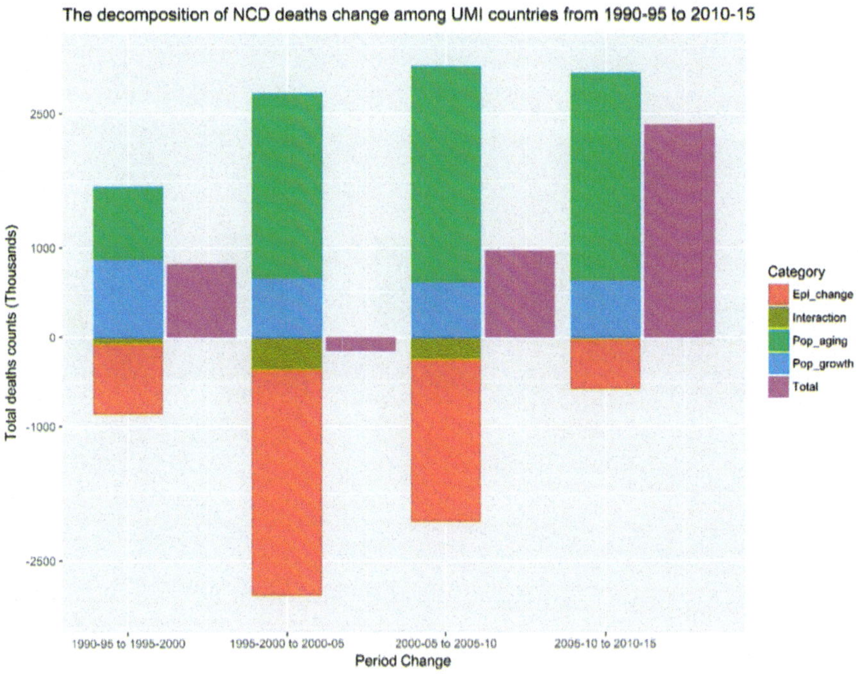

Fig. 9.5a The decomposition of NCD deaths change (age 50–69) for both sexes – UMI

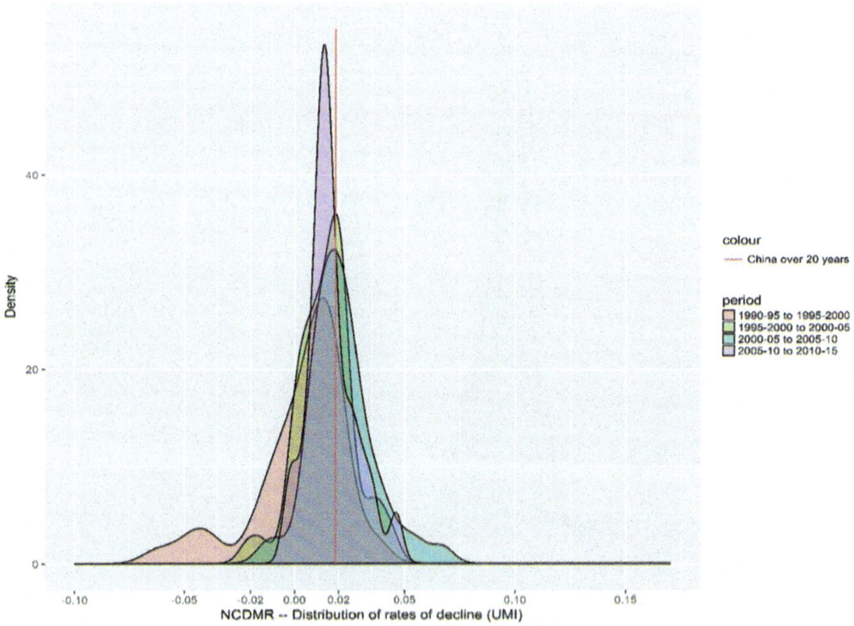

Fig. 9.5b The distribution of the annual rate of decline of premature NCD death rate – UMI

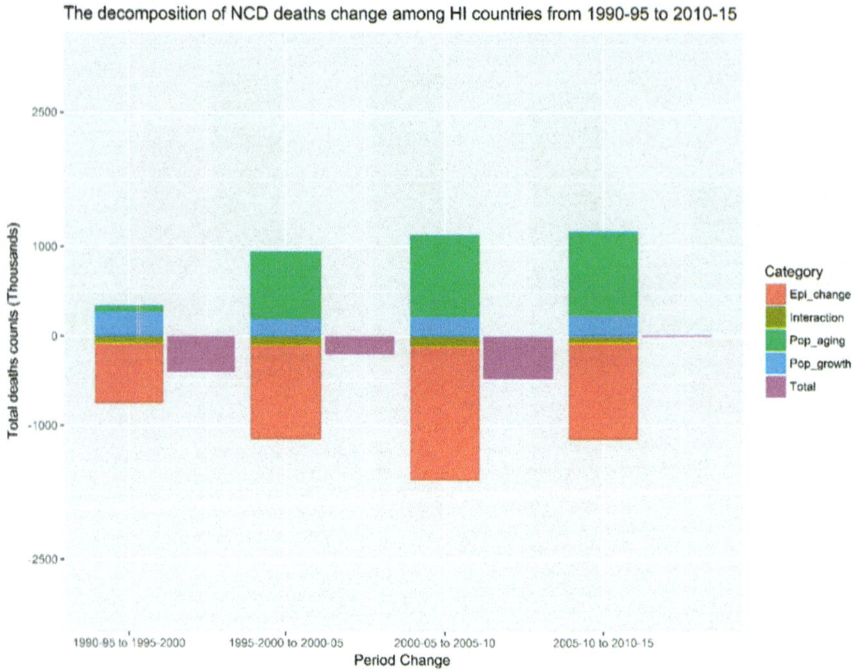

Fig. 9.6a The decomposition of NCD deaths change (age 50–69) for both sexes – HI

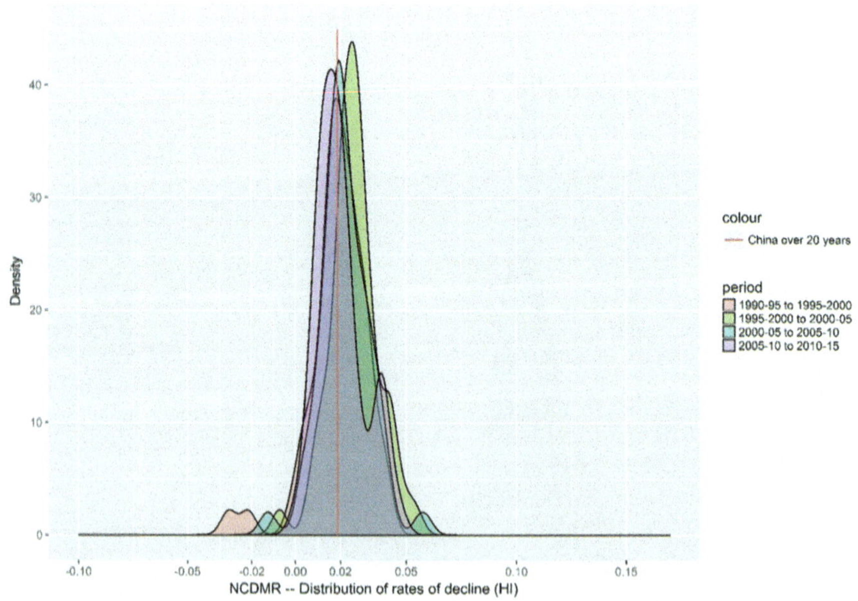

Fig. 9.6b The distribution of the annual rate of decline of premature NCD death rate – HI

the period 1990–95 to the most recent period 2010–15. While the epidemiological change in lower-middle-income countries was decreasing NCD deaths at a faster pace, the epidemiological change peaked from period 1995–2000 to 2000–2005 and declined in recent two periods among upper-middle-income countries. The total change of premature NCD deaths in the peak period was negative, which suggested the significant progress of declining premature NCD mortality in upper-middle-income countries in that period. However, the progress was declining in the following periods and the total premature NCD deaths change were increasing again. In general, except the peak period change, the total premature NCD deaths have been increasing steadily over the last two decades in middle income countries. The most recent period change from 2005–10 to 2010–15 indicated that the increase of NCD deaths occurred at age 50–69 in a 5-year period was around 4 million in middle-income countries.

The premature NCD mortality decline was the most optimistic in high-income countries with the highest annual rate of premature NCD mortality decline (2% per year) from the period 1990–95 to the most recent period 2010–15. While the population growth factor was moderate and stable over time, the population aging has been increasing significantly over time among high-income countries. The epidemiological changing was progressing over the first three period changes with more decline of deaths, but could not keep up with the pace of population aging. Thus the high-income countries have achieved total premature NCD deaths decline for all periods changes except the last one.

In general, demographic factors including population growth and population aging have increased premature NCD deaths, while the epidemiological change has declined premature NCD deaths over all periods in all income groups. While the population growth factor has remained moderate and stable contributing to total NCD deaths increase, the population aging factor has been growing to be a dominant factor driving up global NCD deaths over the last two decades. The decomposition of premature NCD deaths by sex for females and males is shown in Appendix E. Generally, the male and female decomposition were similar to the decomposition for both sexes. Globally, the sex difference of population growth factor was insignificant, but the sex differences of population aging and epidemiological change were evident. Men generally have had more total NCD deaths (age 50–69) increase than women with more NCD death increase driven by population aging and more NCD deaths decrease driven by epidemiological change over the four period changes. Similar patterns are observed in LMICs, but men in high-income countries have had more NCD deaths decrease than women. The peak period change in upper-middle-income countries were mostly contributed by the epidemiological change among men. More analyses of sex differences of premature NCD death rates are included in the next part of this section.

Determinants

Domain A – Health Inequity

Sex

Sex difference was evident globally shown from Fig. 9.7 to Fig. 9.9 covering all five periods. The premature NCD death rates among men were generally higher than those among women. Men and women had different levels of exposure and vulnerability to NCD risk factors. Figure 9.7 shows that women had less years of school than men, and the distribution of education attainment was more diverse among women with more countries having very low-level education attainment among women. Figure 9.8 indicates that women had generally higher obesity prevalence than men but the sex difference was small, while Fig. 9.9 shows that men had

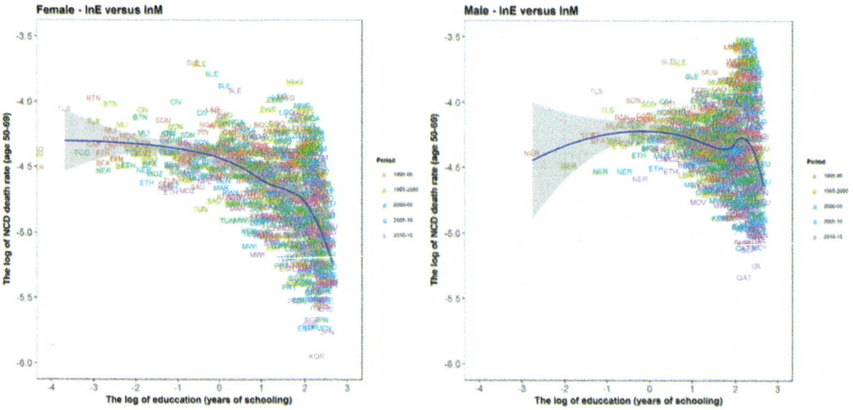

Fig. 9.7 The log of education level versus The log of premature NCD death rate (age 50–69) by sex (all five periods)

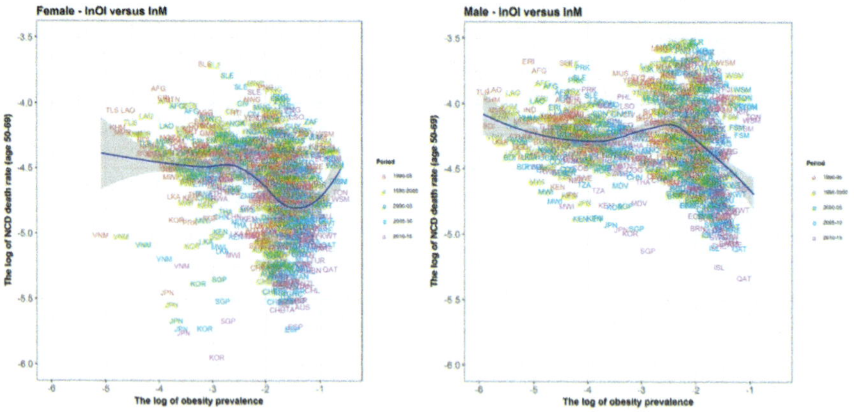

Fig. 9.8 The log of obesity level versus The log of premature NCD death rate (age 50–69) by sex (all five periods)

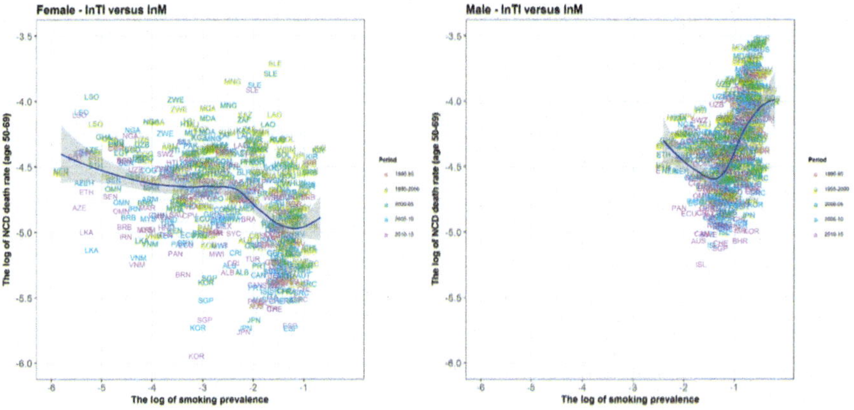

Fig. 9.9 The log of tobacco use level versus The log of premature NCD death rate (age 50–69) by sex (all five periods)

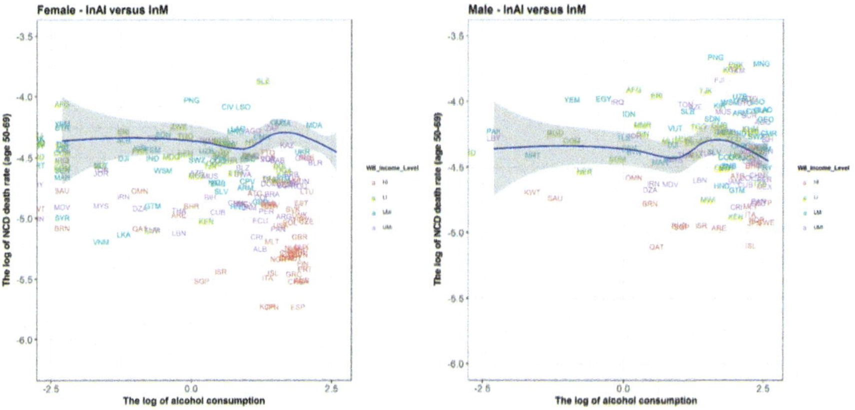

Fig. 9.10 The log of alcohol use level (liters of pure alcohol) versus The log of premature NCD death rate (age 50–69) by sex in the period 2005–10

significantly higher smoking prevalence than women and the distribution among men was very concentrated at high levels. The sex-specific data for alcohol use and physical activity are only available for one period 2005–10 and 2010–15 respectively. Figure 9.10 indicates that men had slightly higher alcohol consumption than women but the difference was insignificant, and no clear positive correlation could be observed between alcohol consumption and premature NCD mortality rate. Figure 9.11 shows that women had higher percentage of insufficient physical activity and the variation was bigger among women. In sum, women had higher risk in terms of lower education, more obesity and insufficient physical activity, while men had higher risk in terms of more tobacco use and alcohol use. The most significant sex

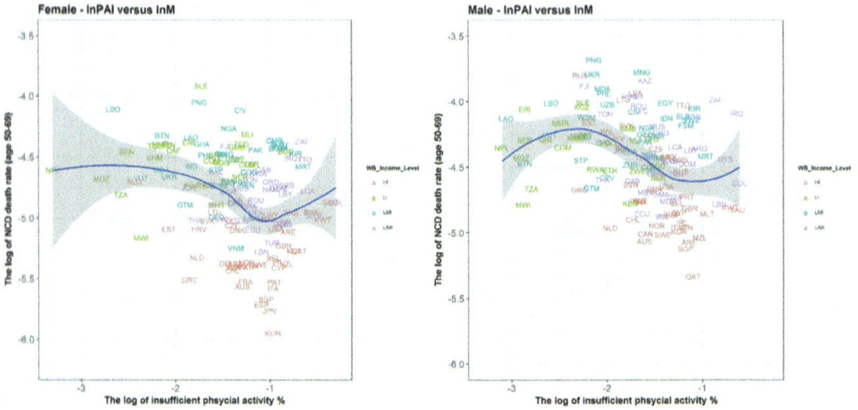

Fig. 9.11 The log of physical activity level (% of insufficient physical activity) versus The log of premature NCD death rate (age 50–69) by sex in the period 2010–15

different was tobacco use. Generally men had higher premature NCD mortality rates than women. The results confirm the hypothesis of evident sex difference from the previous studies.

Income

The previous studies have found that lower income was associated with higher NCD mortality across countries but low-income countries did not have enough evidence. The negative association between GDP per capita and NCD mortality is shown by Model 1 in Table 9.1 but not significant. Figure 9.12 indicates that the relationship was not simply monotonic when low-income countries were included as the lowest low-income countries had relatively low NCD death rates at age 50–69. Thus Model 2 and Model 3 introduce another independent variable – the square of the log of GDP per capita to fit the inversely U-shaped association, which indicates the association between NCD mortality and income is non-linear.

Education

The result for both sexes does not confirm the hypothesis that higher education level was generally associated with lower NCD mortality across countries as the association in all three models – Model 3 even shows a positive correlation, which is contradictory to the hypothesis. This contradictory might be explained by the association between education and risk factors among men in Domain B.

Table 9.1 Fixed-effects models predicting premature NCD death rate at age 50–69 (both sexes)

Premature NCD death rate	Model 1	Model 2	Model 3
ln*Y* GDP per capita	−0.128***	0.391	0.408**
	(0.036)	(0.204)	(0.140)
$(\ln Y)^2$		−0.029*	−0.028***
		(0.011)	(0.008)
ln*E* education attainment	0.052	0.036	0.067**
	(0.034)	(0.034)	(0.021)
ln*OI* obesity level	0.200***	0.129**	0.079*
	(0.040)	(0.048)	(0.035)
ln*TI* tobacco use level	0.010	0.011	
	(0.031)	(0.030)	
ln*AI* alcohol use level	0.042*	0.040*	0.022**
	(0.018)	(0.017)	(0.007)
ln*PHE* public health expenditure per capita	0.020	0.016	−0.004
	(0.021)	(0.020)	(0.012)
N (different due to data availability)	473	473	787
R-squared	0.153	0.169	0.127

Note: *p < 0.05; **p < 0.01; ***p < 0.001

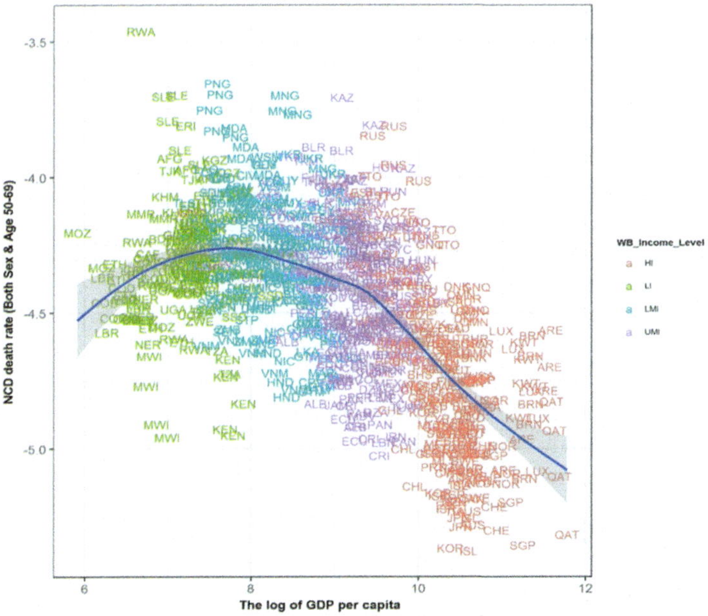

Fig. 9.12 The log of GDP per capita versus The log of premature NCD death rate in all periods

Table 9.2 The fixed-effects models predicting NCD risk factors and public health expenditure

	lnOI	lnTI	lnAI	lnPHE
lnGDP	0.288***	−0.109**	0.351**	0.599***
	(0.023)	(0.046)	(0.136)	(0.063)
lnE	0.291***	0.430***	0.194	0.064
	(0.024)	(0.048)	(0.142)	(0.066)
Observations N	808	480	802	793
R^2	0.309	0.199	0.013	0.127

Note: *$p < 0.05$; **$p < 0.01$; ***$p < 0.001$

Domain B – Social Stratification and Intermediate Determinants

All biological and behavioral factors in Domain B would increase premature NCD death rates according to the literature. Table 9.1 further examines the relationship between the risk factors and NCD mortality. Since social stratification could influence intermediate determinants as stated in the Domain A – sex and SES were associated with different risk factors. The sex differences of risk factors are included in the Domain A1 section above. The fixed effects models in Table 9.2 indicate the relationship between different risk factors and SES.

Obesity

The Fig. 9.8 suggests that women tended to have higher obesity prevalence than men as previous studies found. The model in Table 9.2 indicates that higher GDP per capita and higher education level could predict higher obesity prevalence across countries. The obesity prevalence was generally increasing over the five periods. Higher obesity level was strongly associated with higher NCD mortality in all models regardless of sex. For physical activity and unhealthy diets, there are not enough data to examine the association, but could be reflected from the results of obesity level, since the literature found insufficient physical activity and unhealthy diets were associated with higher risk for obesity. Figure 9.11 suggests that higher income countries tended to have higher percentage of insufficient physical activity, which is in line with previous studies.

Tobacco Use

Figure 9.9 confirms that men had significantly higher smoking prevalence than women as found in the literature. As no clear cross-national association between tobacco use and SES was found in the literature, the result in Table 9.2 suggests that countries with higher education attainment were associated with higher smoking prevalence, and the tobacco use has been declining over time globally, which shows a contradictory negative association between income and smoking prevalence. Tobacco use was strongly associated with high NCD mortality in models for both-sexes and male, but not for females. There was no strong correlation between tobacco use and NCD mortality among women. This result is in line with the finding of significant sex difference of tobacco use levels – the distribution of smoking prevalence among females was a lot more dispersed. The male models also have larger coefficients of tobacco use than female models. Tobacco use data availability is limited, thus Model 3 was introduced with a bigger sample size by excluding the variable indicating smoking prevalence.

Harmful Use of Alcohol

The sex difference of alcohol use between men and women was moderate shown in Fig. 9.10. The model in Table 9.2 indicates that greater GDP per capita was ssociated with higher levels of alcohol consumption. Model 1 and Model 2 confirm that more alcohol consumption was positively associated with higher premature NCD death rate. Once more data were included in Model 3, this association for alcohol use was not significant with a smaller coefficient. The Fig. 9.10 plotting sex-specific alcohol use in 2005–2010 does not suggest strong association with premature NCD mortality either.

Domain C – Structural Determinants

This study used two measurements, time and public health expenditure to quantitatively compare cross-national differentials of structural determinants on premature NCD mortality. The findings in Table 9.2 suggests GDP was an important positive determinant of public health expenditure. Further, the public health expenditure has been increasing over time. All the fixed-effects models in Table 9.1 do not support the hypothesis that premature NCD mortality was negatively associated with high public health expenditure per capita. All the variables included in the models are relevant to predict cross-national premature NCD mortality differentials.

Discussion

Through using decomposition and fixed-effects models, this study quantified the demographic changes and epidemiological changes on premature NCD deaths and examined the major determinants of premature NCD death rates across 183 countries. The findings from decomposition indicate demographic factors including population growth and population aging have driven up premature NCD deaths significantly over last two decades, while epidemiological change has shown the constant progress on declining premature NCD deaths. This global pattern was same for all four World Bank income groups. While population growth factor was moderate to increase NCD deaths at a constant pace over time, population aging has become a dominant factor contributing to NCD deaths increase at an increasing pace. This indicates that the pace of epidemiological change has to keep up with the pace of population aging to decline NCD deaths. To achieve SDG, four income groups face the same biggest challenge imposed by population aging. Low- and lower-middle-income countries need to increase their currently low annual rate of decline, while upper-middle- and high-income countries need to keep the current pace of epidemiological change.

Using the major determinants from UNDP's framework, the fixed-effects models could explain some of NCD mortality differentials across countries. More determinants of NCD mortality differentials remain to be discovered. All the risk factors included in the model were positively associated with GDP per capita except smoking, which means that higher income countries generally had higher prevalence of the risk factors. By controlling these risk factors to predict premature NCD death rates, lower premature NCD mortality rate was not shown to be associated with higher socioeconomic development – higher income, education attainment and more public health investments. Different time trends of risk factors (e.g., increasing obesity and decreasing tobacco use) may indicate the public health interventions on tobacco control have been effective over the last two decades, while not enough effective efforts have been devoted to decrease other risk factors. Higher-income countries had lower premature NCD death rates and higher rates of decline of premature NCD death rates, while lower-income countries had higher premature NCD mortality rates and lower rates of decline from 1990–95 to 2010–15. Thus the equity gap across countries in terms of premature NCD mortality has widened over the last two decades.

This macro-level study is largely limited by the data availability and data quality, especially the limited low-quality data from low-income countries. Low-income countries are very special for cross-national comparison of premature NCD mortality, as life expectancy is very low and the age distribution of premature NCD deaths is different from other groups with more missing data on risk factors. All three measurements analyzed in Appendix C may not be adequate to describe NCD motility pattern in low-income countries. The insignificant epidemiological change among low-income countries and inversely U-shaped association between GDP and NCD deaths rate at age 50–69 globally may also be caused by the limited mortality data available at age 50–69 from low-income countries. As mentioned above, cross-national comparison cannot be conducted on major determinants including physical activity, unhealthy diet and sex-specific alcohol consumption. The current available data may not be appropriate to represent the risk factors. For example, the mean consumption of alcohol at country level does not indicate the distribution and prevalence of harmful use of alcohol. This is also reflected in some models with insignificant association between alcohol consumption and NCD mortality. To better monitor the epidemiology of NCD and develop effective health interventions, collecting more data and improving data quality are crucial, especially in low-income countries with little information available now.

By analyzing premature NCD mortality rates across countries over last two decades, this study identified past progress on epidemiological change, current challenges imposed by population aging, and enlarging equity gaps across countries with different stages of economic development. To achieve SDG, faster declining rate of premature NCD mortality death needs to be achieved to counteract the NCD deaths driven by population aging, and global resources need to be reallocated to close the equity gap. More studies could be done based on the findings from this study to further examine and learn from the past progress. For instance, the epidemiological change peaked from period 1995–2000 to 2000–2005 among upper-middle-income countries with total NCD deaths decline. Further analysis found that China contributed around half of the total epidemiological change. More implications would be found through analyzing what happened during that period in China, whether this progress could happen in other countries through effective interventions, and why the pace of the epidemiological changing slowed down later.

Appendices

Appendix A: The Full List of Non-communicable Diseases Included

GBD cause ID	GBD group	GBD cause name	GBD cause ID	GBD group	GBD cause name
409	**B**	**Non-communicable diseases**			
410	***B.1***	***Neoplasms***	***521***	***B.4***	***Cirrhosis and other chronic liver diseases***
411	B.1.4	Esophageal cancer	***526***	***B.5***	***Digestive diseases***
414	B.1.5	Stomach cancer	527	B.5.1	Peptic ulcer disease
417	B.1.7	Liver cancer	528	B.5.2	Gastritis and duodenitis
423	B.1.10	Larynx cancer	529	B.5.3	Appendicitis
426	B.1.11	Tracheal, bronchus, and lung cancer	530	B.5.4	Paralytic ileus and intestinal obstruction
429	B.1.14	Breast cancer	531	B.5.5	Inguinal, femoral, and abdominal hernia
432	B.1.15	Cervical cancer	532	B.5.6	Inflammatory bowel disease
435	B.1.16	Uterine cancer	533	B.5.7	Vascular intestinal disorders
438	B.1.18	Prostate cancer	534	B.5.8	Gallbladder and biliary diseases
441	B.1.6	Colon and rectum cancer	535	B.5.9	Pancreatitis

(continued)

GBD cause ID	GBD group	GBD cause name	GBD cause ID	GBD group	GBD cause name
444	B.1.1	Lip and oral cavity cancer	541	B.5.10	Other digestive diseases
447	B.1.2	Nasopharynx cancer	***542***	***B.6***	***Neurological disorders***
450	B.1.3	Other pharynx cancer	543	B.6.1	Alzheimer disease and other dementias
453	B.1.8	Gallbladder and biliary tract cancer	544	B.6.2	Parkinson disease
456	B.1.9	Pancreatic cancer	545	B.6.3	Epilepsy
459	B.1.12	Malignant skin melanoma	546	B.6.4	Multiple sclerosis
462	B.1.13	Non-melanoma skin cancer	554	B.6.5	Motor neuron disease
465	B.1.17	Ovarian cancer	557	B.6.9	Other neurological disorders
468	B.1.19	Testicular cancer	***558***	***B.7***	***Mental and substance use disorders***
471	B.1.20	Kidney cancer	559	B.7.1	Schizophrenia
474	B.1.21	Bladder cancer	560	B.7.2	Alcohol use disorders
477	B.1.22	Brain and nervous system cancer	561	B.7.3	Drug use disorders
480	B.1.23	Thyroid cancer	572	B.7.7	Eating disorders
483	B.1.24	Mesothelioma	***586***	***B.8***	***Diabetes, urogenital, blood, and endocrine diseases***
484	B.1.25	Hodgkin lymphoma	587	B.8.1	Diabetes mellitus
485	B.1.26	Non-Hodgkin lymphoma	588	B.8.2	Acute glomerulonephritis
486	B.1.27	Multiple myeloma	589	B.8.3	Chronic kidney disease
487	B.1.28	Leukemia	594	B.8.4	Urinary diseases and male infertility
488	B.1.29	Other neoplasms	603	B.8.5	Gynecological diseases
491	***B.2***	***Cardiovascular diseases***	613	B.8.6	Hemoglobinopathies and hemolytic anemias
492	B.2.1	Rheumatic heart disease	619	B.8.7	Endocrine, metabolic, blood, and immune disorders
493	B.2.2	Ischemic heart disease	***626***	***B.9***	***Musculoskeletal disorders***
494	B.2.3	Cerebrovascular disease	627	B.9.1	Rheumatoid arthritis
495	*B.2.3.1*	*Ischemic stroke*	639	B.9.5	Other musculoskeletal disorders
496	*B.2.3.2*	*Hemorrhagic stroke*	***640***	***B.10***	***Other non-communicable diseases***
498	B.2.4	Hypertensive heart disease	641	B.10.1	Congenital birth defects
499	B.2.5	Cardiomyopathy and myocarditis	653	B.10.2	Skin and subcutaneous diseases

(continued)

GBD cause ID	GBD group	GBD cause name	GBD cause ID	GBD group	GBD cause name
500	B.2.6	Atrial fibrillation and flutter	686	B.10.5	Sudden infant death syndrome
501	B.2.7	Aortic aneurysm			
502	B.2.8	Peripheral artery disease			
503	B.2.9	Endocarditis			
507	B.2.10	Other cardiovascular and circulatory diseases			
508	***B.3***	***Chronic respiratory diseases***			
509	B.3.1	Chronic obstructive pulmonary disease			
510	B.3.2	Pneumoconiosis			
515	B.3.3	Asthma			
516	B.3.4	Interstitial lung disease and pulmonary sarcoidosis			
520	B.3.5	Other chronic respiratory diseases			

Appendix B: The Full List of 183 WHO Member States, ISO Alpha-3 Country Codes, World Bank Income Groups, and World Bank Regions

WHO member states	World Bank income	ISO code	World Bank region
Afghanistan	LI	AFG	South Asia
Albania	UMI	ALB	Europe and Central Asia
Algeria	UMI	DZA	Middle East and North Africa
Angola	UMI	AGO	Sub-Saharan Africa
Antigua and Barbuda	HI	ATG	Latin America and Caribbean
Argentina	UMI	ARG	Latin America and Caribbean
Armenia	LMI	ARM	Europe and Central Asia
Australia	HI	AUS	East Asia and Pacific
Austria	HI	AUT	Europe and Central Asia
Azerbaijan	UMI	AZE	Europe and Central Asia
Bahrain	HI	BHR	Middle East and North Africa
Bangladesh	LI	BGD	South Asia

(continued)

WHO member states	World Bank income	ISO code	World Bank region
Barbados	HI	BRB	Latin America and Caribbean
Belarus	UMI	BLR	Europe and Central Asia
Belgium	HI	BEL	Europe and Central Asia
Belize	UMI	BLZ	Latin America and Caribbean
Benin	LI	BEN	Sub-Saharan Africa
Bhutan	LMI	BTN	South Asia
Bolivia	LMI	BOL	Latin America and Caribbean
Bosnia and Herzegovina	UMI	BIH	Europe and Central Asia
Botswana	UMI	BWA	Sub-Saharan Africa
Brazil	UMI	BRA	Latin America and Caribbean
Brunei	HI	BRN	East Asia and Pacific
Bulgaria	UMI	BGR	Europe and Central Asia
Burkina Faso	LI	BFA	Sub-Saharan Africa
Burundi	LI	BDI	Sub-Saharan Africa
Cambodia	LI	KHM	East Asia and Pacific
Cameroon	LMI	CMR	Sub-Saharan Africa
Canada	HI	CAN	North America
Cape Verde	LMI	CPV	Sub-Saharan Africa
Central African Republic	LI	CAF	Sub-Saharan Africa
Chad	LI	TCD	Sub-Saharan Africa
Chile	HI	CHL	Latin America and Caribbean
China	UMI	CHN	East Asia and Pacific
Colombia	UMI	COL	Latin America and Caribbean
Comoros	LI	COM	Sub-Saharan Africa
Congo	LMI	COG	Sub-Saharan Africa
Costa Rica	UMI	CRI	Latin America and Caribbean
Cote d'Ivoire	LMI	CIV	Sub-Saharan Africa
Croatia	HI	HRV	Europe and Central Asia
Cuba	UMI	CUB	Latin America and Caribbean
Cyprus	HI	CYP	Europe and Central Asia
Czech Republic	HI	CZE	Europe and Central Asia
Democratic Republic of the Congo	LI	COD	Sub-Saharan Africa
Denmark	HI	DNK	Europe and Central Asia
Djibouti	LMI	DJI	Middle East and North Africa

(continued)

WHO member states	World Bank income	ISO code	World Bank region
Dominican Republic	UMI	DOM	Latin America and Caribbean
Ecuador	UMI	ECU	Latin America and Caribbean
Egypt	LMI	EGY	Middle East and North Africa
El Salvador	LMI	SLV	Latin America and Caribbean
Equatorial Guinea	HI	GNQ	Sub-Saharan Africa
Eritrea	LI	ERI	Sub-Saharan Africa
Estonia	HI	EST	Europe and Central Asia
Ethiopia	LI	ETH	Sub-Saharan Africa
Federated States of Micronesia	LMI	FSM	East Asia and Pacific
Fiji	UMI	FJI	East Asia and Pacific
Finland	HI	FIN	Europe and Central Asia
France	HI	FRA	Europe and Central Asia
Gabon	UMI	GAB	Sub-Saharan Africa
Georgia	LMI	GEO	Europe and Central Asia
Germany	HI	DEU	Europe and Central Asia
Ghana	LMI	GHA	Sub-Saharan Africa
Greece	HI	GRC	Europe and Central Asia
Grenada	UMI	GRD	Latin America and Caribbean
Guatemala	LMI	GTM	Latin America and Caribbean
Guinea	LI	GIN	Sub-Saharan Africa
Guinea-Bissau	LI	GNB	Sub-Saharan Africa
Guyana	LMI	GUY	Latin America and Caribbean
Haiti	LI	HTI	Latin America and Caribbean
Honduras	LMI	HND	Latin America and Caribbean
Hungary	UMI	HUN	Europe and Central Asia
Iceland	HI	ISL	Europe and Central Asia
India	LMI	IND	South Asia
Indonesia	LMI	IDN	East Asia and Pacific
Iran	UMI	IRN	Middle East and North Africa
Iraq	UMI	IRQ	Middle East and North Africa
Ireland	HI	IRL	Europe and Central Asia
Israel	HI	ISR	Middle East and North Africa

(continued)

WHO member states	World Bank income	ISO code	World Bank region
Italy	HI	ITA	Europe and Central Asia
Jamaica	UMI	JAM	Latin America and Caribbean
Japan	HI	JPN	East Asia and Pacific
Jordan	UMI	JOR	Middle East and North Africa
Kazakhstan	UMI	KAZ	Europe and Central Asia
Kenya	LI	KEN	Sub-Saharan Africa
Kiribati	LMI	KIR	East Asia and Pacific
Kuwait	HI	KWT	Middle East and North Africa
Kyrgyzstan	LI	KGZ	Europe and Central Asia
Laos	LMI	LAO	East Asia and Pacific
Latvia	HI	LVA	Europe and Central Asia
Lebanon	UMI	LBN	Middle East and North Africa
Lesotho	LMI	LSO	Sub-Saharan Africa
Liberia	LI	LBR	Sub-Saharan Africa
Libya	UMI	LBY	Middle East and North Africa
Lithuania	HI	LTU	Europe and Central Asia
Luxembourg	HI	LUX	Europe and Central Asia
Madagascar	LI	MDG	Sub-Saharan Africa
Malawi	LI	MWI	Sub-Saharan Africa
Malaysia	UMI	MYS	East Asia and Pacific
Maldives	UMI	MDV	South Asia
Mali	LI	MLI	Sub-Saharan Africa
Malta	HI	MLT	Middle East and North Africa
Mauritania	LMI	MRT	Sub-Saharan Africa
Mauritius	UMI	MUS	Sub-Saharan Africa
Mexico	UMI	MEX	Latin America and Caribbean
Moldova	LMI	MDA	Europe and Central Asia
Mongolia	LMI	MNG	East Asia and Pacific
Montenegro	UMI	MNE	Europe and Central Asia
Morocco	LMI	MAR	Middle East and North Africa
Mozambique	LI	MOZ	Sub-Saharan Africa
Myanmar	LI	MMR	East Asia and Pacific
Namibia	UMI	NAM	Sub-Saharan Africa
Nepal	LI	NPL	South Asia
Netherlands	HI	NLD	Europe and Central Asia
New Zealand	HI	NZL	East Asia and Pacific

(continued)

WHO member states	World Bank income	ISO code	World Bank region
Nicaragua	LMI	NIC	Latin America and Caribbean
Niger	LI	NER	Sub-Saharan Africa
Nigeria	LMI	NGA	Sub-Saharan Africa
North Korea	LI	PRK	East Asia and Pacific
Norway	HI	NOR	Europe and Central Asia
Oman	HI	OMN	Middle East and North Africa
Pakistan	LMI	PAK	South Asia
Palestine	LMI	PSE	Middle East and North Africa
Panama	UMI	PAN	Latin America and Caribbean
Papua New Guinea	LMI	PNG	East Asia and Pacific
Paraguay	LMI	PRY	Latin America and Caribbean
Peru	UMI	PER	Latin America and Caribbean
Philippines	LMI	PHL	East Asia and Pacific
Poland	HI	POL	Europe and Central Asia
Portugal	HI	PRT	Europe and Central Asia
Qatar	HI	QAT	Middle East and North Africa
Romania	UMI	ROU	Europe and Central Asia
Russia	HI	RUS	Europe and Central Asia
Rwanda	LI	RWA	Sub-Saharan Africa
Saint Lucia	UMI	LCA	Latin America and Caribbean
Saint Vincent and the Grenadines	UMI	VCT	Latin America and Caribbean
Samoa	LMI	WSM	East Asia and Pacific
Sao Tome and Principe	LMI	STP	Sub-Saharan Africa
Saudi Arabia	HI	SAU	Middle East and North Africa
Senegal	LMI	SEN	Sub-Saharan Africa
Serbia	UMI	SRB	Europe and Central Asia
Seychelles	UMI	SYC	Sub-Saharan Africa
Sierra Leone	LI	SLE	Sub-Saharan Africa
Singapore	HI	SGP	East Asia and Pacific
Slovakia	HI	SVK	Europe and Central Asia
Slovenia	HI	SVN	Europe and Central Asia
Solomon Islands	LMI	SLB	East Asia and Pacific
Somalia	LI	SOM	Sub-Saharan Africa
South Africa	UMI	ZAF	Sub-Saharan Africa

(continued)

WHO member states	World Bank income	ISO code	World Bank region
South Korea	HI	KOR	East Asia and Pacific
South Sudan	LI	SSD	Sub-Saharan Africa
Spain	HI	ESP	Europe and Central Asia
Sri Lanka	LMI	LKA	South Asia
Sudan	LMI	SDN	Sub-Saharan Africa
Suriname	UMI	SUR	Latin America and Caribbean
Swaziland	LMI	SWZ	Sub-Saharan Africa
Sweden	HI	SWE	Europe and Central Asia
Switzerland	HI	CHE	Europe and Central Asia
Syria	LMI	SYR	Middle East and North Africa
Tajikistan	LI	TJK	Europe and Central Asia
Tanzania	LI	TZA	Sub-Saharan Africa
Thailand	UMI	THA	East Asia and Pacific
The Bahamas	HI	BHS	Latin America and Caribbean
The Gambia	LI	GMB	Sub-Saharan Africa
Timor-Leste	LMI	TLS	East Asia and Pacific
Togo	LI	TGO	Sub-Saharan Africa
Tonga	UMI	TON	East Asia and Pacific
Trinidad and Tobago	HI	TTO	Latin America and Caribbean
Tunisia	UMI	TUN	Middle East and North Africa
Turkey	UMI	TUR	Europe and Central Asia
Turkmenistan	UMI	TKM	Europe and Central Asia
Uganda	LI	UGA	Sub-Saharan Africa
Ukraine	LMI	UKR	Europe and Central Asia
United Arab Emirates	HI	ARE	Middle East and North Africa
United Kingdom	HI	GBR	Europe and Central Asia
United States	HI	USA	North America
Uruguay	HI	URY	Latin America and Caribbean
Uzbekistan	LMI	UZB	Europe and Central Asia
Vanuatu	LMI	VUT	East Asia and Pacific
Venezuela	UMI	VEN	Latin America and Caribbean
Vietnam	LMI	VNM	East Asia and Pacific
Yemen	LMI	YEM	Middle East and North Africa
Zambia	LMI	ZMB	Sub-Saharan Africa
Zimbabwe	LI	ZWE	Sub-Saharan Africa

Appendix C: The Age Distribution of Premature NCD Deaths

As Fig. 9.13 shows, more than 50% of premature NCD deaths occurred between age 30–69 in more than 170 countries over the five periods. The percentage of premature NCD deaths among all NCD deaths was associated with country's income level as high-income countries had the lowest percentage of premature NCD deaths while low-income countries generally had more than 50% of NCD deaths occurred prematurely. Higher percentage of premature NCD deaths occurred at age 30–69 was associated with lower level of premature percentage. In the other words, in the countries with lower percentage of premature NCD deaths among all NCD deaths, most of the premature NCD deaths happened in later ages. The similar pattern is observed in Fig. 9.14 too, but around 30 low- and lower-middle-income countries having less than 50% of premature NCD deaths occurred at age 50–69. This indicates these lower-income countries had a different pattern of age distribution of premature NCD deaths from most other countries. Figure 9.15 shows the percentage of premature NCD deaths occurred at age 30–49 was lower than 30% for most countries, and there was no linear relationship between two percentages. Some lower-income countries having lower percentage of premature NCD deaths occurred beyond age 30 had higher death rate caused by congenital birth defects (categorized as other NCDs in Appendix A) in newborns (GBD 2015). Thus, this study cannot capture these NCD deaths occurred at early ages for these countries. Since the proportion of working age group 30–49 does not necessarily indicate population

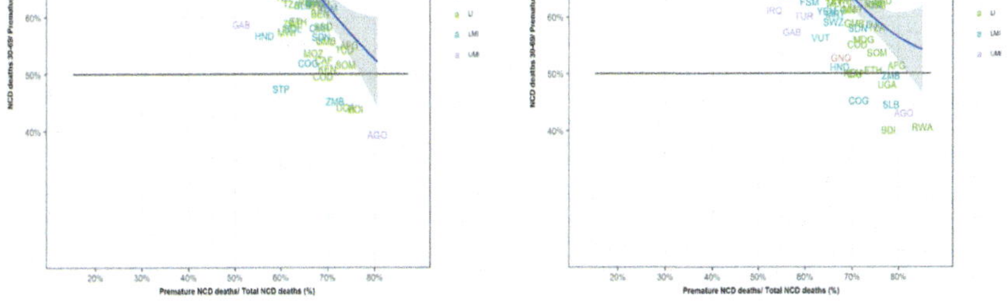

Fig. 9.13 The percentage of NCD deaths between age 30–69 among all premature NCD deaths versus The precentage of premature NCD deaths among all NCD deaths

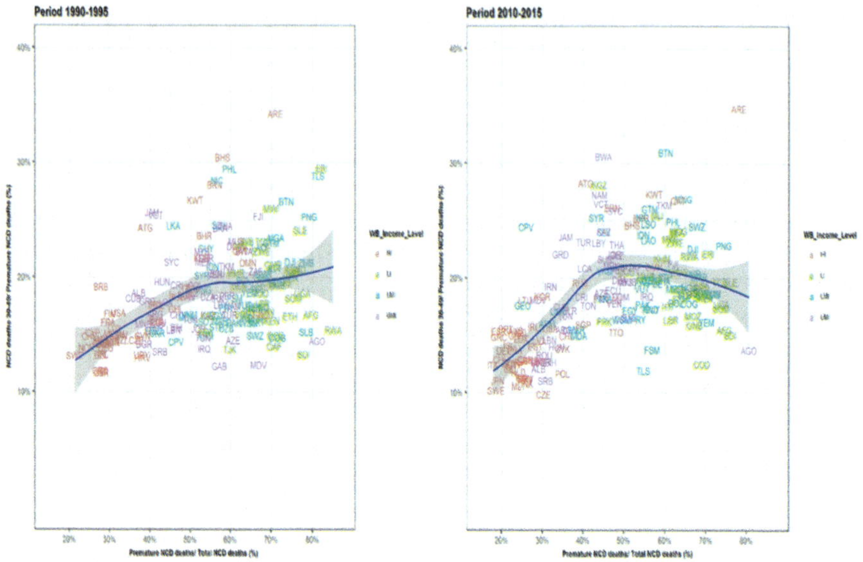

Fig. 9.14 The percentage of NCD deaths between age 50–69 among all premature NCD deaths versus The precentage of premature NCD deaths among all NCD deaths

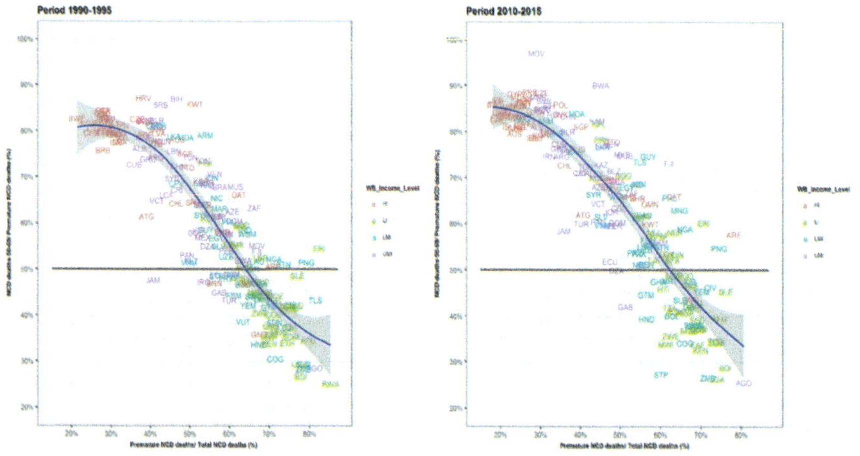

Fig. 9.15 The percentage of NCD deaths between age 30–49 among all premature NCD deaths versus The precentage of premature NCD deaths among all NCD deaths

aging and the percentage of NCD deaths occurred in age 30–49 was low and relatively stable across countries, focusing on the age group 50–69 took most of premature NCD deaths into account and could show the general trends for all premature NCD deaths in most countries.

Appendix D: A Comparison of Three Premature NCD Mortality Measurements

The first measurement is age-specific (50–69) NCD death rate from the first part of the analysis. The other two measurements are the probability of dying because of NCD between age 50 and 69, and the probability of dying because of NCD between age 30 and 69. The probability of dying because of NCD in the presence of causes was calculated from the UN lifetables (UN 2015) and GBD cause-of-death percentage (GBD 2015) for the two age groups. All three measurements are for both sexes.

As Fig. 9.16 shows, the two measurements of probability of dying in two age groups were significantly correlated with the Pearson's r over 0.9 in all five periods. The probability of dying was higher between age 30 and 69, but the percentage of NCD among all causes of deaths were lower in age group 30–69. Thus the probability of dying because of NCD at two age groups were very close estimates. In Fig. 9.17, similar

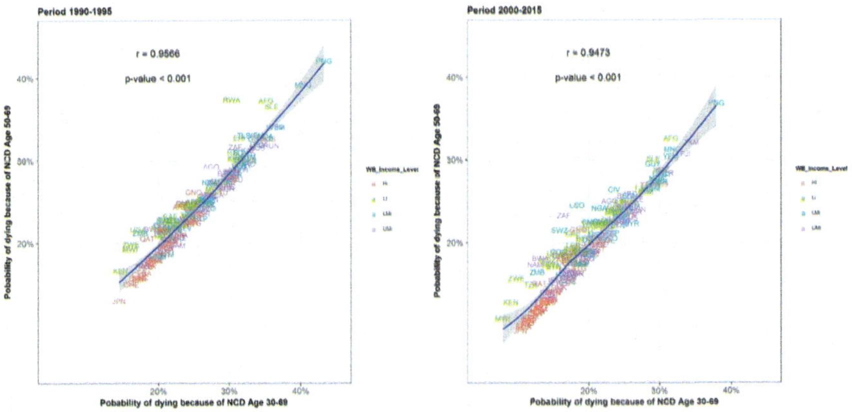

Fig. 9.16 The probability of dying because of NCD at age 50–69 versus The probability of dying because of NCD at age 30–69

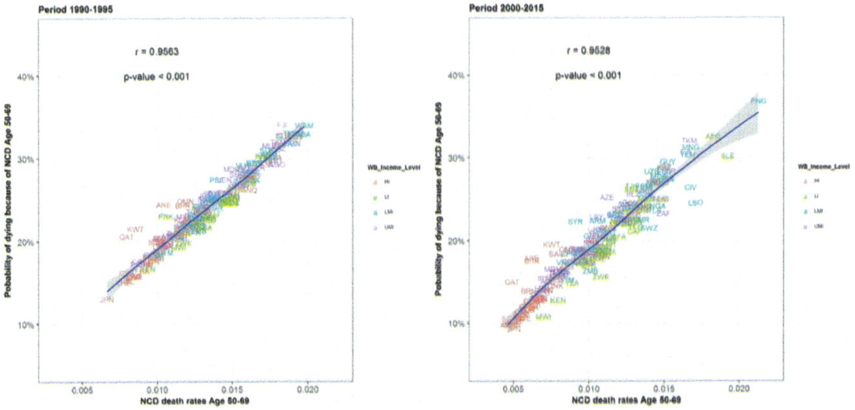

Fig. 9.17 The probability of dying because of NCD at age 50–69 versus The age-specific (50–69) NCD death rate

patterns could be observed for the comparison between the probability of dying because of NCD at age 50–69 and the age-specific (50–69) NCD death rate with the Pearson's r over 0.9 in all five periods too. Although the scale of NCD death rates at age 50–69 were different from the probability of dying, NCD death rates at age 50–69 were as good as age-standardized measurements (the probability of dying) to indicate cross-national premature NCD mortality differentials The fixed-effects models using the three measures as dependent variables does not show different results in terms of the direction and the significance of the association with independent variables. Thus, the results section only presents the models using the NCD death rates as the dependent variable.

Appendix E: The Decomposition of NCD Deaths (Age 50–69) Change by Sex (Figs. 9.18, 9.19, 9.20, 9.21 and 9.22)

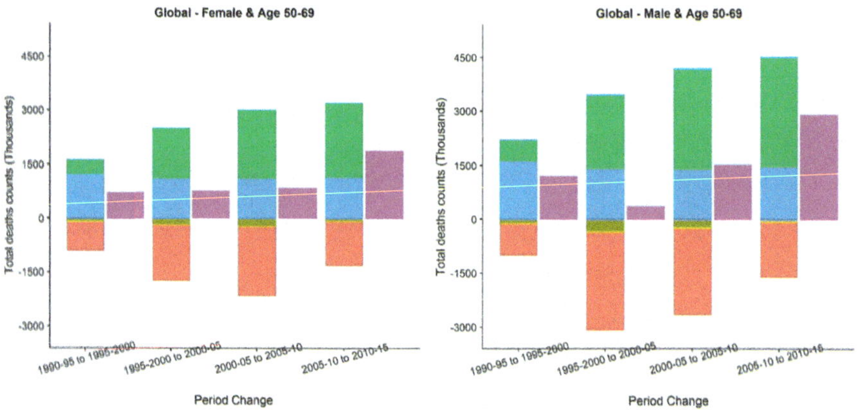

Fig. 9.18 The decomposition of premature NCD deaths at age 50–69 by sex – global

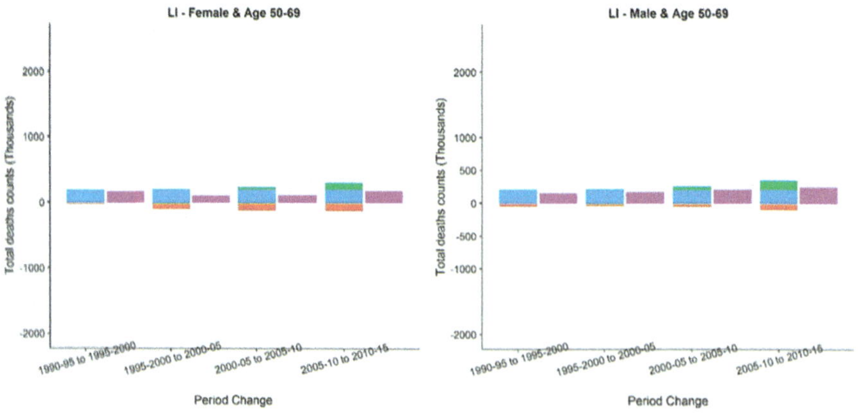

Fig. 9.19 The decomposition of premature NCD deaths at age 50–69 by sex – LI

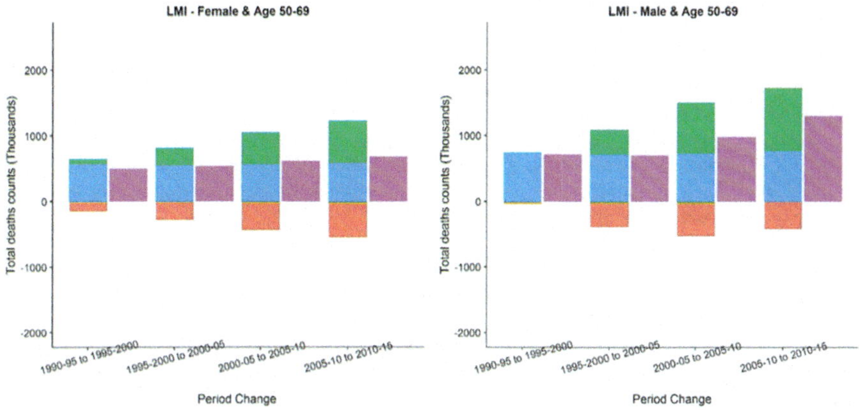

Fig. 9.20 The decomposition of premature NCD deaths at age 50–69 by sex – LMI

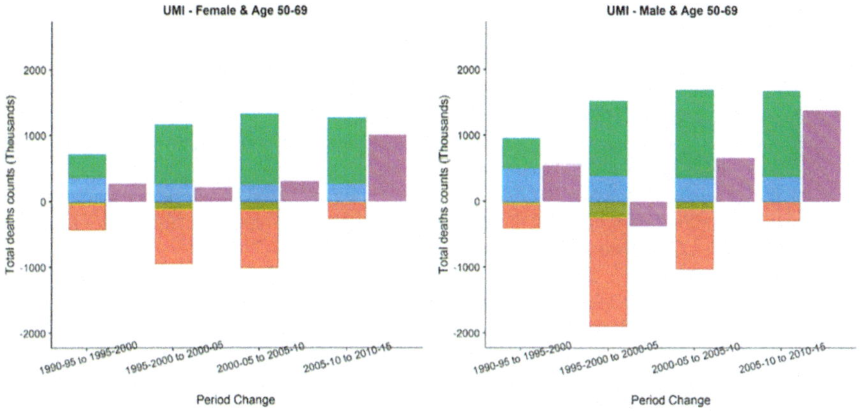

Fig. 9.21 The decomposition of premature NCD deaths at age 50–69 by sex – UMI

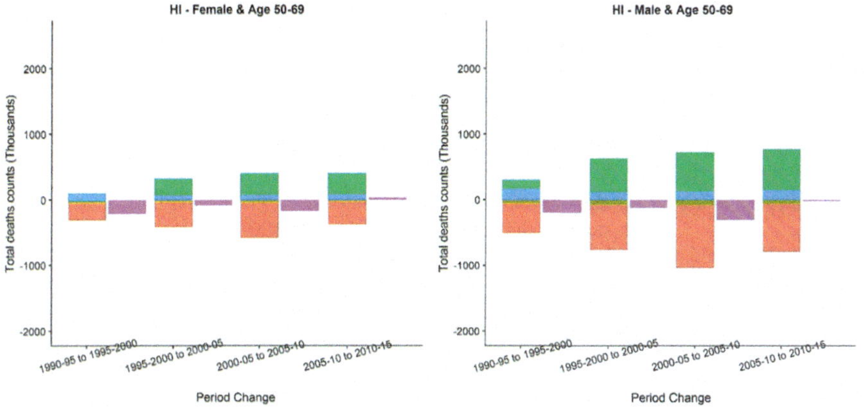

Fig. 9.22 The decomposition of premature NCD deaths at age 50–69 by sex –HI

References

Addy, C. L., Wilson, D. K., Kirtland, K. A., Ainsworth, B. E., Sharpe, P., & Kimsey, D. (2004). Associations of perceived social and physical environmental supports with physical activity and walking behavior. *American Journal of Public Health, 94*(3), 440–443.

Allen, L., Williams, J., Townsend, N., Mikkelsen, B., Roberts, N., Foster, C., & Wickramasinghe, K. (2017). Socioeconomic status and non-communicable disease behavioural risk factors in low-income and lower-middle-income countries: A systematic review. *The Lancet Global Health, 5*(3), e277–e289.

Bloom, D. E., Cafiero, E., Jané-Llopis, E., Abrahams-Gessel, S., Bloom, L. R., Fathima, S., et al. (2012). *The global economic burden of noncommunicable diseases* (No. 8712). Program on the Global Demography of Aging.

Bosdriesz, J. R., Mehmedovic, S., Witvliet, M. I., & Kunst, A. E. (2014). Socioeconomic inequalities in smoking in low and mid income countries: Positive gradients among women? *International Journal for Equity in Health, 13*(1), 14.

Cecchini, M., Sassi, F., Lauer, J. A., Lee, Y. Y., Guajardo-Barron, V., & Chisholm, D. (2010). Tackling of unhealthy diets, physical inactivity, and obesity: Health effects and cost-effectiveness. *The Lancet, 376*(9754), 1775–1784.

Chaker, L., Falla, A., van der Lee, S. J., Muka, T., Imo, D., Jaspers, L., et al. (2015). The global impact of non-communicable diseases on macro-economic productivity: A systematic review. *European Journal of Epidemiology, 30*(5), 357–395.

Daar, A. S., Singer, P. A., Persad, D. L., Pramming, S. K., Matthews, D. R., Beaglehole, R., et al. (2007). Grand challenges in chronic non-communicable diseases. *Nature, 450*(7169), 494–496.

Dauchet, L., Amouyel, P., Hercberg, S., & Dallongeville, J. (2006). Fruit and vegetable consumption and risk of coronary heart disease: A meta-analysis of cohort studies. *The Journal of Nutrition, 136*(10), 2588–2593.

Eriksen, M., Mackay, J., & Ross, H. (2013). *The tobacco atlas* (4th ed.). American Cancer Society.

Evans, D. B., Tandon, A., Murray, C. J., & Lauer, J. A. (2001). Comparative efficiency of national health systems: Cross national econometric analysis. *BMJ, 323*(7308), 307–310.

Gajalakshmi, C. K., Jha, P., Ranson, K., Nguyen, S., & Mundial, B. (2000). Global patterns of smoking and smoking-attributable mortality. In *Tobacco control in developing countries* (pp. 9–39). Oxford University Press.

Giles-Corti, B., & Donovan, R. J. (2002). Socioeconomic status differences in recreational physical activity levels and real and perceived access to a supportive physical environment. *Preventive Medicine, 35*(6), 601–611.

Global Burden of Disease Study (GBD). (2015). *Results*. Seattle: Institute for Health Metrics and Evaluation (IHME). Available at: http://ghdx.healthdata.org/gbd-results-tool. Accessed on 27 May 2017.

Goyal, A., Bhatt, D. L., Steg, P. G., Gersh, B. J., Alberts, M. J., Ohman, E. M., et al. (2010). Attained educational level and incident atherothrombotic events in low-and middle-income compared with high-income countries. *Circulation, 122*(12), 1167–1175.

Groth, M. V., Fagt, S., & Brondsted, L. (2001). Social determinants of dietary habits in Denmark. *European Journal of Clinical Nutrition, 55*(11), 959.

Guthold, R., Ono, T., Strong, K. L., Chatterji, S., & Morabia, A. (2008). Worldwide variability in physical inactivity: A 51-country survey. *American Journal of Preventive Medicine, 34*(6), 486–494.

He, F. J., Nowson, C. A., & MacGregor, G. A. (2006). Fruit and vegetable consumption and stroke: Meta-analysis of cohort studies. *The Lancet, 367*(9507), 320–326.

Hiscock, R., Bauld, L., Amos, A., Fidler, J. A., & Munafò, M. (2012). Socioeconomic status and smoking: A review. *Annals of the New York Academy of Sciences, 1248*(1), 107–123.

Hitchman, S. C., & Fong, G. T. (2011). Gender empowerment and female-to-male smoking prevalence ratios. *Bulletin of the World Health Organization, 89*(3), 195–202.

Hitiris, T., & Posnett, J. (1992). The determinants and effects of health expenditure in developed countries. *Journal of Health Economics, 11*(2), 173–181.

Institute for Health Metrics and Evaluation (IHME). (2015). *GBD compare*. Seattle: IHME, University of Washington. Available at: http://vizhub.healthdata.org/gbd-compare. Accessed on 27 May 2017.

Kelly, B. B., & Fuster, V. (Eds.). (2010). *Epidemiology of cardiovascular disease. Promoting cardiovascular health in the developing world: A critical challenge to achieve global health.* National Academies Press.

Lim, S. S., Vos, T., Flaxman, A. D., Danaei, G., Shibuya, K., Adair-Rohani, H., et al. (2013). A comparative risk assessment of burden of disease and injury attributable to 67 risk factors and risk factor clusters in 21 regions, 1990–2010: A systematic analysis for the Global Burden of Disease Study 2010. *The Lancet, 380*(9859), 2224–2260.

Lu, C., Schneider, M. T., Gubbins, P., Leach-Kemon, K., Jamison, D., & Murray, C. J. (2010). Public financing of health in developing countries: A cross-national systematic analysis. *The Lancet, 375*(9723), 1375–1387.

Marmot, M., et al. (2008). *Closing the gap in a generation: Health equity through action on the social determinants of health: Final report of the Commission on Social Determinants of Health (CSDH)*. Geneva: World Health Organization (WHO).

Mathers, C. D., & Loncar, D. (2006). Projections of global mortality and burden of disease from 2002 to 2030. *PLoS Medicine, 3*(11), e442.

Monteiro, C. A., Conde, W. L., Lu, B., & Popkin, B. M. (2004). Obesity and inequities in health in the developing world. *International Journal of Obesity and Related Metabolic Disorders, 28*(9), 1181–1186.

Pampel, F. (2010). Divergent patterns of smoking across high-income nations. In *International differences in mortality at older ages: Dimensions and sources* (pp. 132–163). Washington, DC: The National Academies Press.

Popkin, B. M., Adair, L. S., & Ng, S. W. (2012). Global nutrition transition and the pandemic of obesity in developing countries. *Nutrition Reviews, 70*(1), 3–21.

Rehm, J., Baliunas, D., Borges, G. L., Graham, K., Irving, H., Kehoe, T., et al. (2010). The relation between different dimensions of alcohol consumption and burden of disease: An overview. *Addiction, 105*(5), 817–843.

Roth, G. A., Forouzanfar, M. H., Moran, A. E., Barber, R., Nguyen, G., Feigin, V. L., et al. (2015). Demographic and epidemiologic drivers of global cardiovascular mortality. *New England Journal of Medicine, 372*(14), 1333–1341.

Schmidt, L. A., Mäkelä, P., Rehm, J., & Room, R. (2010). Alcohol: Equity and social determinants. In *Equity, social determinants and public health programmes* (p. 11). Geneva: World Health Organization.

Singh, G. K., Siahpush, M., & Kogan, M. D. (2010). Neighborhood socioeconomic conditions, built environments, and childhood obesity. *Health Affairs, 29*(3), 503–512.

Solar, O., & Irwin, A. (2007). *A conceptual framework for action on the social determinants of health* (Discussion Paper Series on Social Determinants of Health, 2). Geneva: World Health Organization (WHO). Available at: http://apps.who.int/iris/bitstream/10665/44489/1/9789241500852_eng.pdf?ua=1&ua=1. Accessed on 27 May 2017.

Sommer, I., Griebler, U., Mahlknecht, P., Thaler, K., Bouskill, K., Gartlehner, G., & Mendis, S. (2015). Socioeconomic inequalities in non-communicable diseases and their risk factors: An overview of systematic reviews. *BMC Public Health, 15*(1), 914.

Subramanian, S. V., Perkins, J. M., Özaltin, E., & Smith, G. D. (2011). Weight of nations: A socioeconomic analysis of women in low-to middle-income countries. *The American Journal of Clinical Nutrition, 93*(2), 413–421.

Swinburn, B. A., Sacks, G., Hall, K. D., McPherson, K., Finegood, D. T., Moodie, M. L., & Gortmaker, S. L. (2011). The global obesity pandemic: Shaped by global drivers and local environments. *The Lancet, 378*(9793), 804–814.

The Sustainable Development Solutions Network (SDSN). (2016). *The indicator report target 3.4.* Available at: http://indicators.report/targets/3-4/. Accessed on 20 May 2017.

United Nations, Department of Economic and Social Affairs, Population Division. (2015). *World population prospects: The 2015 revision, DVD edition*. New York: United Nations (UN).
United Nations Development Programme (2013). *Discussion paper – Addressing the social determinants of noncommunicable diseases*. New York: United Nations Development Programme (UNDP). Available at: http://www.undp.org/content/undp/en/home/librarypage/hiv-aids/discussion-paper%2D%2Daddressing-the-social-determinants-of-noncommu.html. Accessed on 27 May 2017.
Verguet, S., Ogbuoji, O., Qi, J., Olson, Z. D., Yamey, G., Nugent, R. A., et al. (2017). Annual rates of decline in child, maternal, tuberculosis and NCD mortality across 109 countries of low and middle income from 1990 to 2015. *The Lancet*. (forthcoming).
Wilsnack, S. C., Wilsnack, R. W., & Kantor, L. W. (2014). Focus on: Women and the costs of alcohol use. *Alcohol Research: Current Reviews, 35*(2), 219.
Wittgenstein Centre for Demography and Global Human Capital (2015). Wittgenstein Centre Data Explorer Version 1.2. Available at: www.wittgensteincentre.org/dataexplorer. Accessed on 27 May 2017.
World Bank (2016). World development indicators. Available at: http://databank.worldbank.org/data/home.aspx. Accessed on 27 May 2017.
World Health Organization. (2009). *Women and health: today's evidence tomorrow's agenda*. Geneva: World Health Organization (WHO).
World Health Organization. (2010). *Global recommendations on physical activity for health*. Geneva: World Health Organization (WHO).
World Health Organization. (2014a). *Global status report on alcohol and health*. Geneva: World Health Organization (WHO). Available at: http://www.who.int/substance_abuse/publications/global_alcohol_report/msb_gsr_2014_1.pdf. Accessed on 27 May 2017.
World Health Organization. (2014b). *Global status report on noncommunicable diseases 2014*. Geneva: World Health Organization (WHO). Available at: http://apps.who.int/iris/bitstream/10665/148114/1/9789241564854_eng.pdf?ua=1. Accessed on 27 May 2017.
World Health Organization. (2017a). Global health expenditure database. Available at: http://www.who.int/health-accounts/ghed/en/. Accessed on 27 May 2017.
World Health Organization. (2017b). Global health observatory data repository. Available at: http://apps.who.int/gho/data/node.main.A867?lang=en. Accessed on 27 May 2017.
World Health Organization. (2018). *Premature NCD deaths from Global Health Observatory (GHO) data*. World Health Organization (WHO). Available at: https://www.who.int/gho/ncd/mortality_morbidity/ncd_premature/en/. Accessed on 30 Aug 2020.

Part III
Working Life: Issues and Patterns

Chapter 10
Expected Consumer Surplus from Medicaid in a Prototypical Working-Age Household

Carlos A. Ulibarri

> *"The art of dealing with any problem at the theoretical, empirical, or applied economic level is to oversimplify in an artful and useful way versus oversimplifying in a dumb, crazy way...".*
>
> – Arnold Harberger.

Introduction

Medicaid and the Children's Health Insurance Program (Medicaid) constitute a joint federal-state entitlement program, providing health care coverage and medical benefits to low income individuals, including families with children, pregnant women, the elderly and people with disabilities (Gruber 2000). State Medicaid programs generally subsidize the demand for health insurance across these populations by not requiring premiums or cost-sharing, thereby impacting the welfare of millions of individuals and families across the United States through the provision of health care benefits and lower medical/insurance costs. The impact of Medicaid programs seems most significant in states with the highest uninsured and poverty rates (Sommers and Oellerich 2013). New Mexico is a case in point.

In 2018 Medicaid provides health insurance to nearly one in five Americans, approximating 73 million individuals. New Mexico's Medicaid program currently covers 850,000 beneficiaries, approaching 40% of the state's population.[1] These numbers reflect expanded coverage of low income working-age adults (19–64; under 138% FPL) under the Patient Protection and Affordable Care Act (ACA). New Mexico (NM) became an expansion State (ES) on January 1, 2014. At the time, the

[1]New Mexico's Medicaid program is referred to as Centennial Care, administered by the Medical Assistance Division of the Human Services Department.

C. A. Ulibarri (✉)
Medical Assistance Division, New Mexico Human Services Department, Santa Fe, NM, USA
e-mail: Carlos.Ulibarri@state.nm.us

B. Jivetti, Md. N. Hoque (eds.), *Population Change and Public Policy*, Applied Demography Series 11, https://doi.org/10.1007/978-3-030-57069-9_10

Medicaid-insured populations were approximately 600,000 in NM and 61.7 million in the US. Currently New Mexico is among the leading states in terms of Medicaid participation by the state population (KFF 2019).[2]

Medicaid growth under the ACA offers a 'natural experiment' for examining Medicaid participation and household welfare – both at the national- and state-levels. This paper estimates a simple random utility model (RUM) of a prototypical household participating in Medicaid with extended income eligibility limits. The value-based framework is used to compute expected consumer surplus from Medicaid, and examine welfare implications of changes in health status, labor force participation and medical out-of-pocket costs.

The study is organized as follows. Section 10.2 reviews literature related to eligibility and participation in Medicaid programs. Section 10.3 describes the RUM framework and variables explaining the participation choice. Section 10.4 discusses the empirical application and findings of the study. Section 10.5 ends with concluding remarks.

Literature

Why do households participate in Medicaid programs? Surely a primary rationale is the mix of medical benefits gained from subsidized insurance coverage; access to medical providers, coverage of pharmaceutical products, and use of health care services such as vision and dental care. These subsidized benefits promote health status and increase the consumers' purchasing power over all other goods (Sommers et al. 2017; Blavin et al. 2018). This argument is the focal point of the study, and applies to other entitlement goods, such as the Supplemental Nutrition Assistance Program (SNAP).

Financial eligibility for Medicaid and other public assistance programs is determined according to the official poverty measure (OPM) of the US Census, developed in 1963.[3] The OPM uses pre-tax cash income as the measure of resources available to meet basic consumption needs, long-defined by the real cost of a minimum food plan multiplied by a factor of 3. The metric is widely critiqued for inadequacies in representing the contemporary needs of individuals and families or the resources available to them (Citro and Michaels 1995; Burtless and Siegel 2001). A more comprehensive poverty measure is given by the Supplemental Poverty Measure (SPM), described in Dalakar (2017).

---

[2]At the time of writing New Mexico is one of 31 expansion states. The timeline and status of State Medicaid Expansion decisions is available in "Status of State Action on the Medicaid Expansion Decision," State Health Facts, Kaiser Family Foundation (2019).

[3]The focus of the present study is on Medicaid/CHIP eligibility of non-disabled working-age and children populations under the ACA. For current status see, "Where are States Today? Medicaid and CHIP Eligibility Levels for Children, Pregnant Women, and Adults." Kaiser Family Foundation (2019).

The Supplemental measure accounts for certain in-kind benefits and transfer payments received by consumer units, thus recognizing household resources and needs more broadly, e.g. food, housing, utilities, medical needs, clothing, and a small amount for other needs. Total household resources to meet these needs include cash incomes from private and public sources, and transfer payments from public assistance programs. The SPM deducts specific expenses from household resources in accounting for poverty status, including medical/insurance costs, social security payroll taxes, child support payments, work-related expenses and child care costs.

In 2016 New Mexico was among 20 states where poverty, as measured by the SPM, was lower than indicated by the official measure: 15.7% compared to 19.2% (Fox 2018). Compared to the official measure, a relatively lower SPM reflects substantial transfer payments to households from means-tested public assistance programs. The SPM accounts healthcare costs as a basic living expense and subtracts these costs from household resources. Fox examines the impact of this deduction on poverty measurement using a counterfactual comparison, finding the hypothetical omission of medical/insurance costs lowers the supplemental poverty rate by 3.6% in 2015 and 3.3% in 2016. The experimental result manifests the major impact of medical out-of-pocket costs on poverty status and economic welfare.

The relationship between poverty metrics and healthcare costs/benefits is instrumental in explaining the Medicaid participation choice. While the official measure sets the actual poverty threshold for Medicaid eligibility, accounting for supplemental poverty reveals the impact of medical out of pocket costs on household budgets (Caswell and Short 2011). To reiterate, the SPM subtracts medical out-of-pocket costs from household resources, thus treating these expenses like a tax on households. Conversely, in-kind health benefits are instrumental in restoring family welfare, like a subsidy to households.

Note that Medicaid is *not* among the public assistance programs included in the SPM, nor is the value of health benefits. Filling this void, the 'health-inclusive' poverty measure (HIPM) developed by Remler et al. (2017) incorporates Medicaid insurance benefits to household resources, providing a more complete accounting of the impact of public assistance on household welfare. The HIPM equates the per-capita value of Medicaid benefits to the second-tier Silver Plan. In 2014 the individual Silver Plan was worth $6350, approximating the average Medicaid spending per full-benefit enrollee: $6396 for the United States, $6026 for New Mexico (KFF 2019).[4]

Applying the HIPM, Remler et al. find Medicaid reduced the health-inclusive poverty rate by 3.8% in 2014, close to the 3.6% reduction reported by Fox using the SPM for data year 2015. These poverty-reducing effects exceed those reported by Sommers and Oellerich (2013) for data year 2011, i.e. reductions of 1% for children, 2.2% among disabled adults, and 0.7% among elderly individuals without a

[4]Imputed values of in-kind benefits are time-specific and reflect household needs. Here, the costs of a vanilla market-based Silver Plan reflect average individual medical needs in 2014. These costs are noted to provide a sense of parity relative to Medicaid per capita costs.

disability. Notably, Sommers and Oellerich do not consider health care benefits from Medicaid coverage, i.e. "whatever gains in health care access or heath status the program confers."

Remler et al. also report Medicaid lowered the poverty rate of Hispanics by 6.1% compared to 1.8% for non-Hispanic whites, and by 5.3% for children under 18. Indeed, the poverty-reducing impact of Medicaid on children is larger than all other means-tested programs, *combined*. What's more, a Medicaid-beneficiary poverty rate (limited to the sub-population receiving benefits) shows the entitlement program cut this rate by 17.1% – relating most directly to households in ACA expansion states. These findings provide new perspective on the poverty-reducing effects of Medicaid in relation to demographic characteristics.

The costs and benefits of being in Medicaid are central to the present study, bringing-up the question: do Medicaid-covered households have lower medical/insurance costs relative to being uninsured or having private-insured coverage? Sommers and Oellerich measure cost-savings from Medicaid using random-matching, inferring cost-savings of $495 per beneficiary.[5] Baicker et al. (2013) report cost-savings of $215 per beneficiary from the 'Oregon Experiment' –a randomized study of Oregon's Medicaid expansion targeting low-income uninsured nonelderly adults. Majerol et al. (2016) conduct an observational study of 2014 Consumer Expenditure Survey data, finding Medicaid cost-savings of $244 and $525 over uninsured and private-insured households. And Glied et al. (2017) report Medicaid families in expansion states saved $387 per year in average healthcare costs compared to families in non-expansion states. Differences in study design, data, and analytical approaches preclude direct comparisons of Medicaid cost-savings across these studies. Nonetheless, they provide guidance in identifying pecuniary gains from Medicaid programs.

The value-based framework includes selected demographic characteristics identified in the study of Medicaid-covered households. Analysis of the 2014 Consumer Expenditure Survey by Foster and Rojas (2018) finds 84% of families in means-tested programs participated in Medicaid. The average family had 4 members, 2.1 children under 18, and 1.3 wage earners. The average age of the reference person was 37, with 57% having up to a high school education. Medicaid beneficiaries were more likely to be homeowners, and their transportation costs tend to occupy a larger share of family budgets. Meanwhile, the average family income/expenditures were $32,963/$36,450, and their healthcare costs were lower and occupied a smaller budget-share than those of non-beneficiaries ($1278, 3.5% vs $1961, 5.4%).

A fundamental assumption in the RUM framework is that Medicaid participation depends on anticipated health benefits from coverage: increased access to providers, increased utilization of medical services and better health status. The US Government Accounting Office (GAO 2018) reports evidence of increased health benefits in

[5]The study assumes a sample of Medicaid-covered individuals lose coverage and either become uninsured or take-up private coverage. The sample population is then randomly matched to a non-Medicaid population (uninsured or private-insured) to infer cost-savings from Medicaid

states expanding Medicaid eligibility under the ACA.[6] In sum, Medicaid participants in expansion states report fewer financial barriers to care, having a usual place of care, using more services and meeting their medical needs. Baicker et al. report further evidence of health benefits for the nonelderly adult population targeted in Oregon's Medicaid expansion.

The present study tests causal linkages relating to health-status, labor force participation and medical/insurance costs. Regarding health-status, Gruber notes Medicaid coverage may be a function of health status, and vice versa, leading to endogeneity bias in estimating the effects of health on insurance. The focus of the present study warrants Medicaid participation be specified as a function of health-status. Consequently, health status is treated as an exogenous determinant of participation in the RUM framework, ignoring changes in health outcomes resulting from taking-up insurance (Sommers et al. 2017; Levy and Meltzer 2008; Meer and Rosen 2004).

Endogeneity bias could also occur in estimating the causal effect of labor force status on Medicaid participation, or through a collinear association between labor force participation and health status. In these regards the empirical analysis does not consider potential impacts of Medicaid coverage may on household labor supply. Thus, labor force status is treated in the simplest form possible – as an independent exogenous predictor of participation in the RUM framework, focusing on its main effects.[7]

Finally, Medicaid expansion under the ACA is a unique testbed for examining participation from extending income eligibility limits (FPL thresholds). The Urban Institute (2018) reports eligibility limits in expansion states for parents and other adults increased from 61% of FPL in 2013 ($23,550 in a 4-person household) to the 138% threshold in 2016 ($33,465 in a 4-person household). Meanwhile, the median eligibility threshold for parents residing in non-expansion states remained below 50% of FPL in 2016 ($12,125 in a 4-person household). The coverage gap in per-capita poverty status has pivotal implications for participation/welfare and is examined in Sect. 10.4 across study units on a one-by-one basis.

This selective review of health/financial costs and benefits from Medicaid offers empirical guidance for the present study, where the working hypothesis is that households participate in Medicaid to maximize CS. The following section presents a simple value-based model of household behavior to test this hypothesis and the impact on household welfare.

[6]The GAO study draws data from the National Health Institute Survey (NHIS), compiled by the National Center for Health Statistics (NCHS).

[7]Following Bound (1991) studies of endogenous labor supply tend to be sensitive to the measures of health used; in particular self-reported health status. This seems less complicating in the present study, which treats both variates exogenously, as well as education. Bound (1991, 2000) provides helpful discussions of measurement errors associated with subjective- and objective-measured health variables.

RUM Framework

Consider an 'all-or-nothing' consumer choice over an entitlement good. Suppose the consumers' income level is sufficiently low to qualify for a price subsidy covering the cost of the good. An exact welfare measure of the subsidy is obtained applying a quasi-linear utility function: $U(x_1, x_2) = U(x_1) + x_2$,where x_1 denotes a binary choice over the entitlement good and x_2 is a numeraire commodity representing all-other-goods. The budget set containing prices and income (p, M) is defined by $B(p, M) = \{(x_1, x_2) \in \mathbb{R}^2_+ | \, p_1 x_1 + x_2 \leq M\}$, with $p_2 = \$1$.

The consumers' choice problem is formulated by solving the budget constraint for the numeraire commodity and substituting like-terms in the utility function, yielding:

$$\max_{x_1=(0,1)} \{U(x_1) + M - p_1 x_1\}.$$

Choosing $x_1^* = 1$ follows from the necessary condition $U(1) + M - p_1 > U(0) + M$. So under the common assumption $U(0) = 0$, the choice implies the utility from the entitlement good exceeds its' price, $U(1) > p_1$. For comparison, let p_r denote the consumers' *reservation price* – the highest price the consumer is willing-to-pay for the entitlement good. Consequently, the reservation price sets the consumers' demand for the entitlement good, $U(1) = p_r$. The optimal choices are then described by $x_1^* = 1$ and $x_2^* = M - p_1 \forall p_1 \leq p_r$, yielding the indirect utility function, $IUF = U(1) + M - p_1$.

Quasi-linear utility simplifies the analysis of welfare change by imposing restrictions on consumer choices i.e. constant marginal utility of income, and zero-income effects from price changes. Under these restrictions the CS measure of welfare change is equivalent to the compensating variation (CV) and equivalent variation (EV) welfare measures (Varian). Furthermore, the CS measure in the all-or-nothing choice reflects the consumers' reservation price. This argument is put in the following proposition.

Proposition Under the simplifying assumption of quasi-linear preferences over an all-or-nothing choice, the consumers' reservation price reflects consumer surplus from a price-subsidy covering the cost of the good.

Proof we wish to show $CS = p_r$ for the non-zero solution to the all-or-nothing choice $(x_1^* = 1)$. This solution is only feasible in the price interval $[0, p_r]$. Hence, the definite integral of the unit constant yields the result: $CS = \int_0^{p_r} x_1^* \, ds = p_r$.

The heuristic description of welfare gains from subsidization shows a non-zero demand for the entitlement good depends on the consumer's reservation price and meeting income eligibility requirements for subsidization. The RUM framework explains this all-or-nothing choice assuming study units assign utility values to Medicaid coverage. Participation is modeled as a binary dependent variable (y), to be explained by attributes of the choice alternative and characteristics describing the study units, i.e. the structural component explaining choice behavior. All other aspects of the binary choice are assumed to be unobservable and represented in the random component of the model (Train 2009).

The RUM uses a binary Logit specification, giving a probability estimate of participation from a rational-choice perspective, and a log-sum measure of the expected change in CS. Consider the equation $U_{i,\,m} = V_{i,\,m} + \varepsilon_{i,\,m}$, where $U_{i,\,m}$ is the utility value assigned by individual i for alternative m, explained by the structural and random components $V_{i,\,m}$ and $\varepsilon_{i,\,m}$. The structural component is assumed to be a linear function of attributes and characteristics explaining choice behavior.

The study unit chooses alternative m if it provides higher utility than any other alternative, that is iff $U_{i,\,m} > U_{i,\,j}$, $\forall\, m \neq j$. In the binary choice ($m = 0, 1$) the probability of choosing one alternative over the other is equivalent to the probability that the chosen alternative provides higher utility, thus $P(y_i = 1) \rightarrow P(U_{i,\,1} > U_{i,\,0})$. In order to specify the random component of the RUM it is useful to rewrite this probability as $P(U_{i,\,1} > U_{i,\,0}) = P(V_{i,\,1} + \varepsilon_{i,\,1} > V_{i,\,0} + \varepsilon_{i,\,0}) = P(\varepsilon_{i,\,0} - \varepsilon_{i,\,1} < V_{i,\,1} - V_{i,\,0})$. The common practice assumes the random disturbances ($\varepsilon_{i,\,m}$) follow the Type 1 Extreme Value Distribution, so that the difference in random terms ($\varepsilon_{i,\,1} - \varepsilon_{i,\,0}$) is distributed according to the logistic distribution.

Following Train, if each $\varepsilon_{i,\,m}$ is iid extreme value and utility is linear in income then expected consumer surplus can be calculated using Small and Rosen's (1981) formula: $E(CS) = \frac{1}{\gamma}\left[\ln\left(e^{V_{i,0}} + e^{V_{i,1}}\right)\right] + C$, where γ is the marginal utility of income and C is an unknown constant. Welfare gains/losses from a policy change or a change in the attributes of a choice alternative are measured by $\Delta E(CS) = \frac{1}{\gamma} \times \left[\ln\left(e^{V^f_{i,0}} + e^{V^f_{i,1}}\right) - \ln\left(e^{V^0_{i,0}} + e^{V^0_{i,1}}\right)\right]$, where the superscripts refer to before (*0*) and after (f) the policy/attribute change. Finally, note that the probability of household i participation is given by $P_i = e^{V_{i,1}}/\left(e^{V_{i,0}} + e^{V_{i,1}}\right)$, where $V_{i,\,1}$ and $V_{i,\,0}$ are the *conditional* indirect utility components with and without Medicaid coverage. The next section describes the empirical application of the model and findings of the study.

Empirical Analysis

Logit models are estimated by the method of Maximum Likelihood using pooled data (2013–2018) from the Current Population Survey's Annual Social and Economic Supplement (CPS ASEC).[8] The binary dependent variable of the study is Medicaid participation. Samples of nonelderly populations under sixty-five are

[8]The CPS ASEC study data were drawn from the Integrated Public Use Microdata Series, University of Minnesota Population Center. See Sarah Flood, Miriam King, Renae Rodgers Steven Ruggles and J. Robert Warren. Integrated Public Use Microdata Series, Current Population Survey: Version 6.0. Minneapolis, MN: IPUMS 2018. https://doi.org/10.18128/D030.V6.0

drawn for the US and NM study units applying household sampling weights. The US sample is studied in two strata, representing residency in Medicaid expansion and non-expansion states. The explanatory variables used in the model construction are identified below and described more fully in the proceeding subsection.

Choice behavior depends on population characteristics, including: age, marital status, children in households, educational attainment, labor force participation and health status. Health status ranges from excellent-to-poor, with poor as the reference. Other explanatory variables include SPM poverty status, out-of-pocket medical/insurance costs, SNAP benefits and total household resources. Total household resources is included as a budget measure (defined by SPM), along with family income below \$37,527.[9] Imposing this cap on the income distributions of the sample populations yields mean-values approximating \$20,000 – just shy of the 100% FPL threshold for a 3-person household in 2016 (\$20,090).

The Logit models are estimated using 'alternative specific constants' (ASC terms). The ASC terms align the estimated participation rates with reported rates for nonelderly populations under 100 and 200% of the FPL in 2016 (KFF 2019). The reported rates are computed for insurance units covered by Medicaid relative to being insured or uninsured. Consequently, the two ASC terms reflect insurance coverage under alternate income eligibility limits.

Table 10.1 reports the Logit estimates for the sample populations. The number of valid cases in the sample models are sufficient to draw inferences over the study units. Wald F-statistics for the fitted models reject the null hypothesis that all of their coefficients are equal to zero ($p < 0.000$). The McFadden-R^2 values are consistent across the models, ranging between 0.2564 (NM) and 0.3174 (non-ES models). The smaller number of valid cases in the NM models is notable (5665 eligible), nonetheless the McFadden R^2 values approximate those of the larger samples and are supported by the group F-statistic.

All but one of the coverage predictors are statistically significant across the models ($p < 0.01$). The exception is marital-status (married with spouse present) in the NM model. The predictors also have consistent signs in all of the models with the exception of SPM poverty status. This predictor is binary (above, below the SPM poverty threshold), with negative signs in the NM and non-ES models and positive signs in the US and ES models. This sign pattern reflects dichotomous impacts of public assistance on household poverty status, as measured by the SPM.

States receiving significant transfer payments for public assistance programs tend to have lower SPM poverty rates compared to their official poverty measures. This characteristic is common to the NM and non-ES study units, where income support and non-cash benefits are impactful in lifting households out of poverty. For these study units not being in poverty (based on the SPM predictor) increases the probability of being in Medicaid. The opposite prediction occurs in the US and ES models,

[9]The broad SPM definition of household resources: cash income plus in-kind benefits (SNAP and housing assistance), minus tax liabilities, work/child expenses, child support payments, and medical out-of-pocket costs.

Table 10.1 Logit estimates for prototypical households-US, Non-ES, ES and NM sample populations

	US sample			ES sample			Non-IS sample			NM sample		
Valid cases	266,674			163,798			102,876			5665		
Adj. Wald F-stat. All coeff. = 0. (p-value)	4578.9 (p < 0.000)			2772.2(p < 0.000)			1828.1 (p < 0.000)			93.5 (p < 0.000)		
McFadden R-sq.	0.2782			0.2681			0.3174			0.2564		
Coefficient sig. Level p < 0.01	**B est.**	**B * Mean**	**Marg. eff.**	**B est.**	**B * Mean**	**Marg. eff.**	**B est.**	**B * Mean**	**Marg. eff.**	**B est.**	**B * Mean**	**Marg. eff.**
Age(< 65)	−0.0126	−0.3913	−0.30%	−0.0090	−0.2820	−0.31%	−0.0208	−0.6392	−0.26%	−0.0136	−0.4061	−0.32%
Education years	−0.0675	−0.5238	−1.59%	−0.0630	−0.4945	−1.7%	−0.0740	−0.5649	−1.39%	−0.0696	−0.5301	−1.69%
Marital status(d:l) - married, spouse present	−0.1763	−0.0358	na	−0.0566	−0.0112	na	−0.3236	−0.0682	na	−0.1294	−0.0269	na
Spm no. of children(d:0) -no children	−0.2609	−0.1192	na	−0.3220	−0.1486	na	−0.2809	−0.1264	na	−0.2875	−0.1221	na
Spm snap amnt.	1.0530	0.3906	24.85%	1.0555	0.3946	26.0%	1.1690	0.4288	21.64%	0.9565	0.4227	26.31%
Spm poverty status	0.1540	0.0764	na	0.1716	0.0885	na	−0.0907	−0.0425	na	−0.3045	−0.1357	na
Medical out of pocket costs	−7.10E-05	−0.1352	−3.19%	−1.00E-04	−0.1818	−4.48%	−3.30E-05	−0.0671	−1.38%	−1.52E-04	−0.1891	−4.72%
Labor force participation (d:2) - yes	−0.7550	−0.3033	na	−0.5874	−0.2346	na	−1.0844	−0.4391	na	−0.4934	−0.1957	na
Health(d:l) - excellent	−1.1427	−0.3365	na	−1.1021	−0.3226	na	−1.3165	−0.3909	na	−1.4286	−0.4776	na
Health(d:2)-very good	−1.0316	−0.2919	na	−0.9953	−0.2784	na	−1.1878	−0.3416	na	−1.1314	−0.2908	na
Health(d:3) - good	−0.7261	−0.1920	na	−0.6813	−0.1840	na	−0.8936	−0.2290	na	−1.0305	−0.2797	na
Health(d:4) - fair	−0.2326	−0.0252	na	−0.1948	−0.0211	na	−0.3120	−0.0338	na	−0.4879	−0.0448	na
Spm total household resources	7.30E-05	1.4257	33.65%	7.50E-05	1.4761	35.2%	5.70E-05	1.0925	29.30%	5.50E-05	1.1012	35.62%
Spm total family income < $37,527)	−6.10E-05	−1.2109	−28.57%	−6.20E-05	−1.2344	−29.9%	−5.50E-05	−1.1068	−24.89%	−5.60E-05	−1.0539	−30.25%

(continued)

Table 10.1 (continued)

	US sample			ES sample			Non-IS sample			NM sample		
Constant	1.1895			1.1986			1.6282			2.2780		
ASC: 200% of FPL	11.1895 + 0.522			1. 1986 + 0.48			1.6282 + 0.53			2.27803 + 0.44		
ASC: 100% otFPL	1.1895 + 0.928			1.1986 + 0.91			16,281 + 0.95			2.27803 + 0.895		
Coefficient sig. Level $p > 0.10$												
Marital status(d:l) - married, spouse present										$p = 0.1447$		

Notes: marginal effect is the change in the est. probability of being in Medicaid from a unit change in a predictor variable, dPr/dX = B(Pr) (1-Pr),where Bis coeff. Est. for predictor X where public assistance programs appear to be less impactful. Consequently, SPM poverty status is a "mixed pickle" when it comes to predicting Medicaid participation.

The coefficient estimates for the household income variable relate to sample distributions drawn from truncated data, i.e. family income distributions with an upper limit of $37,527. The estimated coefficients under this constraint show the expected inverse relationship between income and Medicaid participation. The income coefficients have dual roles: first, as predictors of Medicaid participation, and secondly, as estimates of the marginal utility of household income in the computation of expected consumer surplus.

Medical out-of-pocket costs and their 'marginal effects' on Medicaid participation are a focal point of the study. Without exception, the parameter estimates show higher costs discourage low-income households from seeking Medicaid coverage, esp. study units with extended FPL thresholds under the ACA (ES and NM). The welfare implications of this cost-avoiding behavior are discussed further below, along with health status and labor force participation.

The empirical relationship between health status and Medicaid participation is complex and studied here on the crudest level, using self-assessed survey responses. The health status variable is categorical, providing a five-point ranking with 'poor health' being the lowest rank and serving as the reference. Linking health status to Medicaid participation is revealing. The findings show significantly higher probability of participation by households in poor health compared to other states-of-health. This finding is common to the sample populations, supporting the health-based rationale for why households participate in Medicaid programs.

Several socio-demographic variables lower the chance of Medicaid participation, esp. educational attainment and labor force participation. Educational attainment has a special place in healthcare economics as a proxy for human capital, originating in the 'demand for health' model pioneered by Grossman (1972, 2015). As he notes, "the positive relationship between more schooling and better health is one of the most fundamental ones in health economics." To be clear, this relationship is not considered in the present study. The focus here is simply on the independent causal effect of education on Medicaid participation. The results are consistent across study units, showing education has a negative impact on participation.

Meanwhile, isolating the causal effect of labor force status on Medicaid participation is complicated by potential collinearity with health status, e.g. non-participation due to poor health. Given the possible endogeneity of health status described above, the non-independence between health and labor supply is concerning. Here again, a simple approach is taken insofar as limiting the estimation to main effects – not interaction. Under this assumption labor supply follows education in having consistent negative causal effects on participation, esp. the non-ES study unit.

To illustrate the results thus far, Fig. 10.1a, b, c show predicted probabilities of Medicaid participation for a US household over a range of family incomes and household resources, i.e. ± 1 standard deviation from their sample means.

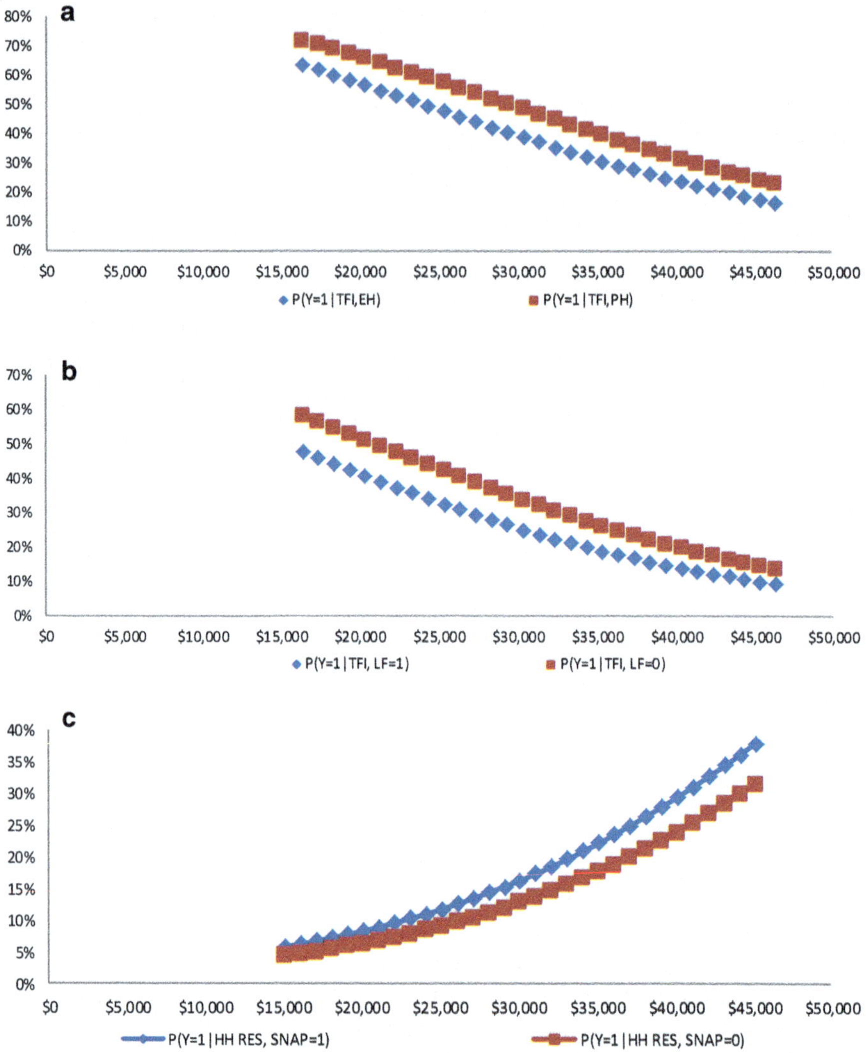

Fig. 10.1 Predicted Probabilities of Medicaid Coverage –US sample Household (2013–2018) (**a**) Variation of total family income (±1 SD from mean) and self-reported health status (*EH* excellent, *PH* poor) (**b**) Variation in total family income (±1 SD from mean) and labor force participation (yes, no) (**c**) Variation in SPM household resources (1 ± SD from mean) and receipt of SNAP benefits (yes/no)

The two-dimensional graphs provide images of the causal effects of heath-status (excellent, poor), labor force participation (yes, no), and SNAP benefits (yes, no).

Figures 10.1a, b depict the inverse relationship between Medicaid participation and family income, with parametric variations in other predictors: (a) shows the higher chance of being in Medicaid if households report poor health compared to excellent health; and (b) shows the higher chance of being in Medicaid if households are not in the labor force. Meanwhile Fig. 10.1c shows Medicaid participation increasing with higher levels of household resources and the impact of receiving SNAP benefits.

In review, the Logit estimates provide useful predictors of Medicaid participation for all of the sample populations and give support to the working hypothesis of the study: subsidized medical coverage promotes health and increases household purchasing power over all other goods. The welfare implications behind this argument are explored below.

Expected Consumer Surplus

CS for a binary choice is illustrated using the heuristic example from Sect. 10.3. Figure 10.2a depicts the consumer's choice space, given income M^0 and prices $p_1$and p_2. Figure 10.2b illustrates CS relative to the consumer's reservation price p_r.

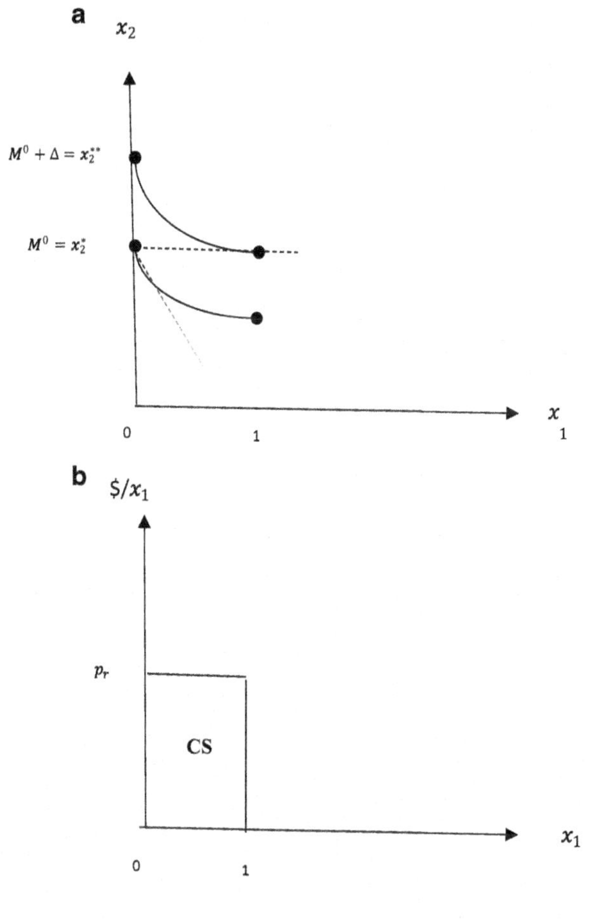

Fig. 10.2 a and b Consumer Surplus from subsidizing a discrete good

If either the consumer's income or reservation price is less than the market price (M^0 or $p_r < p_1$) then the optimal choice is $x_1^* = 0, x_2^* = M$. With subsidization ($p_1 = 0$) the consumer prefers $x_1^* = 1, x_2^* = M^0$. For the example given, there are no income effects from price changes. Consequently, the CS, CV and EV welfare metrics are equivalent measures of the change in indirect utility, represented here by the income-increment $M^1 - M^0 = \Delta$. In the application to follow the CS measures are computed using Small and Rosen's formula and parameter estimates for the various study units.

Panel (a) of Table 10.2 reports estimates of the 'willingness-to-pay' (WTP) for lower medical out-of-pocket costs – a standard economic welfare compensation measure. The estimates are computed using alternate coefficient ratios: either $-(\beta_{moop}/\beta_{spmfaminc})$, or $-(\beta_{moop}/\beta_{spmhhres})$. Both ratios measure WTP to restore utility to original levels following a loss. Based on family income, the non-ES household is willing-to-pay \$0.60 out of family income per dollar saved compared to \$1.61 for the ES household – note the relatively higher marginal value of medical benefits in ES study units. Correspondingly, US and NM households have WTP values of \$1.16 and \$2.71. Using household resources as the payment instrument has

Table 10.2 Predicted Medicaid participation and expected consumer surplus, 100% and 200% of FPL

Sample study units →		US	ES	Non-ES	NM
Panel A: Willingness-to-pay per dollar of medical cost-savings					
	wtp	\$1.16	\$1.61	\$0.60	\$2.71
Panel B: ASC 200% FPL eligible					
	Status-quo				
	pr	51.0%	56.1%	40.9%	62.0%
	cs	\$11,688	\$13,271	\$9549	\$17,279
	Poor health				
	pr	70.8%	74.1%	65.1%	83.0%
	cs	\$20,170	\$21,781	\$19,164	\$31,596
	Poor health and not in the labor force				
	pr	76.6%	78.3%	74.4%	85.5%
	cs	\$23,838	\$24,666	\$24,742	\$34,541
Panel C: ASC 100% FPL eligible					
	Status-quo				
	pr	61.0%	66.2%	51.3%	72.0%
	cs	\$15,416	\$17,519	\$13,064	\$22,734
	Poor health				
	pr	78.4%	81.5%	74.0%	88.5%
	cs	\$25,144	\$27,185	\$24,485	\$38,572
	Poor health and not in the labor force				
	pr	83.1%	84.7%	81.5%	90.3%
	cs	\$29,164	\$30,332	\$30,705	\$41,697

Notes: Panel A, WTP values are invariant to ASC. Panels B and C, status quo definitions apply ASC 200% and 100% FPL, and mean values of Predictor variables a modest impact on the welfare compensation amounts: \$0.58 (non-ES), \$1.33 (ES), \$0.97 (US), and \$2.76 (NM).

Panels (b) and (c) of Table 10.2 report the effects of health and labor force status on participation and CS for the 100 and 200% FPL models. The CS values are computed using family income as the *numeraire*. Status quo participation and CS levels are computed at the mean-values of the predictors. Poor health significantly increases participation and CS relative to the status quo; especially for study units with family incomes below 100% FPL.

Further, study units in poor health and not in the labor force are the most likely to participate in Medicaid and have the highest gains in expected CS. For example, a prototypical US household under 200% FPL, in poor health and not in the labor force, has a 76.6% chance of participating in Medicaid, thus gaining \$23,838 in CS. Meanwhile, a similar household under 100% FPL has an 83.1% chance of participating in Medicaid and gaining \$29,164 in CS. By comparison, the average annual premium for a private sector employee enrolled in family coverage was \$18,687 in 2017.

What can we conclude from these findings; how do they contribute to the Medicaid expansion literature; do they have policy applications?

Concluding Remarks

Medicaid is vital to millions of lower income working-age households across the US. The general hypothesis posed at the onset of this study is that households participate in Medicaid to improve their welfare. Correspondingly, the study rationalizes the participation choice of prototypical households and estimate their expected consumer surplus from program participation. The study data (CPS-ASEC, 2013–2018) cover Medicaid expansion under the ACA, permitting a natural experiment of participation choices in expansion and non-expansion states. Several conclusions are reached from the utility theory considered here.

Overall, the expected gain in consumer surplus from Medicaid participation is highest for Medicaid-eligible households in expansion states; esp. those reporting poor health, being unmarried with children, having fewer years of schooling, not being in the labor force and experiencing ‘deep poverty,’ i.e. less than one-half of the OPM poverty line. Further, households in expansion states have a significantly higher marginal value in avoiding out-of-pocket costs than households in non-expansion states. These findings underscore the significance of household demography in welfare analysis of Medicaid coverage.

Further, the study links welfare compensation to medical out-of-pocket costs and the probability of coverage. The expectation of lower out-of-pocket costs increases the probability of Medicaid participation and household welfare – above all, for the lowest income households in ACA expansion states. These findings are intuitive given the growth of subsidized coverage, and the reported impacts of Medicaid in reducing poverty.

The study contributes to Medicaid expansion literature by applying a simple RUM framework to evaluate welfare effects from variations in health status, labor force participation, and medical out-of-pocket costs – in particular, how changes in these predictors impact healthcare coverage and welfare under varying FPL limits. The study also uses the Supplemental Poverty Measure in explaining differences in participation across ACA expansion and non-expansion states. Finally, the 'willingness-to-pay' estimates provide a valuation metric for guiding policy decisions that impact out-of-pocket medical costs. In these regards, the tools of applied welfare analysis used here may help advance more efficient allocation and funding of healthcare resources.

Acknowledgements and Disclaimer I wish to acknowledge Secretaries Brent Earnest and David Scrace, and Directors Julie Weinberg, Nancy Leslie-Smith and Nicole Comeaux in leading the implementation and administration of the Affordable Care Act through the New Mexico Human Services Department. I also extend my thanks to the University of New Mexico Department of Geospatial and Population Studies for including this work in the second Biennial Population and Public Policy Conference, in Albuquerque, New Mexico. Finally, I am grateful to Jason Sanchez, Jaclyn Herrera, Shane Shariff and anonymous referees for reviews and suggestions in the development of this work. That said, I assume sole responsibility for the hypothesis, method of analysis and reported findings of the study.

References

Baicker, K., et al. (2013). The Oregon experiment – Effects of Medicaid on clinical outcomes. *The New England Journal of Medicine, 368*(18), 1713–1722.

Blavin, F., Karpman, M., Kenney, G. M., & Sommers, B. D. (2018). Medicaid versus marketplace coverage for near-poor adults: Effects on out-of-pocket spending and coverage. *Health Affairs, 37*(2), 299–307.

Bound. (1991). Self-reported versus objective measures of health in retirement models. *The Journal of Human Resources, 26*(1), 106–138.

Bound. (2000). *Measurement error in survey data*. Report 00–450, Population Studies Center, University of Michigan.

Burtless, G., & Siegel, S. (2001). *Medical spending, health insurance, and measurement of American poverty*. The Brookings Institutiton, Washington, D.C.

Caswell, K. L., & Short, K. S. (2011). *Medical out-of-pocket spending among the uninsured: Differential spending & the supplemental poverty measure* (Working paper 2011–24). U.S. Census Bureau, Social, Economic, and Housing Statistics Division.

Citro, C. F., & Michaels, R. T. (Eds.). (1995). *Measuring poverty: A new approach*. National Academy Press, Washington, D.C.

Dalakar, J. (2017). *The supplemental poverty measure: Its core concepts, development and use*. Congressional Research Service, Washington D.C.

Foster, A. C., & Rojas A. (2018). *Program participation and spending patterns of families receiving government means-tested assistance*. Monthly Labor Review, U.S. Bureau of Labor and Statistics.

Fox, L. (2018). *The supplemental poverty measure: 2017*. United States Census Bureau, Current Population Reports, Washington, D.C. (pp. 60–265).

Glied, S., Ougni, C., & Russo T. (2017). *How medicaid expansion affected out-of-pocket health care spending for low-income families*. Issues Brief, The Common Wealth Fund.

Grossman, M. (1972). On the concept of health capital and the demand for health. *The Journal of Political Economy, 80*(2), 223–255.

Grossman, M. (2015). *The relationship between health and schooling: What's new?* (Working paper 21609). National Bureau of Economic Research.

Gruber, J. (2000). "*Medicaid*," (Working paper 7829). National Bureau of Economic Research.

Harberger, A. C., & Just, R. (2012). A conversation with Arnold Harberger. *The Annual Review of Resource Economics, 4*, 1–26.

Kaiser Family Foundation. (2019). *Where Are States Today? Medicaid and CHIP Eligibility Levels for Children, Pregnant Women, and Adults.*

Levy, H., & Meltzer, D. (2008). The impact of health insurance on health. *Annual Review of Public Health, 29*, 399–409.

Majerol, M., Tolbert, J., & Damico, A. (2016). *Health care spending among low-income households with and without Medicaid.* Kaiser Family Foundation.

Meer, J., & Rosen, H. S. (2004). Insurance and the utilization of medical services. *Social Science & Medicine, 58*, 163–1632.

Remler, D. K., Korenman, S. D., & Hyson, R. T. (2017). Estimating the effects of health insurance and other social programs on poverty under the affordable care act. *Health Affairs, 36*(10), 1828–1837.

Small, K. A., & Rosen, H. S. (1981). Applied welfare economics with discrete choice models. *Econometrica, 49*, 105–130.

Sommers, B. D., & Oellerich, D. (2013). The poverty-reducing effects of Medicaid. *Journal of Health Economics, 32*(5), 816–832.

Sommers, B. D., Gwande, A. A., & Baicker, K. (2017). Health insurance coverage and health – What the recent evidence tells us. *New England Journal of Medicine, 377*, 586–593.

Train, K. E. (2009). *Discrete choice methods with simulation* (2nd ed.). Cambridge, MA: Cambridge University Press.

United States Government Accounting Office. (2018). *Medicaid expansion and access to care.* GAO-18-607, Washington D.C.

Chapter 11
Is the Indian Labor Market Biased Against Women? Yes!

Soumyajit Chakraborty, Alok K. Bohara, Melissa Binder, and Richard Santos

Introduction

Gary Becker's *The Economics of Discrimination* (1957) identifies the existence of racial prejudices and discrimination in a competitive labor market. Not only racial, there exists discrimination by sex; and gender division of labor is prevalent all over the world (Bielby and Baron 1986). In writings of neoclassical economists, discrimination is sometimes also motivated by a continuing occupational sex segregation. Limited access to the labor market due to family obligations has worked against the liberal notion of 'feminine role' in the labor market (Mincer and Polachek 1974). Reskin (1993) argues that practice of childrearing has forced women to accept such jobs which do not penalize any skill depreciation from paid leave and have possibilities of re-entering.

In the pre-industrialization age, men 'specialized' to work in the labor market while women worked in the home, which led to a separation of domestic and childcare work from other types of work. This separation made two things possible: first, it made children and women vulnerable and economically dependent on men; and second, it waived off men's participation in household chores. Thus, the economic role of household activities was taken out of consideration and the notion of 'housewife' evolved. Since industrialization made household activities less burdensome, women began to participate in economic activities in a changed market

---

Disclaimer: The original article entitled "*Working and Getting Paid, But Eventually Less Than Men: By How Much in India?*" by Soumyajit Chakraborty, Alok K. Bohara, Melissa Binder, and Richard Santos is being submitted exclusively to the International Association of Applied Demography for publication consideration into an edited book titled "Population Change and Public Policy" by Springer International Publishing.

S. Chakraborty (✉) · A. K. Bohara · M. Binder · R. Santos
Department of Economics, University of New Mexico, Albuquerque, NM, USA
e-mail: soumyajitchakraborty@unm.edu; bohara@unm.edu; mbinder@unm.edu; santos@unm.edu

B. Jivetti, Md. N. Hoque (eds.), *Population Change and Public Policy*, Applied Demography Series 11, https://doi.org/10.1007/978-3-030-57069-9_11

scenario. The twentieth century thus saw a changing gender division of labor and the demand for white-collar female workers increased. Meanwhile, effective contraception methods also made it possible for women to identify themselves beyond their reproductive role and allowed them to focus on the economic role as well. But, a patriarchal system, which was established much before capitalism, allowed men to develop hierarchical control of the family. This led to economic inequality in terms of opportunities and choices in the labor market (Hartmann 1976).

Formal analysis of labor market outcomes involves an application of human capital theory (Schultz 1961; Becker 1962). Investment in human capital plays a significant role in terms of employment opportunities. For empirical testing, Mincer (1974) provides an earnings equation in relation to certain parameters of human capital e.g., age, education, and experience. Especially in a male-biased society like that of India's, access to human capital and continuation to higher studies is a stand-alone problem. But, even achieving equal human capital does not ensure equitable labor market treatment due to the pervasive nature of institutional discrimination. Thus, it is important to identify the extent of the gender gap in earnings and factors attributing to that differential. Additionally, different socioreligious groups (formed out of castes and religions) are used to address the intra-group gender gaps.

The remainder of this study is organized as follows. Section 11.2 provides a review of the Indian labor market, and renders some descriptive statistics derived from the National Sample Survey (NSS) data. Section 11.3 gives a brief overview of empirical methodologies. Findings are presented in Sect. 11.4. Finally, the paper draws concluding remarks with reference to government policies in Sect. 11.5.

Review of the Indian Labor Market and Descriptive Statistics

In between 1993 to 2004, the Indian labor market has seen an almost twofold increase in earnings disparities (Mukherjee and Majumder 2011). There is a possible scope of occupational segregation playing a decisive role in gender gap of earnings. Chattopadhyay et al. (2013) find that majority of the Indian women work mostly in two industries - food and apparel. A study by Uppal (2007) shows prominent indication of 'feminization of poverty' which implies women workers are concentrated in ill–/un-paid occupations. The intra-occupation gender gap is more than 60% of the total gender gap (i.e., inter-occupations gender gap is about 40%) and there exists considerable earnings gap for almost each occupation (Chakraborty 2016).

The National Sample Survey (NSS) provides Indian workers' micro-data with individual-level information on human capital, industry, occupation, and other demographic characteristics. This study analyzes the working-age (18–60 years) samples weighted for the population. To incorporate the essence of joblessness, we consider both primary activity and up to the first secondary activity as an individual's occupational engagement. NSS also provides information on workers' castes and

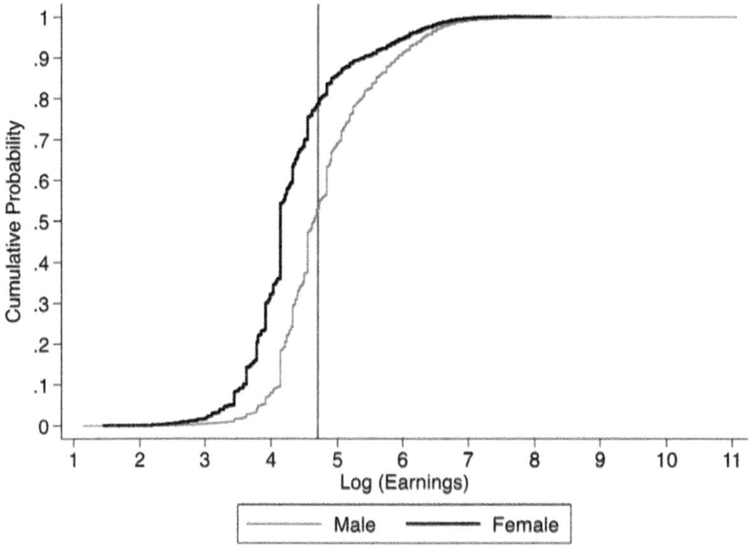

Fig. 11.1 Cumulative distribution of earnings for male and female workers. (Source: Authors' calculation from the 68th round NSS data)

religions separately. We merge that information in forming socioreligious groups and classify the workers into six categories: Scheduled Tribes (STs), Scheduled Castes (SCs), Other "Backward" Castes (OBCs), "High"-Caste Hindus (HHs), Muslims and Other Religions. The NSS data indicate that a representative female worker earns less than her male counterpart, irrespective of socioreligious groups (see Figs. 11.1 and 11.2). Specifically, about 78% of female workers earn less than the overall mean earnings level (vertical reference line).

As shown in the summary statistics (see Table 11.1), though an average female worker has 2 years of less educational attainment than an average male worker, the former is more experienced than the latter. It is clear that there is a mismatch of human capital endowment. In terms of wage, a representative female worker earns almost \$61 less than a male one. While 70% of male workers are engaged in non-agricultural activities, about a half of the total female workers are still employed in agricultural sector. This indicates a visible gender division of labor. Now, when the sex earnings ratio $\left(W_f/W_m\right)$ is calculated, we see a major variation across states, industries, and occupations (see Fig. 11.3 and Table 11.2).

First, the relationship between earnings and human capital & other income-generating characteristics is estimated by following the Mincerian earnings equation (Mincer 1974). Then we estimate the overall gender gap in earnings; and also, the intra-socioreligious gender gaps. Decomposing these gaps demand for both parametric and non-parametric techniques, which is discussed in the following section.

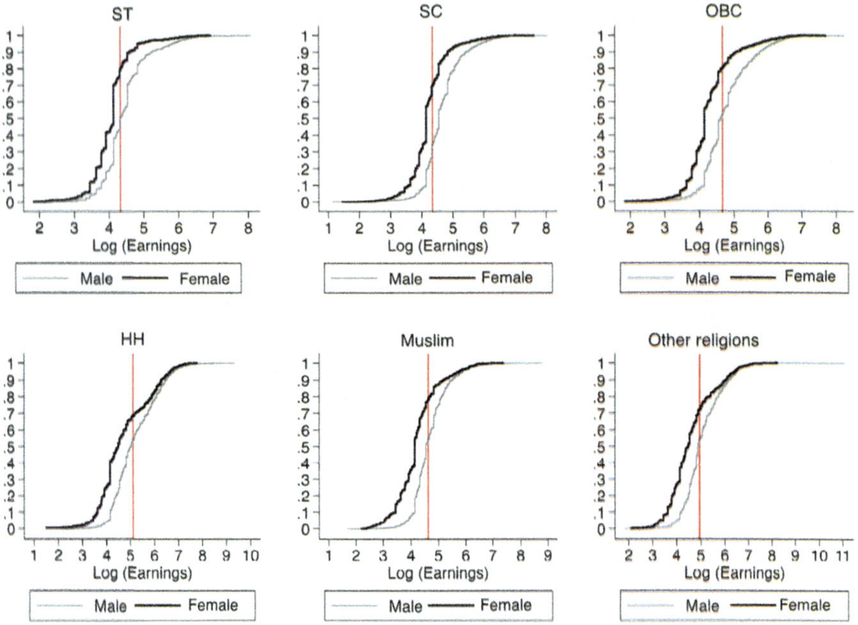

Fig. 11.2 Cumulative distribution of intra-socioreligious earnings for male and female workers. (Source: Authors' calculation from the 68th round NSS data)

Table 11.1 Summary statistics of selected variables

	Pooled		Female		Male	
Variables	Mean	S.E.	Mean	S.E.	Mean	S.E.
Age	35.9	0.04	36.8	0.09	35.6	0.05
Female (%)	0.2					
Household size	4.7	0.01	4.4	0.02	4.8	0.01
Educational attainment	6.2	0.02	4.6	0.04	6.6	0.02
Experience	23.9	0.05	26.0	0.10	23.2	0.05
Experience groups (base: 0–10 years)						
11–20 years (%)	28.1		23.1		29.5	
21–30 years (%)	26.0		26.8		25.8	
> 30 years (%)	30.4		37.2		28.4	
Monthly earnings	160.3	0.84	113.5	1.32	174.2	1.00
Region (Base: Rural)						
Urban (%)	34.03		28.28		35.74	
Marital status (Base: Otherwise)						
Married (%)	76.4		71.26		77.94	
Broad employment sector (Base: Agriculture)						
Non-agriculture (%)	65.37		50.18		69.89	
Observations	69,583		15,005		54,578	

Source: Authors' calculation from the 68th round NSS data

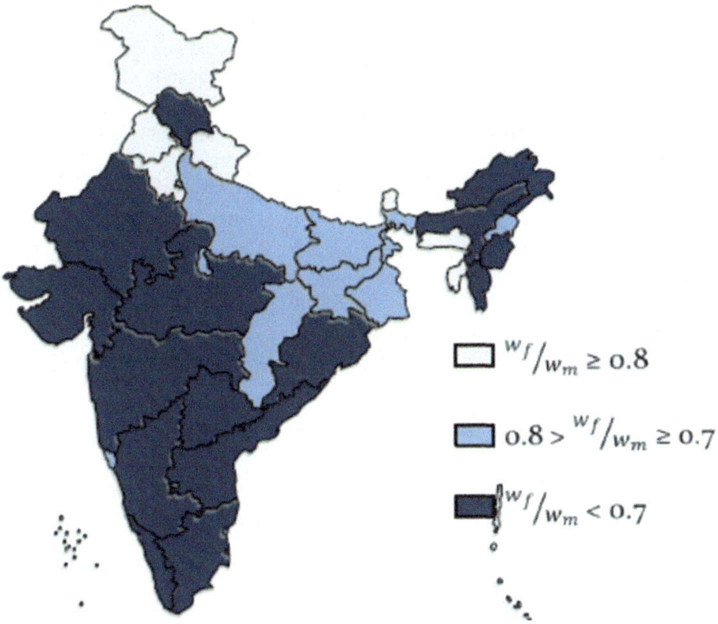

Fig. 11.3 Sex earnings ratio by states. (Source: Authors' calculation from the 68th round NSS data)

Methodology

We can interpret the gender gap in earnings in different ways based on the techniques of decomposition. First, we follow the parametric three-fold Blinder-Oaxaca decomposition (Blinder 1973; Oaxaca 1973). According to this method, we can decompose the gender gap into three parts: (1) due to endowment e.g., human capital and other income-generating characteristics (*E*); (2) due to coefficients e.g., discrimination and other unobservable factors (*C*); and (3) due to interaction between endowment and coefficients (*I*). Thus, if *D* represents the gender gap in earnings, then $D = E + C + I$. But as the parametric decomposition fails to incorporate any possible likelihood of historical discrimination, we also decompose the gender gap in earnings following a non-parametric matching technique. This method intrinsically assumes a different probability distribution for a group of representative workers based on their identical human capital and other income-generating characteristics. Matching techniques have become crucial to causal effects after the introduction of propensity score (Rosenbaum and Rubin 1983) in experimental design of evaluation studies. On the other hand, Heckman et al. (1997), and Heckman and Todd (2009) have familiarized the role of propensity score matching in non-experimental designs as well. After a few semi-parametric methods developed by DiNardo et al. (1996), and Donald et al. (2000), Barsky et al. (2002)

Table 11.2 Sex earnings ratio by industries and occupations

Selected National Industrial Classification (NIC)	w_f/w_m	Selected National Classification of Occupations (NCO)	w_f/w_m
Arts, entertainment and recreation	1.70	Physical, mathematical and engineering science professionals	1.04
Real estate activities	1.34	General managers	0.90
Electricity, gas, steam and air conditioning supply	1.23	Corporate managers	0.90
Transportation and storage	1.20	Office clerks	0.90
Administrative and support service activities	1.19	Legislators and senior officials	0.85
Information and communication	1.04	Models, sales persons and demonstrators	0.85
Professional, scientific and technical activities	0.92	Subsistence agricultural and fishery workers	0.82
Financial and insurance activities	0.87	Teaching professionals	0.79
Wholesale and retail trade; repair of motor vehicles and motorcycles	0.84	Customer services clerks	0.76
Public administration and defense; compulsory social security	0.75	Agricultural, fishery and related laborers	0.76
Agriculture, forestry and fishing	0.73	Laborers in mining, construction, manufacturing and transport	0.72
Accommodation and food service activities	0.70	Metal, machinery and related trades workers	0.70
Construction	0.67	Precision, handicraft, printing and related trades workers	0.68
Education	0.67	Drivers and mobile-plant operators	0.67
Human health and social work activities	0.66	Life science and health professionals	0.66
Activities of households as employers; undifferentiated goods and services producing activities of households for own use	0.53	Market oriented skilled agricultural and fishery workers	0.64
		Extraction and building trades workers	0.56
Manufacturing	0.48	Stationary plant and related operators	0.53
Mining and quarrying	0.38	Sales and services elementary occupations	0.49
Water supply; sewerage, waste management and remediation activities	0.36	Personal and protective service workers	0.40

Source: Authors' calculation from the 68th round NSS data

examines the racial wealth gap with a non-parametric decomposition technique. Advancing this literature, Ñopo (2008) proposes a rigorous decomposition method in non-parametric setup while measuring the Peruvian earnings gap. Using the Ñopo method, Karki and Bohara (2014) shows that *non-Dalit* workers have historically built some favorable labor market characteristics which help them to earn more than *Dalits* in a typical caste-ridden and patriarchic society. This is a four-fold decomposition method based on one-to-many likelihood score matching, where Δ represents the gender gap in earnings, and $\Delta = \Delta_0 + \Delta_x + \Delta_m + \Delta_f$. Here, Δ_0 and Δ_x are equivalent to Blinder-Oaxaca's C and E, respectively. Additionally, Δ_m and Δ_f are calculated over the differences within common support. Specifically, Δ_m indicates a component of gender gap due to the difference within two groups of male workers: one, whose likelihood distribution match with that of female workers; and two, whose do not match. On the contrary, Δ_f indicates a component of gender gap due to the difference within two groups of female workers matching or not matching with that of male workers. Typically, as female workers are considered as the reference group, Δ_m is expected to be positive and Δ_f to be negative. Alternatively, Δ_m and Δ_f will be zero if the likelihood distribution of characteristics for male and female workers match exactly with each other.

This study uses both Blinder-Oaxaca and Ñopo decomposition to estimate the extent of different contributing factors of the gender gap in earnings. Human capital characteristics are used for matching purposes with a set of covariates including labor market characteristics.

Results

Mincerian Earnings Equation

First, Empirical estimation for a representative worker i can be formulated as follows: $Y_i = \beta_0 + \beta_1 HC_i + \beta_2 X_i + \gamma_i + \varepsilon_i$; where Y_i is the log (earnings), HC_i is the human capital, X_i is the set of covariates, γ_i denotes state/industry/occupation fixed effects. Results in Table 11.3 confirm that age, educational attainment, and experience play positive roles to workers' earnings. Specifically, an extra year of education may result into a significant 3% increase in earnings. Also, if a worker is 10 years more experienced than another, (s)he is likely to earn 9% higher than the latter. The coefficients of experience-squared support the theory of diminishing marginal productivity suggesting an inverted-U shape relationship between earnings and experience.

In terms of gender differential in earnings, an average female worker tends to earn significantly less (32% lesser, column 6 of Table 11.3) than her male counterpart. Again, the role of socioreligious groups are visible as "high" caste Hindus (HHs) earn almost 17% higher (column 6 of Table 11.3) than the tribal workers (STs). There prevail earnings differentials according to region and broad employment sectors too. As expected, having a Lewisian two-sector developing economy, rural and agricultural workers earn significantly less than those in urban and non-agricultural sectors, respectively.

Table 11.3 Ordinary Least Squares (OLS) results

	Dependent variable: Log (earnings).					
Variables	(1)	(2)	(3)	(4)	(5)	(6)
Human capital						
Educational attainment	0.0956***	0.0681***	0.0521***	0.0456***	0.0358***	0.0306***
	(0.000940)	(0.00377)	(0.00361)	(0.00449)	(0.00456)	(0.00452)
Experience	0.0424***	0.0128**	0.00581	0.000910	0.000638	0.00891
	(0.00147)	(0.00548)	(0.00527)	(0.00749)	(0.00733)	(0.00692)
Experience squared	−0.000525***	−0.000445***	−0.000378***	−0.000384***	−0.000356***	−0.000362***
	(0.0000265)	(0.0000292)	(0.0000284)	(0.0000293)	(0.0000295)	(0.0000285)
Other characteristics						
Gender (Base: Male)						
Female		−0.324***	−0.297***	−0.322***	−0.298***	−0.322***
		(0.0103)	(0.0103)	(0.0149)	(0.0133)	(0.0136)
Age		0.0247***	0.0278***	0.0325***	0.0296***	0.0213***
		(0.00523)	(0.00500)	(0.00723)	(0.00744)	(0.00681)
Socioreligious groups (Base: ST)						
SC		0.109***	0.0748***	0.0176	0.0929***	0.102***
		(0.0167)	(0.0158)	(0.0360)	(0.0333)	(0.0329)
OBC		0.114***	0.0735***	0.00595	0.0985***	0.0953***
		(0.0160)	(0.0152)	(0.0338)	(0.0316)	(0.0305)
HH		0.287***	0.199***	0.159***	0.215***	0.172***
		(0.0187)	(0.0179)	(0.0413)	(0.0394)	(0.0384)
Muslim		0.175***	0.0827***	0.0257	0.120***	0.101***
		(0.0184)	(0.0176)	(0.0387)	(0.0342)	(0.0334)
Other religions		0.278***	0.218***	0.0989**	0.239***	0.218***
		(0.0218)	(0.0210)	(0.0407)	(0.0368)	(0.0358)

Marital status (Base: Otherwise)						
Married		0.0392***	0.0723***	0.0768***	0.0650***	0.0624***
		(0.0121)	(0.0117)	(0.0135)	(0.0132)	(0.0120)
Household size		−0.0155***	−0.0135***	−0.00911***	−0.0137***	−0.0100***
		(0.00199)	(0.00195)	(0.00322)	(0.00340)	(0.00312)
Region (Base: Rural)						
Urban			0.212***	0.208***	0.224***	0.180***
			(0.00972)	(0.0174)	(0.0200)	(0.0184)
Broad employment sector (Base: Agriculture)						
Non-agriculture			0.210***	0.186***		0.0943***
			(0.00974)	(0.0163)		(0.0327)
State fixed effects				*Yes*		*Yes*
Industry fixed effects					*Yes*	*Yes*
Occupation fixed effects						*Yes*
Districts clustered				*Yes*	*Yes*	*Yes*
Constant	3.470***	3.369***	3.277***	3.325***	3.540***	3.630***
	(0.0201)	(0.0491)	(0.0476)	(0.0762)	(0.0824)	(0.0758)
Observations	69,583	69,583	69,583	69,583	69,583	69,583
R-squared	0.339	0.382	0.418	0.445	0.465	0.489

Robust standard errors in parentheses
***$p < 0.01$, **$p < 0.05$, *$p < 0.1$
Source: Authors' calculation from the 68th round NSS data

Blinder-Oaxaca Decomposition

From the OLS results, we adopt the specification in column (6) and perform the decomposition analysis. Table 11.4 records the overall and intra-socioreligious gender gap in earnings (column D). Then these gaps are decomposed into (1) gap due to endowment (column E), (2) gap due to coefficients (column C), and (3) interaction between them (column I).

The overall/pooled gender gap in earnings is 0.48 or 48%, of which 37% (0.178) can be explained due to differences in human capital endowment. On the other hand, 70% (0.335) of the gender gap is attributed to institutional discrimination and other unobservables. Specifically, discrimination accounts for 70–100% of the intra-socioreligious gender gap, except that for STs (see Fig. 11.4). This implies that among ST workers, gender gap in earnings is mainly due to difference in human capital. Or, it might be the case that tribal workers, either men or women, are seen as a homogeneous entity of labor source from an employer's perspective. It is interesting to note that these ST workers are heavily engaged in ill-paid occupations.

Ñopo Decomposition

Table 11.5 presents the overall and intra-socioreligious gender gap in earnings (column Δ). Then these gaps are decomposed into four summands: (1) unexplained (column Δ_0), (2) explained-I (column Δ_m), (3) explained-II (column Δ_f), and (4) explained-III (column Δ_x). Two columns i.e., Δ_x and Δ_0 are analogous to the columns E and C in Sect. 11.4.2. Non-parametric results are technically different from that of parametric decomposition and yield much lower values for gender gap in earnings.

Table 11.4 Gender gap in earnings (Blinder-Oaxaca decomposition)

Groups	Gender gap (D)	Due to endowment (E)	Due to coefficients (C)	Due to interaction (I)
Pooled	0.477***	0.178***	0.335***	−0.0357***
	(0.00708)	(0.00674)	(0.00618)	(0.00583)
ST	0.306**	0.279***	0.0945	−0.0670***
SC	0.357***	0.202***	0.250***	−0.0946***
OBC	0.596***	0.180***	0.452***	−0.0352***
HH	0.689***	0.0983***	0.653***	−0.0628***
Muslim	0.604***	0.0236	0.626***	−0.0463*
Other religions	0.396*	0.0454**	0.409*	−0.0581***

Standard errors in parentheses

***$p < 0.01$, **$p < 0.05$, *$p < 0.1$

Source: Authors' calculation from the 68th round NSS data

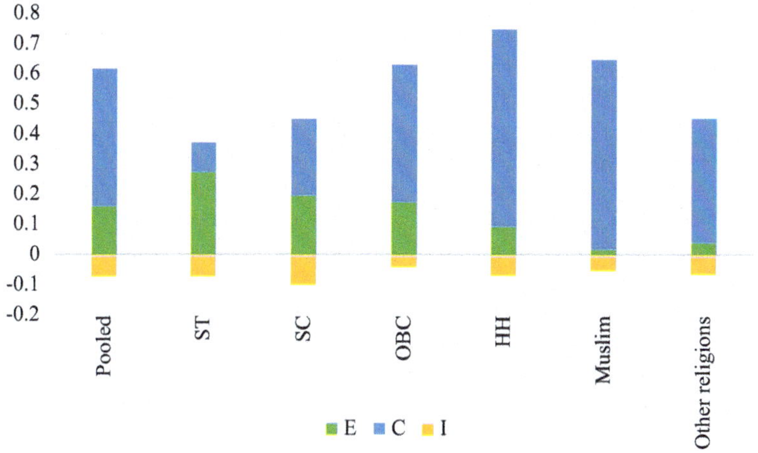

Fig. 11.4 Gender gap in earnings (Blinder-Oaxaca decomposition). (Source: Authors' calculation from the 68th round NSS data)

Table 11.5 Gender gap in earnings (Ñopo decomposition)

Groups	Gender gap (Δ)	Unexplained (Δ_o)	Explained-I (Δ_m)	Explained-II (Δ_f)	Explained-III (Δ_x)
Pooled	0.110	0.101	0.036	−0.024	−0.003
ST	0.093	0.075	0.030	−0.012	−0.000
SC	0.103	0.097	0.024	−0.016	−0.001
OBC	0.113	0.102	0.033	−0.020	−0.003
HH	0.096	0.093	0.045	−0.035	−0.007
Muslim	0.110	0.121	0.022	−0.031	−0.001
Other religions	0.099	0.101	0.042	−0.036	−0.008

Source: Authors' calculation from the 68th round NSS data

The overall/pooled gender gap in earnings is 0.11 or 11%, of which 91% (0.101) is due to labor market discrimination and other unobservables. In fact, for all intra-socioreligious gender gaps, discrimination is the highest contributor. While all Δ_m coefficients are positive, Δ_f column throughout shows negative coefficient values. In other words, historically, male workers have earned some favorable labor market characteristics, which is continuously helping them to earn much more than female workers with identical human capital attributes. This accounts for 32% (0.036) of the overall gender gap in earnings. On the contrary, historical discrimination has worked against women who participate in the labor market, and 21% (−0.024) of the gender gap is due to this. We might think of specific institutional factors such as labor union which is likely to be male dominant and male segregated industries are likely to have strong union presence (Chakraborty 2016).

According to the stacked column graph (see Fig. 11.5) of Ñopo decomposition, the overall and intra-socioreligious gender gaps in earnings are mostly due to

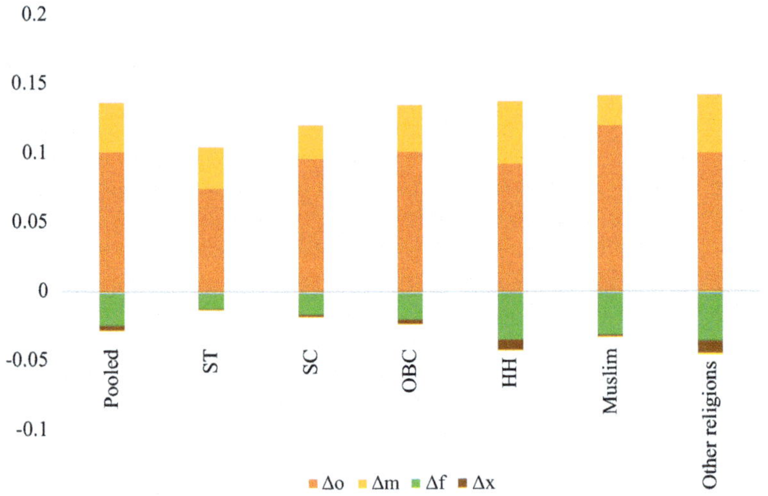

Fig. 11.5 Gender gap in earnings (Ñopo decomposition). (Source: Authors' calculation from the 68th round NSS data)

institutional discriminatory factors. And then, the favorable male attributes and non-favorable female attributes play a crucial role in explaining these gaps. This is expected in a patriarchal setup that regardless of socioreligious background, men continue to be considered "superior" in labor market participation.

Concluding Remarks

This study extensively examines if the Indian labor market faces gender discrimination and whether it contributes to the gender gap in earnings. Additionally, we also incorporate the regressive caste system and religions into our analysis. Both Blinder-Oaxaca and Ñopo methods of decomposition confirm an active role of gender discrimination irrespective of castes and religions. While the first technique suggests that employers consider tribal workers as a homogeneous pool of labor supply, the nonparametric technique attests to the fact that institutional background and historical discrimination play significant roles in widening the overall and intra-socioreligious gender gaps.

Though the Indian government focused to dampen economic inequality through abolition of discriminatory practices at the workplace, it was surely not enough. The 2011–12 NSS data suggests that the top 10% of income ladder earns about 40% of total earnings and the bottom 50% earn a mere 20% of that. On top of that, there are disparities involving sectoral differences such as rural versus urban, agricultural versus non-agricultural, etc. Therefore, policymakers should consider a strategy of "inclusive growth" that attempts to reduce human capital differential in terms of opportunities. Skill-enhancing training programs, flexible work environment with affordable quality childcare, and model anti-discriminatory employment generation programs are a few ways to promote women empowerment. The government needs to assess its performance in achieving gender equality and set agendas in the Sustainable Development Goals (SDGs). Otherwise, employers will always be prejudiced against certain group of workers (Becker 1957), in this case- Indian women.

Acknowledgments The authors like to acknowledge Professor Guillermina Jasso for the Student Paper Award at 2019 Biennial Population and Public Policy Conference. We also wish to thank Dr. Amanda Kerr, Dr. Soumi R. Chowdhury, and Dr. Kira Villa for their valuable feedback toward the first draft of the manuscript. Our sincere thanks to the anonymous reviewer whose comments and suggestions have significantly improved the manuscript. Finally, we extend our thanks to two small funding- J. Raymond Stuart Award in Economics, and Graduate & Professional Student Association (GPSA) Student Research Grant for the data purchase and conference participations.

References

Barsky, R., Bound, J., Charles, K. K., & Lupton, J. P. (2002). Accounting for the black–white wealth gap: A nonparametric approach. *Journal of the American Statistical Association, 97*(459), 663–673.

Becker, G. S. (1957). *The economics of discrimination*. Chicago: University of Chicago Press.

Becker, G. S. (1962). Investment in human capital: A theoretical analysis. *Journal of Political Economy, 70*(5, Part 2), 9–49.

Bielby, W. T., & Baron, J. N. (1986). Men and women at work: Sex segregation and statistical discrimination. *American Journal of Sociology, 91*(4), 759–799.

Blinder, A. S. (1973). Wage discrimination: Reduced form and structural estimates. *Journal of Human Resources, 84*, 436–455.

Chakraborty, S. (2016). Occupational sex segregation: An inquiry into Indian Job Market. *Journal of Regional Development and Planning, 5*(2), 1.

Chattopadhyay, M., Chakraborty, S., & Anker, R. (2013). Sex segregation in India's formal manufacturing sector. *International Labour Review, 152*(1), 43–58.

DiNardo, J., Fortin, N. M., & Lemieux, T. (1996). Labor market institutions and the distribution of wages, 1973-1992: A semiparametric approach. *Econometrica, 64*(5), 1001–1044.

Donald, S. G., Green, D. A., & Paarsch, H. J. (2000). Differences in wage distributions between Canada and the United States: An application of a flexible estimator of distribution functions in the presence of covariates. *The Review of Economic Studies, 67*(4), 609–633.

Hartmann, H. (1976). Capitalism, patriarchy, and job segregation by sex. *Signs: Journal of Women in Culture and Society, 1*(3, Part 2), 137–169.

Heckman, J. J., & Todd, P. E. (2009). A note on adapting propensity score matching and selection models to choice based samples. *The Econometrics Journal, 12*(suppl_1), S230–S234.

Heckman, J. J., Ichimura, H., & Todd, P. E. (1997). Matching as an econometric evaluation estimator: Evidence from evaluating a job training programme. *The Review of Economic Studies, 64*(4), 605–654.

Karki, M., & Bohara, A. K. (2014). Evidence of earnings inequality based on caste in Nepal. *The Developing Economies, 52*(3), 262–286.

Mincer, J. (1974). *Schooling, experience, and earnings*. Human behavior & social institutions No. 2. National Bureau of Economic Research (NBER). New York.

Mincer, J., & Polachek, S. (1974). Family investments in human capital: Earnings of women. *Journal of Political Economy, 82*(2, Part 2), S76–S108.
Mukherjee, D., & Majumder, R. (2011). Occupational pattern, wage rates and earning disparities in India: A decomposition analysis. *Indian Economic Review, 46*, 131–152.
Ñopo, H. (2008). Matching as a tool to decompose wage gaps. *The Review of Economics and Statistics, 90*(2), 290–299.
Oaxaca, R. (1973). Male-female wage differentials in urban labor markets. *International Economic Review, 14*, 693–709.
Reskin, B. (1993). Sex segregation in the workplace. *Annual Review of Sociology, 19*(1), 241–270.
Rosenbaum, P. R., & Rubin, D. B. (1983). The central role of the propensity score in observational studies for causal effects. *Biometrika, 70*(1), 41–55.
Schultz, T. W. (1961). Investment in human capital. *The American Economic Review, 51*, 1–17.
Uppal, A. (2007). *Gender discrimination in quality of employment and wages in unorganized manufacturing sector of India*. Paper presented in gender issues and Empowerment of Women Platinum Jubilee of Indian Statistical Institute, 1–2 February 2007.

Chapter 12
The Role of Flexibility at Work on Residential Location: From the Work-Life Balance Perspective

Jeongsoo Kim

Introduction

The development of Information and Communication Technology (ICT) enables people to telecommute in the more flexible working environment: saving commuting time and finding job opportunities in distant places (Gareis 2002; Choo et al. 2005; Ory and Mokhtarian 2006; Jiang 2008; Zhu 2013). The recent trends of remote working and telecommuting in the 'new-urban era' benefits employees to overcome both distance and time constraints of residential location (Kim et al. 2015) so that housing costs become affordable and work-life can be compatible (Lund and Mokhtarian 1994; Shen 2000).

Since the early 2000s, people questioned with a pragmatic approach, "does it matter where you work?" (Hill et al. 2003). Even from the early 2000s, about 24% of 65 million employed adults used personal computers at home to perform their jobs in the United States (Buzar et al. 2005). As telecommuters are increasing, urban planners and economists begin to pay attention to ICT advance and urbanization (Zhu 2013). Even though existing empirical studies contribute to the functional efficiency of telecommuting (Ettema 2010; Moos and Skaburskis 2007; Kelly et al. 2011; Hill et al. 2003), how the flexibility at work transform residential locations, which potentially contributes to people's well-being, is relatively limitedly studied.

The concept of work-life balance in this study denotes flexibility in a working schedule, which allows workers to spend more time with families (Odle-Dusseau et al. 2012). So, the context of work-life balance in this study does not extend toward either better working conditions or better earnings. Also, a distant home location is

J. Kim (✉)
Department of Demography, University of Texas at San Antonio, San Antonio, TX, USA
e-mail: Jeongsoo.kim@utsa.edu

B. Jivetti, Md. N. Hoque (eds.), *Population Change and Public Policy*, Applied Demography Series 11, https://doi.org/10.1007/978-3-030-57069-9_12

irrelevant to housing values or educational benefits that probably become an obstacle for the young generation who have K-12 enrolled children (Black 1999). Thus, this study focuses more on the relationship between recent trends of remote working with a flexible working environment.

This study aims to examine how flexible working correlates with the distance between home to work in the recent decade in the United States. This paper hypothesizes that the transition of working conditions influences residential locations, which increases the distance of home-work. This study used the 2009 and 2017 National Household Travel Survey (NHTS) at the individual level. Also, the American Community Survey (ACS) is applied to illustrate how geographic characteristics reflect the trends of residential-relocation both at the state and the MSA level.

Literature Review

Urbanization in the United States, beginning with a concentration of skyscrapers and dense population since the nineteenth century, created both the costly transportation and job opportunities. Then, decentralization followed the dense population as the development of cheaper and faster automobiles and the advent of telephone communication (Levy 2009). Also, demographic transitions and the popularity of small families with high economic activities in urban areas become catalysts for reurbanization that people prefer living in metropolitan areas among the young generation. Indeed, the transforming age structure in the U.S. alleviated the shrinkage of larger central cities from 2000 to 2010 despite the fast deurbanization (Levy 2016).

The concept of telecommuting starts from futuristic imagination. Toffler (1980) describes that people could work in 'electronic cottages,' and Gates et al. (1995) illustrate that people commute on the 'information highway.' Florida (2002) suggests that the economic geography of human capital based on the diversity of talent is associated with the high technology industries and regional incomes. Specifically, the knowledge-based workers in STEM (Science, Technology, Engineering, and Math) show that growing tendency to reside in suburban areas and the metropolitan fringes, and reluctantly to choose high priced areas (Frenkel et al. 2013). Also, as the core human resources at work are changing to critical thinking and practical communication skills; thereby, the concept of work-life balance gets centered in the workplace (Stepp 2013). Indeed, telecommuting supports the balanced work-family atmosphere compared to the cases of traditional office-commuting workers (Hill et al. 2003).

Flexible working arrangements such as home-based working, part-time working, job sharing, and flexible working schedules befit not only the personal needs of employees but also the demands of transnational companies (Gratton 2010). Advanced ICT provides workers with more freedom from the constraints of fixed and central workplaces substantially, so workers perform routine tasks across remote places. Specifically, telecommuting allows employees to work in various locations, including at home, which also bolsters a flexible working arrangement (Bailey and Kurland 2002). In Hong Kong, the telecommuting penetration among IT industry workers has increased to 25% since 2000, and a similar trend emerges in other East Asian countries (Leung and Zhang 2017). Also, the flexible working policies contribute to higher attentiveness to employers and lower stress of employees that eventually attains the compatible work-family outcomes (Halpern 2005).

Telecommuting releases workers from physical locations and time-constraints, thereby leading to the urban sprawl and associated with more vehicle-miles of travel than when without the options of telecommuting (Ory and Mokhtarian 2006). Telecommuting influences the lifestyle patters through spatial, sequential, and temporal changes: an increase in the distance between home and work, a decrease in commuting time, travel to more distant leisure places (Kim et al. 2012). Also, telecommuters prefer remote places to live, which tendency is stronger in the second-tier and third-tier cities than in the primary central cities (Tayyaran et al. 2003). However, the influence of telecommuting on living locations varies by the characteristics of occupation (Janelle 1995; Kim et al. 2012). Also, differences in daily trips and residential preferences correlate with the various motivations of telecommuting (Ory and Mokhtarian 2006).

Besides, microeconomic theories address the relationship between residential areas and work locations through transportation costs and land prices. (Alonso 1964; Wingo 1961). The commuting distance has a significant effect on the decision of residential location consistently (Tillema et al. 2010). Also, road users adaptively consider road pricing so that expensive road users are more likely to change their route choice, travel time, and frequencies. (Banister 2002). At the intra-metropolitan level, the residential location choice of knowledge-workers depends on socio-economic factors, housing affordability, and commute time (Frenkel et al. 2013). Recently, development in ICT contributes to cultural and informative resources in urban planning of new cities (Mitchell 1996; Graham 2004; Landry 2007). In particular, capital cities are the places where a substantial part of new productive activities are dominant (Cohen 2006).

In sum, advanced ICT actualizes flexible working conditions. This change improves work-life balance for existing employees and provides new job opportunities in a regional economy. In the following sections, we visualize the trends of distance to work at different regional levels. Then, this study estimates the distant home locations from work through conducting a nested multinomial logit model.

Data and Methods

NHTS and ACS

The Federal Highway Administration of the U.S. Department of Transportation has been collecting travel behavior data based on probability random sampling surveys since 1969. NHTS encompasses the changing trends of travel regarding household, person, each vehicle, and day trips. 2017 survey documents the demographic, attitudinal, and behavioral characteristics for all members of 129,112 households. NHTS data which represents all Americans by modifying demographic weight fit for analysis at the national and regional levels. The 2017 NHTS is the recent survey in this series of every 8 years at the Federal, state, and local levels. One of the trendy changes in the 2017 NHTS compared to the 2009 NHTS is to reflect the prevalence of online purchases and its impact on travel behaviors. Although the online purchase is not directly relevant to 'remote working and residential locations,' an increasing trend of ICT development indirectly influences new jobs and travel behaviors.

Also, the American Community Survey (ACS) provides geographic information from states to counties and the census tracks, supported by the Census Bureau database. The Core Based Statistical Areas (CBSAs) in ACS include the county or counties-equivalent entities associated with one core (urbanized areas or urban clusters) of at least 10,000 population. CBSAs became effective in 2003 and referred collectively to metropolitan statistical areas and micropolitan statistical areas. Specifically, the NHTS 2017 addresses Metropolitan Statistical Areas (MSAs) in CBSAs where has a population of more than 50,000.

Analytical Methods

This study examines whether the increase in the distance of home-work is associated with flexible working through GIS techniques and the nested multinomial logit models. GIS mapping illustrates changes in the averaged distance of home-work at the state level between 2009 and 2017. NHTS provides both States and Metropolitan Statistical Areas (MSAs) information, which is also fit for CBSAs in ACS geographic information. Even though MSAs in CBSAs are only partial geographic areas limitedly showing the changes in metropolitan areas, MSAs could reveal the characteristics of the urban commuters' travel behaviors. First, we thus describe the averaged distance of home-work at the state level, then narrow down the scope to the MSA level for the analysis of metropolitan areas.

The outcomes this study aims to estimate are the physical distances between home and work. We thus categorize the responses with six variables by distant orders (Table 12.2). The dependent variables consist of six categories by the distances of home-work: less than 1 miles of 'very close,' less than 3 miles of 'close,' less than 10 miles of 'fair,' less than 100 miles of 'far,' more than 100 miles of 'remote work.' Also, predictors of interest in this study are the extent of flexibility at work, which is measured by having 'an option to work at home' and 'flexibility in working schedule,' which allows workers to change working time by themselves in commute (Table 12.1). Changes toward flexible working are measured as interaction terms between 2009 and 2017, which measures the extent of transitions in flexible working during the period.

On top of the geographical method, this study conducts individual-level analysis by using nested multinomial logit regression models. Each model tests the

Table 12.1 Increasing share of flexible options at individuals' work

Flexibility in work		2009	2017
Option to work at home	No option	103,679	88,239
		85.8%	82.8%
	Have options	17,165	18,341
		14.2%	17.2%
	n	120,844	106,580
		100.0%	100.0%
Flexibility in work schedule	Fixed working time	68,422	64,093
		56.7%	52.0%
	Flexible working	52,317	59,187
		43.3%	48.0%
	n	120,739	123,280
		100.0%	100.0%

Source: NHTS (National Household Travel Survey)

association between the distance of home-work and the flexibility in the workplace by controlling for demographic, economic, and occupational characteristics of individual workers. These models consider demographic covariates: age, education, race/ethnicity, sex, geographic areas, and population size. Also, the following nested models examine the distance of home-work by controlling for economic factors: income and housing types, multiple occupations, and working status. Finally, this study conducts a regression analysis by using the interaction terms between years and the flexibility at work: 'fixed schedule' in 2009, 'flexible schedule' in 2017, 'must-attend' in 2009, and 'flexible attendance' in 2017.

The purpose of this study is to find whether and how flexibility at work correlates with the residential location. First, changes in the averaged distances of home-work at the state level illustrate how people commute in 2017 compared to 2009. Also, the MSA level analysis suggests that how the distance of home-work has changed in urban areas amid the increasing trend of flexible working. Last, the nested multinomial logit models explore the changes in the distances of home-work in the time-varying method, considering four categories of predictors for each model: demographic factors, economic status, occupational status, and flexibility in working.

Results

Figure 12.1 shows the changes in the commuting distance by states between 2009 and 2017. Overall, from 3 to 10 miles of increase in the commuting distance during the recent decade. Specifically, the state of Nevada shows about 54 miles of increase in the commuting distance during the recent decade. Some respondents in Nevada 2017 show that the distance of home-work ranges from 273 to 2578 miles. Because

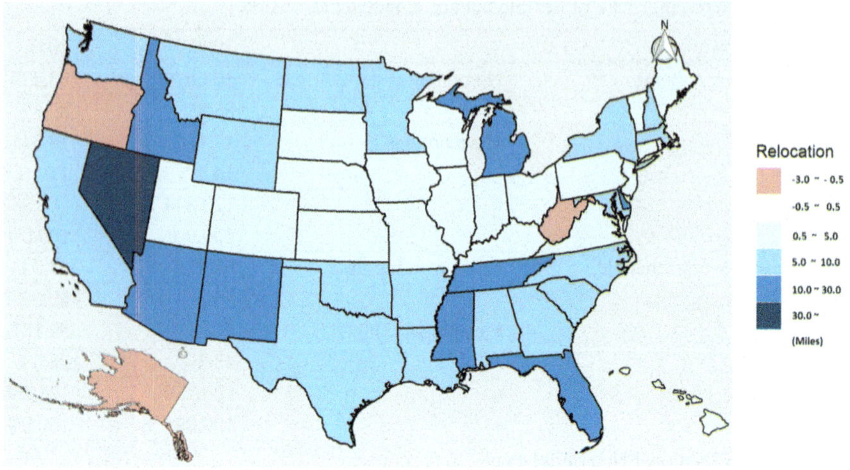

Fig. 12.1 Increase in the distance of home-work by States during 2009–2017. (Source: ACS 2016 and NHTS 2009 & 2017)

we measured the home-work distance on average, workers who commute from more than 200 miles away from work in Nevada are assumed to the remote work or telecommute. Even though a few states exhibit a decrease in commuting distances such as D.C, the state of Virginia, Nebraska, and Alaska, overall, there have been 7.5 miles of increase in the commuting distance in the U.S at the state level during 2009–2017.

Figure 12.2 illustrates the differential of commuting distance at the MSAs level between 2009 and 2017. Specifically, Las Vegas-Paradise MSA exhibits 65 miles increase during the period, which is not the typical case among the traditional commuters. Also, Miami-Fort Lauderdale-Miami Beach MSA, Nashville-Davidson-Murfreesboro MSA, and New Orleans-Metairie-Kenner MSA have increased about 30 miles of commuting distance during the same period. Overall, about 18 miles of increase in commuting distance on average among MSAs. Presumably, the trends of flexible-working and telecommuting influence the residential locations among urban residents.

The focus of this study is how the flexibility at work correlates with a residential location in the United States. The flexibility consists of two conditions; if workers have 'options to work at home' and if workers can 'flexibly change their working schedules' (Table 12.1). Even though the percentage of 'working at home' has decreased in 2017 (Table 12.2), both the share of workers with 'options to work at home' (Table 12.1) and the share of 'remote working' (Table 12.2) have increased. The percentage of 'flexible working' increased to 48% in 2017 from 43% in 2009 (Table 12.1).

The new types of commute include 'working at home' as well as 'working remotely.' On the other hand, the traditional commute denotes the distance within 100 miles of commute except for 'working at home.' The distance of home-work

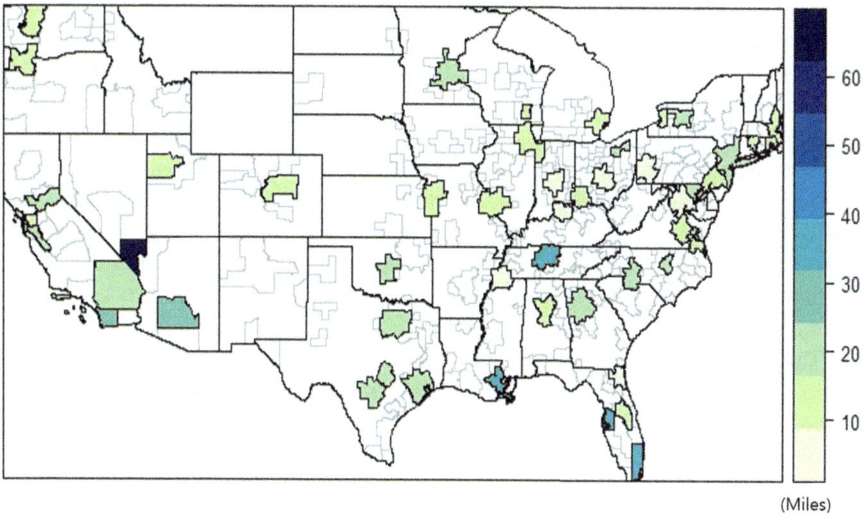

Fig. 12.2 Differences in the commuting distance in MSAs during 2009–2017. (Source: ACS 2016 and NHTS 2017)

Table 12.2 Increasing the commuting distance of individuals during 2001–2017

Distance of home-work		2001	2009	2017
New commuting pattern	Working at home	1166	1042	728
	0 miles	1.6%	0.9%	0.7%
	Remote working	203	473	1829
	101–1000 miles	0.3%	0.4%	1.7%
Traditional commuting pattern	Very close	8007	8549	5559
	0–1 miles	11.2%	7.2%	5.2%
	Close	10,574	14,792	14,151
	1–3 miles	14.8%	12.5%	13.3%
	Fair	20,400	34,620	36,185
	3–10 miles	28.6%	29.2%	33.9%
	Far	31,016	59,213	48,192
	10–100 miles	43.5%	49.9%	45.2%
	n	71,366	118,689	106,644
		100.0%	100.0%	100.0%

Source: NHTS (National Household Travel Survey)

shows a change that workers' residential location has outstretched during the last two decades (Table 12.2). Interestingly, the percentage of remote working, defined as more than 100 miles of home-work distance, has increased from 0.3% in 2001 to 1.7% in 2017 (Table 12.2). More importantly, among the traditional commuters, 'less than 3 miles of commute' decreased since 2001 while 'longer than 10 miles' of home-work commute increased during the same period (Table 12.2).

Table 12.3 shows the likelihood of distant commute by nested multinomial logit models. Model 1 shows how sociodemographic factors correlate with the commute distance in the United States at the individual level. The forties live in the longest distance among all age groups while the older ages over the forties tend to live closer to work. In educational attainment, bachelor-degree holders outstretched their residence to the longest distance from their work. Compared to people below high school graduates, people with higher educational attainment live in remoter places from their work, while females are more likely to live closer to their work than male workers.

This study considers geographic characteristics in the United States by the nine regions of census categories because geographic constraints may influence the distance of home-work. Compared to the Pacific region, Middle Atlantic, West Northcentral, and some Mountain areas show closer residential locations to work. On the other hand, Southcentral and South Atlantic areas are more likely to live in distant places from work. Also, this study incorporates the metropolitan areas by population size. Compared to rural areas, the more populated cities are more likely to show a longer commute except for 'MSAs less than 250,000 population.'

Model 2 includes financial factors such as income level and housing types. Compared to the lowest income quintile, the higher income quintiles are more likely to live in distant places from work. Specifically, the highest income quintiles show the remotest living locations among all income groups. Also, compared to homeowners, renters, and other housing types of workers are more likely to live closer to work. Indeed, homeowners seem to have larger freedom in decisions of home location.

Model 3 addresses working conditions through whether people have multiple jobs or work with part-time status. People with multiple jobs are more likely to live closer to workplaces. Likewise, people who work in part-time positions are more likely to live closer to workplaces compared to workers with regular jobs. However, the negative relationship between the distance of home-work and multiple jobs is minor, and the statistical significance of the negative relationship disappears when it takes into account flexibility at work (Model 4).

Model 4 focuses on the relationship between flexibility at work and the distance of home-work in a time-variant model. The time interaction includes the explanatory variables of the year of 2009 and 2017. Because Table 12.1 indicates that the share of having 'the option not to attend' is only about 14% in 2009 and 17% in 2017 each, Model 4 identifies that workers with both 'flexibility in working schedule' and 'options to work at home' show an adverse relationship related to the commute distance. However, when it considers the time variance by interaction with years, the implications appear different. First, workers are more likely to live longer distances in 2017 compared to 2009. Even though workers with a 'flexible schedule' has a minor negative relationship with the commute distance, workers who have 'options not to attend' in 2017 are more likely to live in distant places from work.

Table 12.3 The likelihood of distant commute in the U.S. by nested multinomial logit models

Distance to work		Model 1		Model 2		Model 3		Model 4	
		Log odds		Log odds		Log odds		Log odds	
Age	Teens	*Ref*		*Ref*		*Ref*		*Ref*	
	Twenties	0.37	***	0.53	***	0.31	***	0.36	***
	Thirties	0.65	***	0.67	***	0.37	***	0.44	***
	Forties	0.70	***	0.65	***	0.34	***	0.41	***
	Fifties	0.57	***	0.55	***	0.24	***	0.30	***
	Sixties+	0.32	***	0.32	***	0.13	***	0.21	***
Education	Below HighSchool	*Ref*		*Ref*		*Ref*		*Ref*	
	HighSchool	0.28	***	0.15	***	0.11	***	0.09	***
	SomeCollege	0.38	***	0.25	***	0.21	***	0.17	***
	Bachelor	0.48	***	0.27	***	0.19	***	0.13	***
	Graduate	0.32	***	0.12	***	0.05	***	−0.03	***
Race/ethnicity	Whites	*Ref*		*Ref*		*Ref*		*Ref*	
	Black	−0.04	***	0.12	***	0.11	***	0.12	***
	Asian	−0.10	***	−0.12	***	−0.12	***	−0.12	***
	NativeAmerican	0.00	***	0.09	***	0.09	***	0.12	***
	Others	−0.11	***	−0.01	***	−0.03	***	−0.02	***
Sex	Male	*Ref*		*Ref*		*Ref*		*Ref*	
	Female	−0.30	***	−0.24	***	−0.18	***	−0.18	***
Division	Pacific	*Ref*		*Ref*		*Ref*		*Ref*	
	NewEngland	0.17	***	0.01	***	−0.01	***	−0.02	***
	MiddleAtlantic	−0.04	***	−0.08	***	−0.08	***	−0.09	***
	EastNorthCentral	0.16	***	0.07	***	0.07	***	0.06	***
	WestNorthCentral	−0.05	***	−0.13	***	−0.13	***	−0.13	***
	SouthAtlantic	0.37	***	0.32	***	0.31	***	0.31	***
	EastSouthCentral	0.54	***	0.52	***	0.50	***	0.49	***
	WestSouthCentral	0.29	***	0.27	***	0.25	***	0.25	***
	Mountain	0.02	***	−0.06	***	−0.06	***	−0.07	***

(continued)

Table 12.3 (continued)

		Model 1		Model 2		Model 3		Model 4	
Distance to work		Log odds		Log odds		Log odds		Log odds	
Population	Not MSA	*Ref*		*Ref*		*Ref*		*Ref*	
	0 - 250 k	−0.23	***	−0.15	***	−0.15	***	−0.18	***
	250 - 500 k	0.05	***	0.59	***	0.07	***	0.04	***
	500 k - 1000 k	0.14	***	0.20	***	0.20	***	0.16	***
	1000 k - 3000 k	0.28	***	0.29	***	0.30	***	0.27	***
	3000 k +	0.32	***	0.33	***	0.33	***	0.29	***
Income	First quintile			*Ref*		*Ref*		*Ref*	
	Second quintile			0.15	***	0.06	***	0.06	***
	Middle			0.38	***	0.28	***	0.29	***
	Fourth high quintile			0.49	***	0.38	***	0.43	***
	Fifth high quintile			0.51	***	0.40	***	0.49	***
Housing	Homeowner			*Ref*		*Ref*		*Ref*	
	Renter			−0.44	***	−0.46	***	−0.44	***
	Others			−0.42	***	−0.43	***	−0.50	***
Multiple occupation	One					*Ref*		*Ref*	
	More than one					−0.03	***	−0.02	***
Working status	Fulltime					*Ref*		*Ref*	
	Part-time					−0.52	***	−0.50	***
Flexible time	Fixed time							*Ref*	
	FlexibleSchedule							−0.10	***
Flexible attendance	Must-attend							*Ref*	
	Flexible attendance							−0.27	***

Flexibility trends	2009							*Ref*	
	2017							0.24	***
Flexibility trends	2009*fixed time							*Ref*	
	2017*flexible schedule							−0.01	***
	2009*must attend							*Ref*	
	2017*flexible attendance							0.28	***
	Log-likelihood	−313,354,178		−219,170,268		−217,107,460		−216,392,713	

Source: NHTS (2009 & 2017 combined)
Signif. codes: *** 0.001, ** 0.01, * 0.05, · 0.1

In sum, people who age the forties, males, and bachelor-degree holders seem to have more residential freedom. Geographically, southern areas and highly populated MSAs show more distant living locations from work. The homeowners and high-income earners are more likely to live in extended locations from work. Also, workers with precarious part-time positions have less freedom in residential locations. Most importantly, workers who have an option of flexible attendance in 2017 tend to live in distant places from work compared to workers with less flexibility in 2009.

Discussion

Telecommuting enables employees to overcome the constraints of commuting distance with the help of ICT development (Gareis 2002; Choo et al. 2005; Ory and Mokhtarian 2006; Jiang 2008; Zhu 2013). Also, Levy (2016) suggests that demographic changes and the development of the service industry reshape urbanization. This study has the purpose of illustrating the transition of residential location in the context of work-life balance. Amid the transition to flexible working by ICT development, the demand for work-life balance leads to an increase in flexible working, attaining more freedom in residential locations even with an affordable housing cost.

Nonetheless, several limitations exist in this study. First, this paper finds evidence of residential relocation based on only two recent cross-sectional datasets of 2009 and 2017. Second, the people who answered as 'working at home' are ignored because it is still unclear if they work at home with the help of telecommuting or not (e.g., farming or online sales). Third, despite controlling for regional characteristics, the development of public transportation systems in metropolitan areas and transitions in regional industrial characteristics could influence the commute distance. Fourth, The residential locations could rely on just individual preferences, rather than structural changes in working conditions of flexible working: school districts constraints and safe neighborhood. Fifth, this study limitedly considers family member's influence on home location decisions.

Notwithstanding the above limitations, NHTS provides enormous datasets, over 300,000 individual samples for each wave, thereby providing a strong implication regarding the transition in a residential location. Other than large samples, the new dataset of 2017 reflects the most recent changes in lifestyle in the United States. For example, flexibility at work can provide tremendous benefits not only in work efficiencies but also in life-work balance. More importantly, urban and regional planning needs to consider how the flexible lifestyle could contribute to individuals' well-being from the life-work balance perspective.

Conclusion

Telecommuting based on ICT development enables workers to spend less time in commute and provides occupational opportunities regardless of their home locations (Ory and Mokhtarian 2006; Jiang 2008). Likewise, more flexible working environments benefit employees to have more freedom in residential locations (Kim et al. 2015). This paper aims to find evidence of the association between flexible working conditions and distances of commute. Even though remote working is prevalent in specific areas of industries, the development of ICT enables workers to consider their future skillsets towards having more occupational and residential freedom. Indeed, telecommuting and flexible working conditions fit the desire for affordable housing, living in a green environment, and work-life balance (Gratton 2010).

NHTS data suggests that the distances of a commute are increasing among the traditional commuters. (Table 12.2). The results of the nested multinomial logit models show that people who age forties, males, and bachelor-degree holders appear to have more residential freedom than others. In the United States, southern areas and highly populated MSAs show remoter living locations from work. The homeowners and high-income earners are more likely to live in distant places from work. Also, workers with precarious part-time positions have less freedom in residential locations. Last, the workers with an option of a flexible working schedule in 2017 tend to live in remoter places from work compared to those who do not have the flexibility opportunities in 2009.

Telecommuting is a global phenomenon realizing workers' desire to achieve work-life balance and to live in more favorable environments instead of the overcrowded cities (Hamer et al. 1991; Alizadeh 2013; Kim et al. 2015). This study aims to contribute to the compatible work-family and worker's well-being amid the transition of working conditions such as flexible working and telecommuting. Actually, flexible working condition in technologically developed areas provides enormous residential freedom. Because this study observes the prevalence of flexible working only within a decade, a future wave of NHTS 2025 remains to be seen whether this transition of remoteness reinforces or not.

References

Alizadeh, T. (2013). Planning implications of telework: A policy analysis of the Sydney metropolitan strategy. *Australian Planner, 50*(4), 304–315.

Alonso, W. (1964). *Location and land use.* Cambridge, MA: Harvard University Press.

American Community Survey. (2016). Portal site of Official Database. Retrieved Dec 25, 2018, from https://www.census.gov/programs-surveys/acs.

Bailey, D. E., & Kurland, N. B. (2002). A review of telework research: Findings, new directions, and lessons for the study of modern work. *Journal of Organizational Behavior, 23*(SPEC. ISS), 383–400. https://doi.org/10.1002/job.144.

Banister, D. (2002). The integration of road pricing with land use planning. Imprint-Europe, 1–18.

Black, S. E. (1999). Do better schools matter? Parental valuation of elementary education. *The Quarterly Journal of Economics, 114*(2), 577–599.

Buzar, S., Ogden, P. E., & Hall, R. (2005). Households matter: The quiet demography of urban transformation. *Progress in Human Geography, 29*(4), 413–436.

Choo, S., Mokhtarian, P. L., & Salomon, I. (2005). Does telecommuting reduce vehicle-miles traveled? An aggregate time series analysis for the U.S. *Transportation, 32*(1), 37–64.

Cohen, B. (2006). Urbanization in developing countries: Current trends, future projections, and key challenges for sustainability. *Technology in Society, 28*, 63–80. https://doi.org/10.1016/j.techsoc.2005.10.005.

Ettema, D. (2010). The impact of telecommuting on residential relocation and residential preferences. A latent class modeling. *The Journal of Transport and Land Use, 3*(1), 7–24. https://doi.org/10.5198/jtlu.v3i1.61.

Florida, R. (2002). The economic geography of talent. *Annals of the Association of American Geographers, 92*(4), 743–755.

Frenkel, A., Bendit, E., & Kaplan, S. (2013). Residential location choice of knowledge-workers : The role of amenities, workplace, and lifestyle. *Cities, 35*, 33–41. https://doi.org/10.1016/j.cities.2013.06.005.

Gareis, K. (2002, August). The intensity of telework in 2002 in the E.U., Switzerland and the USA. In *International congress new work* (pp. 3–5).

Gates, B., Myhrvold, N., Rinearson, P., & Domonkos, D. (1995). *The road ahead*. London: Viking.

Graham, S. (Ed.). (2004). *The cybercities reader*. Psychology Press.

Gratton, L. (2010). The future of work. *Business Strategy Review, 21*(3), 16–23.

Halpern, D. F. (2005). How time-flexible work policies can reduce stress, improve health, and save money., *168*, 157–168. https://doi.org/10.1002/smi.1049.

Hamer, R., Kroes, E., & Van Ooststroom, H. (1991). Teleworking in the Netherlands: An evaluation of changes in travel behavior. *Transportation, 18*(4), 365–382. https://doi.org/10.1007/BF00186565.

Hill, E. J., Ferris, M., & Märtinson, V. (2003). Does it matter where you work? A comparison of how three work venues (traditional office, virtual office, and home office) influence aspects of work and personal/family life. *Journal of Vocational Behavior, 63*(2), 220–241. https://doi.org/10.1016/S0001-8791(03)00042-3.

Janelle, D. G. (1995). Metropolitan expansion. Telecommuting and transportation. In *The geography of urban transportation*. New York: Guilford Press.

Jiang, Y. (2008). The impact of telecommuting on the journey to work: A two-sample instrumental variables approach. In *Two empirical essays in environmental and urban economics*. University of Maryland: ProQuest LLC.

Kelly, E. L., Moen, P., & Tranby, E. (2011). Changing workplaces to reduce work-family conflict: Schedule control in a white-collar organization. *American Sociological Review, 76*(2), 265–290. https://doi.org/10.1177/0003122411400056.

Kim, S. N., Mokhtarian, P. L., & Ahn, K. H. (2012). The Seoul of Alonso: New perspectives on telecommuting and residential location from South Korea. *Urban Geography, 33*(8), 1163–1191.

Kim, S. N., Choo, S., & Mokhtarian, P. L. (2015). Home-based telecommuting and intra-household interactions in work and non-work travel: A seemingly unrelated censored regression approach. *Transportation Research Part A: Policy and Practice, 80*, 197–214. https://doi.org/10.1016/j.tra.2015.07.018.

Landry, C. (2007, March). Visual scenescape of the city. In *Proceedings of the urban directions: Seeing the city conference, keynote lecture* (Vol. 7).

Lee, J. H., Hancock, M. G., & Hu, M. C. (2014). Towards an effective framework for building smart cities: Lessons from Seoul and San Francisco. *Technological Forecasting and Social Change, 89*, 80–99. https://doi.org/10.1016/j.techfore.2013.08.033.

Leung, L., & Zhang, R. (2017). Mapping ICT use at home and telecommuting practices: A perspective from work/family border theory. *Telematics and Informatics, 34*(1), 385–396. https://doi.org/10.1016/j.tele.2016.06.001.

Levy, J. (2009). *Contemporary urban planning*. Retrieved from http://search.lib.unc.edu/search?R=UNCb5582582

Levy, J. M. (2016). *Contemporary urban planning*. London: Taylor & Francis.

Lund, J. R., & Mokhtarian, P. L. (1994). Telecommuting and residential location: Theory and implications for commute travel in monocentric metropolis. *Transportation Research Record, 1463*, 10–14.

Mitchell, W. J. (1996). *City of bits: Space, place, and the infobahn*. Cambridge, MA: MIT Press.

Mokhtarian, P. L. (1991). Telecommuting and travel: State of the practice, state of the art. *Transportation, 18*(4), 319–342.

Moos, M., & Skaburskis, A. (2007). The characteristics and location of home workers in Montreal, Toronto, and Vancouver. *Urban Studies, 44*(9), 1781–1808. https://doi.org/10.1080/00420980701507639.

National Household Travel Survey. (2009, 2017). Retrieved Dec 25, 2018, from https://nhts.ornl.gov/.

Neirotti, P., De Marco, A., Cagliano, A. C., Mangano, G., & Scorrano, F. (2014). Current trends in smart city initiatives: Some stylised facts. *Cities, 38*, 25–36. https://doi.org/10.1016/j.cities.2013.12.010.

Odle-Dusseau, H. N., Britt, T. W., & Bobko, P. (2012). Work–family balance, well-being, and organizational outcomes: Investigating actual versus desired work/family time discrepancies. *Journal of Business and Psychology, 27*(3), 331–343.

Ory, D. T., & Mokhtarian, P. L. (2006). Which came first, the telecommuting or the residential relocation? An empirical analysis of causality. *Urban Geography, 27*(7), 590–609. https://doi.org/10.2747/0272-3638.27.7.590.

Rathore, M. M., Ahmad, A., Paul, A., & Rho, S. (2016). Urban planning and building smart cities based on the internet of things using big data analytics. *Computer Networks, 101*, 63–80. https://doi.org/10.1016/j.comnet.2015.12.023.

Shen, Q. (2000). New telecommunications and residential location flexibility. *Environment and Planning A, 32*(8), 1445–1463.

Song, Y., & Knaap, G. J. (2003). New urbanism and housing values: A disaggregate assessment. *Journal of Urban Economics, 54*(2), 218–238. https://doi.org/10.1016/S0094-1190(03)00059-7.

Stepp, K. (2013). *H.R. & the workplace of the future*.

Tayyaran, M. R., Khan, A. M., & Anderson, D. A. (2003). Impact of telecommuting and intelligent transportation systems on residential location choice. *Transportation Planning and Technology, 26*(2), 171–193.

Tillema, T., van Wee, B., & Ettema, D. (2010). The influence of (toll-related) travel costs in residential location decisions of households: A stated choice approach. *Transportation Research Part A: Policy and Practice, 44*(10), 785–796. https://doi.org/10.1016/j.tra.2010.07.009.

Toffler, A., & Alvin, T. (1980). *The third wave*. New York: Bantam books.

Wingo, L. (1961). *Transportation and urban land*. Washington: Resources for the Future.

Zhu, P. (2013). Telecommuting, household commute, and location choice. *Urban Studies, 50*(12), 2441–2459. https://doi.org/10.1177/0042098012474520.

Chapter 13
Working Through Tips: Examining Labor Dynamics in Tipped Workplaces

Eli R. Wilson and Davyd Setter

Introduction

In the United States, the prevalence of tipping parallels the growth of the postindustrial capitalist economy. Over five million Americans today earn gratuities across thirty-three different occupations and counting (Lynn et al. 1993). Tipping is also big business. Azar (2011) estimates that customers leave $47 billion dollars annually in gratuities in U.S. restaurants *alone*; by 2020, the economic impact of tips will be closer to $60 billion a year.[1]

Despite the ubiquity and significance of tips in the U.S. economy, social science research on those who work for tips remains narrowly focused on direct exchanges between workers and customers. This body of research illustrates that while workers use a variety of strategies to interact with customers in an attempt to maximize the tips they receive (Davis 1959; Mars and Nicod 1984; Sallaz 2009; Thompson 2015). The actual tip amount left by customers depends on a wide range of psychological, cultural, and contextual factors that are beyond workers' control (Azar 2007; Brewster and Wills 2013). Tipped employees, because of their dependency on gratuities for a large chunk of their earnings, remain vulnerable to verbal and

Disclaimer: The cover page must contain a disclaimer that that identifies the article as being an original work by the named author(s) and as being submitted exclusively to the International Association of Applied Demography for publication consideration into an edited book on Applied Demography by Springer Publishing Company.

[1]Authors' calculations based on Azar (2011).

E. R. Wilson (✉) · D. Setter
Department of Sociology & Criminology, University of New Mexico, Albuquerque, NM, USA
e-mail: erwilson18@unm.edu

B. Jivetti, Md. N. Hoque (eds.), *Population Change and Public Policy*, Applied Demography Series 11, https://doi.org/10.1007/978-3-030-57069-9_13

emotional abuse at the hands of customers who have the option to withhold tips as punishment for unsatisfactory service (Lynn and McCall 2000; Paules 1991).

Neglected in the current scholarship on tipping is the fact that not all types of workers employed in tipped workplaces in the United States receive gratuities in the same proportion, or any at all. A majority of all tipped workers labor alongside non-tipped workers who operate in job capacities that involve little to no direct customer engagement. Whether in hotels, restaurants, spas, or casinos, different types of service workers must coordinate their actions in order to collectively accomplish all the steps of customer service (Sherman 2007; Strauss 1985; Subramanian and Suquet 2017; Wilson forthcoming). Front-of-the-house workers, such as servers in restaurants and front desk clerks in hotels, interact directly with customers. "Support" workers, such as bussers and housekeepers, perform tasks that involve indirect customer interaction. Meanwhile, back-of-the-house workers, such as cooks and cleaners, produce and maintain service goods behind the scenes or when the establishment is closed to the public (Sherman 2007; Star and Strauss 1999). Each of these categories of service work require distinct types of tasks and skills. They also result in structurally unequal access to tips (Adler and Adler 2004; Wilson 2017).

This study examines the broader scope of labor dynamics found in tipped service workplaces in the United States. We detail how tips affect the labor experiences, social relations, and workplace practices of different types of service workers present in these settings. Current empirical investigations of tipped employment undersell the actual proportion of the U.S. workforce that labors alongside, in conjunction with, or in supervision of, tipped workers. Studying the "downstream" implications of tips—what transpires away from customers and after a gratuity is handed over to workers—captures how the labor experiences of those who operate in the same service environment are shaped by tips. Thus, we aim to broaden scholarly inquiry into tipped labor beyond the point-of-service contexts where they are currently situated. In the sections below, we estimate the scope of workers employed in tipped workplaces in the United States, then turn to a case study of the restaurant industry to illustrate how different subsets of workers are impacted by the presence of tipping in their places of employment.

Background on Tipped Employment

The practice of tipping itself is not new. Tipping goes back at least to the nineteenth century in the United States and Britain, with its roots steeped in the provision of small sums of money passed between wealthy clients and those who perform some kind of personal service for them (Azar 2004; Segrave 1998). By the twentieth century, tipping in the United States and Western Europe had become a normalized cultural practice in many customer service settings, such as restaurants, bars, barber shops, hotels, and railroad cars; working for tips became associated with the iconic customer-facing employees in each of these respective labor settings (see Santino 1989; Segrave 1998). That said, not all service-related jobs are tipped. Skilled

Table 13.1 Total number of tipped workers in United States by industry and occupation

Occupation title	# of Workers	% of Tipped workers	Median wage[a]
Workers outside of the food industry	*1,315,200*	*24.4%*	
Barbers, Hairstylists, and Cosmetologists	673,700	12.5%	$12.23
Gaming Services Workers	176,100	3.2%	$10.97
Massage Therapists	160,300	3.0%	$19.92
Taxi Drivers and Chauffeurs	305,100	5.7%	$12.49
Workers within food industry	*4,081,940*	*75.6%*	
Hosts and Hostesses, Restaurant, Lounge, and Coffee Shop	414,540	7.7%	$10.06
Dining Room, Cafeteria Attendants and Bartender Helpers	455,700	8.4%	$10.71
Bartenders	611,200	11.3%	$10.84
Waiters and Waitresses	2,600,500	48.2%	$10.47
Total Tipped Workers	5,397,140		

Source: Bureau of Labor Statistics Data (2016). Defined using Allegretto and Cooper's (2014) category of "Predominantly Tipped Professions"
[a]Median wage does not include tips

service workers, such as doctors and lawyers, do not customarily receive tips from their clients. Nor do fast-food workers, perhaps because the nature of the service exchange appears to be too transactional and less personalized. Tipping remains most common in interactive service contexts that involve a sufficient degree of personalized yet subordinated service being offered.

Currently, over five million Americans are classified as tipped workers spread across 33 discrete occupations (see Lynn et al. 1993). The bulk of all tipped workers are concentrated in the food and beverage industry, with smaller proportions concentrated in three other labor categories: personal appearance (e.g, barbers, stylists), luxury and service work (e.g., masseuses, concierge, gaming), and transportation (e.g., taxi driving, chauffeurs). Broken down by job category (Table 13.1, below), we can see that servers, bartenders, hosts, and other restaurant workers make up a combined 75.6% of all tipped workers. The category of "waiters and waitresses" remain the most common type of tipped job (48.2%), followed by "barbers, hairstylists, and cosmetologists" (12.5%) and bartenders (11.3%). The high proportion of tipped workers employed in food and beverage jobs is directly related to the steady growth of the restaurant industry relative to other leading tipped industries in the United States.

As a specific type of economic exchange, tipping is influenced by both sociocultural norms and psychological factors (Azar 2007; Bearman 2009). The amount a customer tips is only loosely related to the perceived quality of service they receive, which contradicts popular perceptions of tipping as being solely "to ensure prompt service" (Lynn and McCall 2000; Kim et al. 2017). Instead, tipping practices vary based on the social characteristics of customers and workers, as well as their interactional dynamic (Azar 2007; Brewster and Mallinson 2009). Men have been found to tip more than women on average, just as they tend to leave bigger tips for female servers than male servers (Lynn and Latane 1984). White workers tend to receive bigger tips from customers than nonwhite workers (Brewster and Wills 2013). Similarly, various aspects of the service encounter affect how much customers tip. Lightly touching guests on the shoulder, for instance, can lead to bigger tips for workers, as does repeat patronage to an establishment on the part of customers (Azar 2007; Crusco and Wetzel 1984; Lynn 2006; Lynn and McCall 2000).

Context also influences how much any one tipped worker can earn. Higher-end establishments tend to feature more expensive service items than their lower-end counterparts in the same industry. Because customers often calculate their tip based on a customary tip amount—18–20% of the total bill in restaurants and beauty salons, for example—those employed in higher-end service settings often enjoy greater tip-earning opportunities than those in lower-end settings (see Azar 2007). That said, workers situated in any establishment that faces slow customer traffic are subject to less lucrative tip-earning opportunities.

Laboring in direct contact with tipping customers also leaves workers vulnerable to a number of employment factors beyond their control. Tipped workers regularly face sexual advances by customers (especially women, see Cobble 1991; Ehrenreich 2001; Spradley and Mann 1975), fluctuating work schedules (Halpin 2015; Lambert et al. 2012), and base wages at or around the federal minimum wage (Jayaraman 2018). In all but seven states, employers are legally allowed to pay tipped workers a lower hourly wage and use tips as an earnings "credit" that can be applied towards the regular minimum wage ($7.25 at the federal level). Consequently, 18 states have a tipped minimum wage set to the federal minimum of $2.13, while another 27 states feature a tipped minimum wage that is below the regular minimum wage. Critics argue that the "tip credit" system leaves workers vulnerable to employer violations of minimum wage laws. This may include the illegal appropriation of employee tips for business purposes, or failing to pay out wages to meet the minimum wage should tips be insufficient. Such violations are costly and time-consuming for workers to contest in court (Foshay 2016). Facing these precarious aspects of their employment, tipped workers are left reliant on the goodwill of both tipping customers and law-abiding employers.

Inside Tipped Workplaces

Many of the largest and most common tipped workplaces employ a range of workers who operate in distinct organizational capacities (Lopez 2010). These workers are required to execute a sequence of labor tasks for tipping customers that must be coordinated with one another (Subramanian and Suquet 2017). However, not all workers in tipped service workplaces earn tips. According to the Fair Labor Standards Act of 1938, only workers who are *in the direct line of customer service* may receive tips. This means that most individuals who operate behind the scenes and away from customers are legally barred from receiving tips as part of their compensation.[2] Managers and supervisors, too, are exclude from receiving tips so long as their primary duties are oriented towards supervising staff and not engaging in service-related acts on the behalf of customers.[3]

Other types of employees in tipped workplaces may also be allocated a smaller portion of gratuities based on their task, popularly referred to as a "tip-out." Tip-out policies vary per establishment and depend on both managerial decisions and employee input. While managers at each restaurant disseminate information about tip policies, they stress that employees themselves are responsible for handling tips (which are legally the property of employees). As a result, the exact procedure for distributing tips among workers is often a gray area, and the practice of tipping-out certain workers and not others invokes power relations between workers of different rank. As we detail below, those employed in the same workplace not only receive unequal quantities of tips, their social relations with one another are directly shaped by tip practices in the workplace.

Full-service ("table service") restaurants remain paradigmatic settings to study the labor dynamics of tipping in the United States for two primary reasons. First, more than three out of every four tipped workers in this country are located in the food-service industry, and restaurant diners leave 58 billion dollars in tips each year. Second, on a cultural level, restaurants represent the commercial establishment most closely associated with tipping. The normative process of service in U.S. restaurants begins with a customer service interaction at the host stand or table, and ends with a bill containing a line for a tip.

Yet, only one out of every three workers employed in restaurants formally receives tips. This organizational distinction among workers has little to do with skill level or rank (Fine 1996; Haley-Lock and Ewert 2011; Wilson 2017).[4] Full-service restaurant workers can be categorized as tipped, tipped-out, or non-tipped, corresponding to the share of tips they receive after every shift. Workers who directly interact with guests, such as servers and bartenders, garner the majority of gratuities left by diners. Those with indirect contact with diners, such as bussers, bartender helpers ("bar backs"), and hosts, receive a smaller tip-out ranging from a few dollars to roughly half the gratuities that servers will take home.[5] Finally, restaurant workers who labor behind the scenes or away from customers do not receive tips. This last category of workers includes cooks and dishwashers as well as

[2]In 2018, after fieldwork for this research had concluded, the US government amended the Fair Labor Standards Act to allow tips to be shared among all workers in tipped workplaces so long as the employer pays a full minimum wage to these workers (without taking a tip credit).

[3]According to the Fair Labor Standards Act, workers who are not considered part of the line of direct customer service are ineligible to receive tips.

[4]It is important to note that many fast-food and "quick service" establishments do not feature tipping and are thus not the focus of this study. The lack of appreciable tips eliminates the earning disparity between workers, who unilaterally make low wages.

[5]This was true at all three restaurants in this study, and is in line with the tip-out rates described by Gomberg-Muñoz (2011).

Table 13.2 Total number of workers in food industry

Occupation title	# of Workers	% of Industry workers	Median wage[a]
Tipped workers			
Hosts and Hostesses, Restaurant, Lounge, and Coffee Shop	414,540	3.3%	$10.06
Dining Room, Cafeteria Attendants and Bartender Helpers	455,700	3.5%	$10.71
Bartenders	611,200	4.7%	$10.84
Waiters and Waitresses	2,600,500	20.1%	$10.47
Non-tipped workers			
Chefs and Head Cooks	146,500	1.1%	$23.30
Cooks	2,403,000	18.6%	$12.12
Food and Beverage Serving and Related Workers (Support Staff)	5,122,600	39.6%	$10.45
Food Preparation Workers	871,100	6.7%	$11.41
Food Service Managers	308,700	2.4%	$26.08
Total industry workers	12,933,840		

Source: Bureau of Labor Statistics Data (2016)
[a]Median wage does not include tips

tip-exempt managers (floor supervisors and chefs) who are responsible for overseeing the dining room and kitchen operations (Table 13.2).

The resulting earning imbalance between different types of service workers in the same workplace due to tips can be substantial, particularly at higher-end establishments. Despite the fact that restaurants typically pay back of the house workers slightly more than front of the house workers in terms of base wages (Wilson 2019), cooks and kitchen workers earn on average *two to three times less money* than servers and bartenders because of tips (Haley-Lock and Ewert 2011). Employees who are laboring just feet from one another end up divided into economic winners and losers on a nightly basis. Cooks and other back-of-the-house workers remain among the lowest hourly wage earners in the country, often making less than $13 an hour ($25,000 annually)—hardly a livable wage in most US metropolitan centers (ROCU 2014; US Bureau of Labor Statistics 2017). By contrast, front-of-the-house workers in urban restaurants can top $30 an hour in combined tips and wages, sometimes earning as much as six figures annually (Haley-Lock and Ewert 2011).[6] Meanwhile, support workers and hosts take home earnings somewhere in between the prior two groups. Despite being tipped-out—thus earning more money per hour than kitchen workers—their hours are typically shorter each shift. The following pie chart visually depicts an average distribution of tips among workers in full-service restaurants (Chart 13.1):

[6]My sense is that these earning figures for tipped restaurant workers represent the higher range for the industry as a whole. Restaurant workers in non-urban contexts and less trafficked establishments will earn less.

Chart 13.1 Standard tip distribution in a full-service restaurant

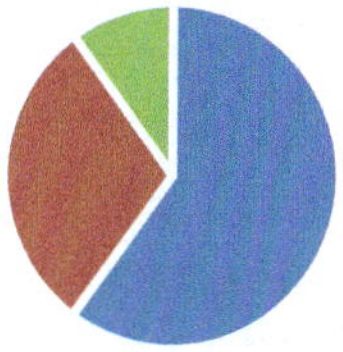

Differential access to large sums of gratuities can lead to tensions between tipped, tipped-out, and non-tipped workers. As Fine (2009) notes about restaurants, tip inequalities between fellow workers complicate traditional organizational and status hierarchies; employee earnings do not reflect the skills and training required for the job. A restaurant manager quoted in Azar (2011) laments that tipped customer service jobs may require only 2 months of training but are compensated at twice the rate of job in the restaurant that requires 2–5 years of experience to get. This disparity impacts not only the interpersonal dynamic within the workplace, but the industry composition as a whole. As several recent reports note, the strong financial incentives for workers to seek tipped jobs has contributed to a shortage of quality cooks available in cities like San Francisco and New York (Ferro 2015).

Tip disparities between types of workers can also exacerbate existing social divides within restaurant labor in which different jobs are coded by race, gender, class, and immigration status (Wilson forthcoming). In many large U.S. cities and in the southwest, back-of-the-house restaurant jobs are racialized as "brown-collar" work associated with unskilled Latino immigrant men (Cantazarite 2000; Gomberg-Muñoz 2011; Waldinger and Lichter 2003). By contrast, front-of-the-house jobs are disproportionately held by class-privileged, white, young men and women, who managers believe to represent desirable social and physical assemblage that will attract customers (Jayaraman 2014; ROCU 2014; Williams and Connell 2010; Wilson 2016). The unequal distribution of tips thus layers onto a socially segmented service workplace. Aside from management, those employed in higher tip-earning positions within restaurants tend to be whiter and more educated than their non-tipped counterparts (ROCU 2014; Wilson forthcoming).

Case Studies: Match, Terroir, the Neighborhood

In order to more closely examine tipped workplaces, we draw on ethnographic data collected by the first author on restaurant workers employed in three full-service restaurants in Los Angeles between 2012–2018. The first author gained employment within each establishment as a restaurant server, working between two and five shifts

Table 13.3 Total number of workers in food industry, by tip classification

Occupation title	# of Workers	% of Industry workers
Tipped workers	3,211,700	24.8%
Tipped out workers	870,240	6.8%
Non-tipped workers	8,851,900	68.4%
Total industry workers	12,933,840	

Source: Bureau of Labor Statistics Data (2016)

a week, and totaling over 500 h at each location. During this time, the first author kept a daily record of the type of shift worked (lunch or dinner), total tips received, and the "section" of tables he was assigned. The ensuing analysis stems from data on observations, interactions, and informal conversations with workers, supervisors, and customers in these workplaces. Additionally, the first author conducted fifty-six, non-random interviews with workers (n = 47), managers (n = 7), and industry professionals (n = 1) that explored the themes of this study. These interviews lasted 30–90 min each and were recorded with interviewees' permission on a handheld electronic device and transcribed immediately afterward.

The three restaurant field sites in this study, referred to as Match, Terroir, and The Neighborhood (pseudonyms), were strategically selected for their broad organizational similarities and subtle differences. All are higher-end (listed on Yelp as 11–60 dollars per person) and full-service establishments in the Los Angeles area that represent medium-sized operations (30–80 total staff). Each has a workforce with demographic similarities to many of their peer restaurants in the city: the majority of back-of-the-house (non-tipped) and support (tipped-out) workers are working-class Latino men, while front-of-the-house workers are college-educated, white men and women (Table 13.3).[7]

Match, Terroir, and The Neighborhood differ with respect to their tip policies and the total amount of gratuities that workers typically earn per shift. At Match, servers and bartenders individually distribute a portion of their tips to bussers, food runners, baristas, bartenders, and hosts, keeping the rest for themselves. The formula for tipping out is based on net sales, with payouts for each server totaling approximately 5% of their sales. For example, a server who sells $1000 worth of food and drink and receives $200 in tips would tip out a total of $50 to various support staff workers and take home $150 in tips. Over the duration of my time at Match, working a combination of breakfast, lunch, and brunch shifts, my average tips per server shift were $153, which averaged out to over $30 an hour in tips and wages combined. By contrast, cooks at Match made a fixed hourly wage of roughly $13 (at the time of study), slightly higher than the national average but also consistent with the wages paid to cooks across all three of the restaurants in this study.

At Terroir and The Neighborhood, tips are pooled and distributed in similar ways. At the end of every shift, servers and bartenders place their tips in a collective pot. Total tips are then tallied and distributed among all dining room employees based on

[7]For more information, please see Wilson (2017).

Table 13.4 Tip distribution at match, Terroir, and the neighborhood (average shift)

	Match (non-pooled)	Terroir (pooled)	Neighborhood (pooled)
Servers	$145	$100	$200
Hosts	$5	$17	$33
Bussers	$20[a]	$50	$100
Bar/Market	$30[b]		

[a]tipped out individually by each server
[b]tip-out earnings in addition to tips earned individually

a set distribution for each worker based on a "point system": six points for servers, three points for bussers and food runners, and one point for hosts. Based on this formula, if servers "walk" with $160 in tips, bussers and food runners would take home $80, and hosts would take home $27. If a worker leaves a shift early or is asked to stay late, their point total can be adjusted up or down accordingly. This policy, however, can be informally negotiated. The actual distribution of tips remains at the discretion of whoever is counting and packaging up the tip money at the end of the shift.

Despite having similar tip pooling systems, average tip earnings per worker differ substantially at Terroir and The Neighborhood (see Table 13.4, above). This is due to differences in customer traffic and staffing practices at each restaurant. Simply put, The Neighborhood has more customers in the restaurant and less tipped workers on the clock than Terroir. My personal tip records from Terroir indicate that servers averaged $96 in tips across lunch and dinner shifts, slightly more during the latter. At The Neighborhood, servers averaged $196 in tips, and both tipped and tipped-out workers had the highest earnings across all three restaurants.

Tips and Worker Tensions

Given the unequal access to tips by different coworkers and the earning imbalances tipping generates, the subject of tips elicits interpersonal tension in the workplace. Tips fray personal relations between workers; they also impede friendships from developing across organizational lines in the first place. Kitchen workers feel it is unfair that they are almost always out-earned by the dining room staff, despite the latter working shorter hours and doing what kitchen workers believe is an easier job. A field note described in Wilson (2019: 7) illustrates this point:

> Rodrigo still has not forgiven Freddy, a Latino food runner a few years older than he, for jokingly waiving a wad of tips in his face and telling him, 'Thank you for making me all this!' Says Rodrigo, "I just wanted to punch him in the face that day. I told him, 'You are a bold man for saying that. Say it again and we are going to have a problem.'

Unresolved tensions between tipped and untipped workers in the same workplace can resurface every time the topic of tips comes up. For Rodrigo, who is a cook, the end of every shift when tips are distributed is a new opportunity for him to get frustrated by the "unfair" earnings between himself and his tipped coworkers. On several occasions, cooks and servers struggled to get along not only because their respective jobs—and social backgrounds—offered few shared experiences, but also because of a tip-fueled economic rift. The meager hourly wages that cooks made pale in comparison to the three-figure cash tips that servers in this study came to expect in *addition* to their base wages (California does not feature a tipped minimum wage). During busy shifts, kitchen workers were aware that their increased efforts to cook or clean will not put more money in their pocket even though it most likely generated additional tips for their front-of-the-house coworkers. Their pointed comments to tipped coworkers in the workplace often reflect this frustration: "where's my tip, huh?"

Tipped-out workers also express frustration over how tips are handled in the workplace. Food runners, hosts, and bussers are chronically dissatisfied with the tip-outs they receive from servers and bartenders. Regardless of how much money they receive in tips, their concern is about *relative* parity in tips. Said one busser: "we work very hard, and they should not be making so much more money than we do." Bussers and food runners must wait for their server colleagues, who are tasked by management to distribute tips at the end of each night in personalized envelopes, to individually hand them their share of tips. This procedure allows servers and bartenders to access tips before anyone else, and make decisions about what to do with "extra" money that cannot be divided up evenly (it usually goes to the servers, or on rare occasion, to the kitchen as described below). Some servers keep the newest 20-dollar bills for themselves and distribute the crumpled ones and fives to the bussers, food runners, and hosts.

Reacting against these types of perceived slights, tipped-out workers occasionally resort to small acts of resistance against their tipped coworkers (see Scott 1985). For example, one busser retaliated by subtly doing less work clearing and resetting tables; a host deliberately avoided seating customers they perceived would spend a lot of money with one server that she disliked. Both efforts disproportionately affect the earnings of the offending server. Tipped-out workers occasionally attempt to lodge complaints with management in an effort to address tip imbalances. However, managers are loath to step in to intervene since tips are considered the property of workers. Approaching management could also backfire should word of the complaint get out. One host who petitioned to change the tip pool by speaking with management quickly became socially isolated in the workplace. She quit 2 months later. Those who make the most in tips—servers and bartenders—often are given jurisdiction to manage tips in the workplace as they see fit.

Under a pooled tip system, workers who do not "pull their weight" can quickly fall out of favor with the rest of the front-of-the-house team. The following is a discussion between three servers, Sally, Rita, and Rudolfo, quoted in Wilson (2019: 10):

> "See?" Sally says in Rita's direction as she exits the restaurant through the back corridor. Sally is clearly frustrated. "Here I am contributing over $400 to the pool and Rita is already gone—and she only brought in $270. And it's like this practically every time!" An hour later, Sally and Rudolfo tally up the pooled tips on the back counter. They glance at each

> other, eyeing their total tips in comparison to Rita's. Privately, they decide to split the "extra" cash tips between the two of them instead of putting it toward the tip pot.
>
> "Shhhh," Sally says, looking at me. "It's only fair. Just look at the numbers!"

Despite a formally pooled tip system, Sally and Rudolfo use their own discretion to put more tips in their pockets and less in Rita's. Tipped workers can often take steps to informally "correct" perceived tip imbalances among their colleagues in ways that, should the latter find out, would constitute grave violations of trust and fairness.

Individualized tip systems hedge against some of these tensions while creating other issues. Workers who keep the tips they make often find themselves in competition with one another for the right to wait on certain groups of diners they think will tip them well. Problems arise when workers overstep unwritten rules in order to get a leg up with their own tips, for instance, by declaring ownership over a table seated in another section while the server for that table is on lunch break, or by not sharing a tip for a large party with other servers who may have assisted with that group.

When tip disparities between workers become too glaring or too unforgivable, they can swell into overt conflict. Yet in general, tipped, tipped-out, and untipped workers alike would rather not incite overt conflict over tips, especially in a labor context where they must work closely with one another to accomplish service. Tensions among tipped and tipped-out workers tend to decrease when tips exceed expectations, or as one server put it, "the money is good anyway, baby!" However, a slow night can draw out simmering feuds between different categories of workers. The way tips are made and distributed in the workplace can mean hundreds of dollars in earning difference each week for a given worker.

Tips and Tenuous Alliances

Tipped workers at Match, Terroir, and The Neighborhood understand that the ability to optimize their tips is about having the right set of opportunities in the workplace. Generating consistent high tips is not just about getting lucky with a few generous diners, it involves strategizing one's approach towards other social and material aspects of the workplace (see Kim et al. 2017; Paules 1991; Sallaz 2009). As a result, despite the tip inequalities that separate them, many tipped workers view their interactions with both tipped-out and non-tipped workers as a source of on-the-job social capital. They attempt to forge tenuous alliances with each.

Front-of-the-house workers negotiate with scheduling managers to secure more favorable shifts. Because the number of workers present on the floor directly affects tips, daily managerial decisions regarding scheduling and "cutting" (releasing workers early) can dramatically swing the tip pool (see Lambert et al. 2012). This makes some traditionally undesirable shifts, such as slow weekdays, unexpectedly lucrative. At Terroir, several of my coworkers expressed more enthusiasm at being scheduled to work the Wednesday night dinner shift, when only three servers were scheduled, than Friday nights, when six servers were scheduled. Tipped workers also favor being scheduled alongside Managers who they know are willing to "short staff" (schedule less workers) and cut labor quickly. For example, one manager at The Neighborhood that was particularly well-liked manager told servers that he was willing to "run with a short staff" (schedule less servers) and would help wait tables if things got too busy—without taking any tips for doing so.

Tipped workers also approach their relations with the schedule managers strategically, based on how this might help them secure more lucrative shifts. At all three restaurants, the scheduling manager draws up both weekly schedules for the front of the house as well as daily "sections" where each server will work. At The Neighborhood, a new server spent the first month strategically arriving early and staying late every day until he felt he had cultivated sufficient cache with management to work on Friday and Saturday nights. At Match, servers understand that some floor sections are more lucrative than others: the corner banquettes always filled up first and with spendy businessmen who preferred the relative seclusion of this area of the restaurant. In an effort to secure this section, one server said she felt keeping strong ties with the scheduling manager was among her top priorities on the job. For tipped workers, strategizing relations that might result in tip-making opportunities each week is half the battle.

Front-of-the-house workers also approach fellow workers with considerable strategic planning in the workplace. Fellow servers give away, trade, and offer to cover shifts. At stake are unequal opportunities to make greater tips: a dinner shift might yield 30% more in tips than lunch that same day. Thus, trading away a coveted Saturday night shift to another worker can build serious goodwill with that individual—as well as a debt that can be evoked at a later date. At Terroir, one worker attempted to switch shifts with the first author in his favor. After these efforts failed, he attempted another strategy: "c'mon Eli, you owe me [from before]!"

Tipped workers also attempt to forge alliances with non-tipped workers in the kitchen. This is because the quicker the food flows to diners, the better the front-of-the-house workers' chances for good tips. Tipped workers would often replenish sodas mid-shift, occasionally buy post-shift beers for the cooks, or less commonly, share a small portion of their tips with them. These efforts differed under Match's system of individualized tips versus the pooled tips at Terroir and The Neighborhood. At the former, workers rely on gestures that could be traced back to them personally, such as by refilling a cook's soda container and handing it back to them, or offering select cooks a five- or ten-dollar bill for extra attention to their food orders that day. One male server named Leroy explains his calculative approach to such gestures, "I don't do it all the time because that would start adding up. But I want to make sure those guys are on my side when I need them."

Under a pooled tip arrangement, front-of-the-house workers at Terroir and The Neighborhood attempt to curry favor from non-tipped workers in ways that could benefit all tipped workers that shift. For instance, after a busy weekend dinner service at The Neighborhood, one in which servers each made nearly $300 in tips, servers figured out whose turn it was to buy "kitchen beers": a six-pack for cooks to sip on while they clean. Other tipped workers may gather together "the juice"—a few extra dollars from the daily tip pool that couldn't be divided evenly—and put it toward a monthly pot intended for kitchen workers. They would label this pot "kitchen tips," and hand it over to the manager on duty for safe keeping. However, since voluntarily distributing leftover tip money to the kitchen relies heavily on coworker goodwill, the procedure could easily be shortchanged. At The Neighborhood, for example, this practice stopped abruptly because the previous installment of kitchen tips had gone missing and servers were highly suspicious that someone had taken the money for themselves. No one bothered to consult any of their kitchen coworkers about the decision to cease passing tips informally their way.

Discussion and Conclusion

The number of tipped workers in the United States has surpassed five million and continues to rise. This means that the day-to-day labor experiences, workplace relations, and employment strategies of a growing number of Americans are impacted by the presence of gratuities in their respective places of employment. Moving the study of tips beyond customer-worker interactions and their tip outcome, this study examines the size, scope, and character of tipped workplaces in the United States.

The unequal distribution of gratuities within tipped workplaces—what we classify as tipped, tipped-out, and un-tipped workers—reflects structural distinctions between different types of service jobs (Adler and Adler 2004; Subramanian and Suquet 2017; Wilson 2019). Tips mediate how different workers interpret, negotiate, and constitute their relations with one another in their place of employment. Tipping can lead to strategic alliances between different subgroups of service workers (see Lewis 2006) just as it can exacerbate chronic workplace tensions. Because of the earning incentive that tips represent in the United States, different kinds of jobs in restaurants come to represent substantially unequal opportunities.

Access to tips affects the ways in which socially distinct employee subgroups approach their jobs. Tips shape vertical labor relations between workers and front-line management. Our findings show that tipped workers become willing accomplices in management-friendly labor schemes such as short-staffing and "cutting" (Lambert et al. 2012) in an effort to personally secure better opportunities to earn tips. These strategies can undercut fellow workers, who tipped workers come to see as just another warm body to split tips between. In this sense, the very logic of tipping, undergirded by self-interested pursuit of additional earning opportunities, threatens the sense of collective unity of service workers in the same workplace, tipped or untipped. The social dynamics of tipped workplaces also reflect the specific labor conditions, norms, and policies where they are found. In settings that generate high amounts of tips, workers compete for shifts that they assume will be more lucrative, and forge alliances with those that can help them secure better earning opportunities. These structural opportunities for tip-making vary across institutional environments with regard to tipping policies and normative practices (see Sallaz 2009). We would therefore expect that in countries with less robust cultural expectations surrounding tipping (Lynn 2006), key aspects of the social environment that employees experience would differ accordingly. For example, it could be that workplace tension between front-of-the-house and back-of-the-house workers is mitigated in the absence of large amounts of tips garnered solely by the former group. However, these tensions may not be eliminated entirely: because the majority of tipped jobs in U.S. restaurants are occupied by young, college-educated whites, and non-tipped jobs tend to be filled by working-class people of color, the unequal earning opportunities that tips engender only *sharpens* employee social divides rather than create them anew.

Beyond restaurants, our study speaks to what goes on inside many other tipped workplaces that maintain a similar array of tipped, tipped-out, and untipped workers. That said, the way these workers will relate to each other will depend on discrete organizational practices and labor market conditions (see Sallaz 2009; Sherman 2007; Subramanian and Suquet 2017). For example, efforts by worker advocacy groups to raise the U.S. federal tipped minimum wage are underway. Should these efforts be successful, more workplaces will feature greater base wage parity between workers—surely a move towards a healthier safety net for this workforce. However, such a change would do little to amend the earning inequality between different categories of workers that we have addressed in this paper.

By centering our examination on the labor dynamics of tipped workplaces, this study highlights the much larger pool of workers who labor in the presence of tips but do not receive them directly. While this is not necessarily new, the sheer number of American workers who are employed in tipped workplaces continues to grow. As the practice of tipping creeps into other industry sectors, such as retail and the so-called "gig economy"—tipped employment appears to be playing an increasingly important role in the U.S. labor market. Areas for future research abound. Tipping is quickly moving beyond traditional, brick-and-mortar workplaces. The prompt for a tip now bookends many service transactions that occur via electronic platforms (e.g., DoorDash, Uber, some trendier retail establishments). The method with which customers tip has also changed: the first author observed that over the course of this research (2012–2018), credit card-based tips have become the overwhelming way in which tipping is conducted, rather than cash. Because credit cards must be processed alongside sales by management and are subject to third-party fees, the distribution of tips today involves an extra layer of managerial oversight. Fittingly, several controversies over company appropriation of worker tips have made national headlines in recent months.[8] As tipped employment extends beyond familiar locales and into the pockets of different kinds of labor actors, scholarly research needs to follow closely with new theorization of this changing employment landscape.

[8]Newman, Andy. "DoorDash Changes Tipping Model After Uproar from Customers." New York Times, July 24, 2019. Retrieved from: https://www.nytimes.com/2019/07/24/nyregion/doordash-tip-policy.html

References

Adler, P., & Adler, P. (2004). *Paradise laborers: Hotel work in the global economy*. Ithaca: Cornell University Press.
Allegretto, S. & Cooper, D. (2014). Twenty-Three Years and Still Waiting For Change: Why It's Time to Give Tipped Workers the Regular Minimum Wage. Economic Policy Institute. Retrieved March 18, 2020 https://www.epi.org/files/2014/EPI-CWED-BP379.pdf.
Azar, O. (2004). The history of tipping—From the sixteenth-century England to United States in the 1910s. *The Journal of Socio-Economics, 33*(6), 745–764.
Azar, O. (2007). The social norm of tipping: A review. *Journal of Applied Social Psychology, 37* (2), 380–402.
Azar, O. (2011). Business strategy and the social norm of tipping. *Journal of Economic Psychology, 32*, 515–525.
Bearman, P. (2009). *Doormen*. Chicago: University of Chicago Press.
Brewster, Z., & Mallinson, C. (2009). Racial differences in restaurant tipping: A labour process perspective. *The Service Industries Journal, 29*(8), 1053–1075.
Brewster, Z., & Wills, J. (2013). The institution of tipping as a source of employee agency: The case of restaurant servers. *Sociological Focus, 46*(3), 193–210.
Bureau of Labor Statistics Data (2016). Overview of BLS Wage Data by Area and Occupation. Retrieved March 18, 2020 https://www.bls.gov/bls/blswage.htm.
Cantazarite, L. (2000). Brown-collar jobs: Occupational segregation and earnings of recent-immigrant Latinos. *Sociological Perspectives, 43*(1), 45–75.
Cobble, D. S. (1991). *Dishing it out: Waitresses and their unions in the twentieth century*. Chicago: University of Illinois Press.
Crusco, A., & Wetzel, C. (1984). The Midas touch: The effects of interpersonal touch on restaurant tipping. *Personality and Social Psychology Bulletin, 10*, 512–517.
Davis, F. (1959). The taxi driver and his fare: Facets of a fleeting relationship. *American Journal of Sociology, 65*(2), 158–165.
Ehrenreich, B. (2001). *Nickel and dimed: On (not) getting by in America*. New York: Metropolitan Books.
Ferro, S. (2015, October 21). There's a shortage of cooks in America. Here's the simple solution. *The Huffington Post*. Retrieved December 30, 2018. http://www.huffingtonpost.com/entry/cook-shortage-restaurant-pay_us_5627b438e4b08589ef4a031d
Fine, G. A. (1996). Justifying work: Occupational rhetorics as resources in restaurant kitchens. *Administrative Science Quarterly, 41*(1), 90–115.
Fine, G. A. (2009). *Kitchens*. Berkeley: University of California Press.
Foshay, K. (2016, August 1). *Wage theft: Millions Stolen, little recovered within the restaurant industry*. KCRW. Retrieved December 30, 2018. http://curious.kcrw.com/2016/08/wage-theft
Gomberg-Muñoz, R. (2011). *Labor and legality: An ethnography of a Mexican immigrant network*. New York: Oxford University Press.
Haley-Lock, A., & Ewert, S. (2011). Waiting for the minimum: US State wages laws, firm strategy, and chain restaurant job quality. *Journal of Industrial Research, 51*(1), 31–48.
Halpin, B. (2015). Subject to change without notice: Mock schedules and flexible employment in the United States. *Social Problems, 62*, 419–438.
Jayaraman, S. (2014). *Behind the kitchen door*. Ithaca: Cornell University Press.
Jayaraman, S. (2018). *Forked: A new standard for American dining*. New York: Oxford University Press.
Kim, P. B., Nemeschansky, B., & Brandt, L. (2017). An exploratory study of determinants for restaurant servers' actual tip earnings: Individual characteristics and work conditions. *Journal of Food Service Business Research, 20*(1), 15–33.
Lambert, S., Haley-Lock, A., & Henly, J. (2012). Schedule flexibility in hourly jobs: Unanticipated consequences and promising directions. *Community, Work & Family, 15*(3), 293–315.

Lewis, J. (2006). 'I'll scratch your Back if You'll scratch mine': The role of reciprocity, power and autonomy in the strip Club. *Canadian Review of Sociology and Anthropology, 43*(3), 297–311.
Lopez, S. (2010). Workers, managers, and customers: Triangles of power in work communities. *Work and Occupations, 37*(3), 251–271.
Lynn, M., Zinkhan, G. M., & Harris, J. (1993). Consumer tipping: A cross-country study. *Journal of Consumer Research, 20*, 478–488.
Lynn, M. W. (2006). Tipping in restaurants and around the globe: An interdisciplinary review. In M. Altman (Ed.), *Handbook of contemporary behavioral economics: Foundations and developments* (pp. 626–643). New York: Sharpe.
Lynn, M., & Latane, B. (1984). The psychology of restaurant tipping. *Journal of Applied Social Psychology, 14*(6), 549–561.
Lynn, M., & McCall, M. (2000). Gratitude and gratuity: A meta-analysis of research on the service-tipping relationship. *Journal of Socio-Economics, 29*, 203–214.
Mars, G., & Nicod, M. (1984). *The world of waiters*. London: George Allen and Unwin.
Paules, G. F. (1991). *Dishing it out: Power and resistance among waitresses in a New Jersey restaurant*. Philadelphia: Temple University Press.
ROCU (Restaurant Opportunities Centers United). (2014). *The great service divide: Occupational segregation and inequality in the US restaurant industry*. Restaurant Opportunities Centers United. Retrieved December 30, 2018 http://rocunited.org/publications/the-great-service-divide-national/
Sallaz, J. (2009). *The labor of luck: Casino capitalism in the United States and South Africa*. Berkeley: University of California Press.
Santino, J. (1989). *Miles of smiles, years of struggle: Stories of Black Pullman porters*. Urbana: University of Illinois Press.
Scott, J. (1985). *Weapons of the weak: Everyday forms of Peasant resistance*. New Haven: Yale University Press.
Segrave, K. (1998). *Tipping: An American social history of gratuities*. Jefferson: McFarland.
Sherman, R. (2007). *Class acts: Service and inequality in Luxury hotels*. Berkeley: University of California Press.
Spradley, J. P., & Mann, B. E. (1975). *The cocktail waitress: Woman's work in a man's world*. New York: John Wiley.
Star, S. L., & Strauss, A. (1999). Layers of silence, arenas of voice: The ecology of visible and invisible work. *Computer Supported Cooperative Work, 8*(1–2), 9–30.
Strauss, A. (1985). Work and the division of labor. *The Sociological Quarterly, 26*(1), 1–19.
Subramanian, D., & Suquet, J.-B. (2017). Unpacking the service triangle: Arranging power relations between frontline occupations. *Work and Occupations, 45*(1), 38–81.
Thompson, A. (2015). Wrangling tips: Entrepreneurial manipulation in fast-food delivery. *Journal of Contemporary Ethnography, 44*(6), 737–765.
US Bureau of Labor Statistics. (2017). *Occupational outlook handbook*. Updated April 13, 2018. Retrieved December 30, 2018. https://www.bls.gov/ooh/food-preparation-and-serving/cooks.htm
Waldinger, R., & Lichter, M. (2003). *How the other half works*. Berkeley: University of California Press.
Williams, C., & Connell, C. (2010). 'Looking good and sounding right': Aesthetic labor and social inequality in the retail industry. *Work and Occupations, 37*(3), 349–377.
Wilson, E. (forthcoming). *Front of the house, Back of the house: Race and inequality in the lives of restaurant workers*. New York: NYU Press.
Wilson, E. R. (2016). Matching up: Producing proximal service in a Los Angeles restaurant. *Research in the Sociology of Work, 29*, 99–124.
Wilson, Eli R. 2017. *Serving across the divide*: *Race, class, and the production of restaurant service*. PhD dissertation, Department of Sociology, University of California, Los Angeles.
Wilson, E. R. (2019). Tip work: Examining the relational dynamics of tipping beyond the service counter. *Symbolic Interaction*. https://doi.org/10.1002/SYMB.413.

Eli R. Wilson is an Assistant Professor of Sociology at the University of New Mexico. His research examines how social inequality is both reproduced and challenged in urban labor markets. His current project examines racialized and gendered career pathways in the craft beer industry. Dr. Wilson's published works can also be found in *Ethnic and Racial Studies*, *Research in the Sociology of Work*, and *The Russell Sage Foundation Journal of the Social Sciences*.

Davyd Setter is a graduate student in sociology at the University of New Mexico. His research interests include social movements and race and ethnicity.

Chapter 14
Immigration Enforcement and Employment in Large Firms: Evidence from County Participation in 287(g)*

Li Zhu, Matthew Hall, and Jordan Matsudaira

Immigrant deportations over the last two decades have grown at an alarming rate, with fewer than 150,000 annual removals in the early 2000s but rising to over 400,000 by 2013. Over this period, demographers estimate that over 4 million immigrants – the vast majority of whom are unauthorized and Latino – have been deported (Gonzalez-Barrera and Krogstad 2014). An important feature of this trend is that local policing agencies played a crucial role in facilitating apprehensions and coordinating with federal agents to deport immigrants from local communities. While post-recession declines in unauthorized migration have led to recent declines in deportations, the risk of apprehension remains high, and hostile rhetoric and intensified approaches for targeting non-criminal migrants heighten the fear and anxiety associated with being undocumented.

A growing body of academic research in economics, demography, sociology, and other fields has aimed at understanding the consequences of mass deportation for migrants and their families, for community relations, and economic productivity. This work has shown, for example, that anti-immigration local policies are associated with declines in non-citizen and Hispanic student populations (Dee and Murphy 2019; Kostandini et al. 2014), as well as measurable impacts on home foreclosure (Rugh and Hall 2016), food insecurity (Potochnick et al. 2017), physical health (Beniflah et al. 2013; Watson 2014; Rhodes et al. 2015), and mental distress (Wang and Kaushal 2018).

L. Zhu · M. Hall (✉)
Policy Analysis & Management, Cornell University, Ithaca, NY, USA
e-mail: mhall@cornell.edu

J. Matsudaira
Economics and Education, Teachers College, Columbia University, New York City, NY, USA

B. Jivetti, Md. N. Hoque (eds.), *Population Change and Public Policy*, Applied Demography Series 11, https://doi.org/10.1007/978-3-030-57069-9_14

As part of the broader motivation for anti-immigration policies and local efforts to facilitate deportation was to reduce the competition that native workers face from immigrant labor, an important question is how enforcement efforts have affected local economies. Several recent studies have explored this topic, generally finding that local enforcement is tied to shrinking labor pools, but higher wages (see Bohn and Santillano 2017; East et al. 2018; Pham and Van 2010). Yet, prior work has not been able to disaggregate these dynamics by racial group or gender, necessary to assess variation in how enforcement is felt and experienced by majority and minority populations.

In this chapter, we exploit variation in the outcomes of county applications to participate in the 287(g) program, one of the primary mechanisms through which local police departments were able to coordinate deportation efforts with Immigration and Customs Enforcement (ICE), in order to generate plausibly-causal estimates of immigration enforcement on the employment trends of large firms. Specifically, we use firm-level data from the Equal Employment Opportunity Commission to document how implementation of 287(g) altered employment for Hispanics and non-Hispanics. We describe these impacts for the total labor pool as well as in specific industries, and also consider patterns of difference between male and female workers. Our results are largely consistent with prior work in that 287(g) was associated with a decline in total employment. Our contribution largely lies in our finding that these impacts were unevenly distributed across racial groups, with 287 (g) reducing Hispanic employment in agriculture & mining, manufacturing, health care/social services, and accommodation/food services, while white employment in wholesale and retail trade, along with administrative support/waste management, is adversely impacted by the anti-immigration policy.

Background

The foreign-born population in the U.S. has grown at an incredible pace over the last several decades, with the nearly 45 million immigrants comprising nearly 1 in 7 of U.S. residents. A defining feature of the foreign-born population is that roughly one-quarter are unauthorized to live and work in the country. The unauthorized population, which totals about 11 million and hails mostly – but not exclusively – from Latin America has transformed communities and labor markets across the country, and represents the primary focus of U.S. immigration policy debates. Their contribution to U.S. labor markets is particularly important, with the 8 million unauthorized workers representing about 5% of the U.S. labor force (see Passel and Cohn 2019). The impacts of immigration, especially unauthorized immigration, on the economic circumstances of native workers has been the source of great scholarly debate (see National Academies of Sciences 2017). Despite a general empirical consensus that immigration has small overall effects on native wages and employment, political arguments and rhetoric strongly imply that low-skill immigrants compete with natives over a shrinking set of quality jobs.

This economic anxiety, along with fears of cultural and social change, has contributed to an intensification of efforts to enforce immigration law and return unauthorized immigrants to their home countries. Accordingly, between 1990 and 2010, the size of the border patrol increased by a factor of five (to over 20,000 agents) and its budget ballooned from $262 million to $3.0 billion. While the impact of these policies may have been to contain (and thereby increase) the unauthorized population residing in the US (Massey et al. 2016), as new flows of Mexican immigration declined, the politics of immigration have increasingly concentrated on removing the resident unauthorized population. As a result, interior enforcement efforts and deportations intensified substantially during the 2000s, with removals peaking at about 400,000 in 2010 (see Meissner et al. 2013).

Given the geographic distribution of unauthorized immigrants, it is not surprising that enforcement efforts have been distributed unevenly across the nation. While California and Texas account for a large share of total removals, the spatial patterns of deportation highlight the local nature of enforcement efforts, with many areas with modest unauthorized populations having elevated rates of apprehension (see Moinester 2019). Part of the explanation for these spatial patterns is the devolution of immigration enforcement to states and local areas (Varsanyi 2010). As federal policymakers failed to legislate comprehensive reform, states and local areas increasingly took it upon themselves to develop policies with the explicit intention of discouraging unauthorized migration (Steil and Vasi 2014; O'Neil 2010). These efforts took various forms, such as housing vacancy regulations, mandates to verify workers' employment eligibility, and laws excluding undocumented students from in-state tuition benefits (see Varsanyi 2010).

An important component of this shift to local control was partnerships between federal immigration officials and local policing agencies (for a detailed description of the various partnerships, see Meissner et al. 2013). In this project, we leverage variation in the 287(g) program, which allowed local areas to enter into agreements with Immigration and Customs Enforcement (ICE) to allow local police officers to enforce aspects of immigration law. More specifically, established as part of immigration reform in 1996 (IIRIRA), the 287(g) program creates partnerships that give officers in participating agencies authority to question, arrest, investigate, and detain individuals suspected of being in violation of immigration law. Authorizing local police to enforce perceived immigration violations expands the scope of the enforcement system substantially, and allows potential immigrants to be detained during routine police duties, including traffic stops (Coleman 2012). In some instances, local police established checkpoints in Latino neighborhoods ostensibly designed to regulate traffic, but became, in essence, an internal border stop for verifying immigration status (Stuesse and Coleman 2014).

These partnerships between ICE and local areas grew in number and intensity during the 2000s. In response to the 9/11 terrorist attacks, several states entered into 287(g) in order to address growing concerns about transportation security and the vetting of terror suspects. Yet, as nationalist and anti-immigrant movements built force, the 287(g) program became viewed as an effective tool for reducing

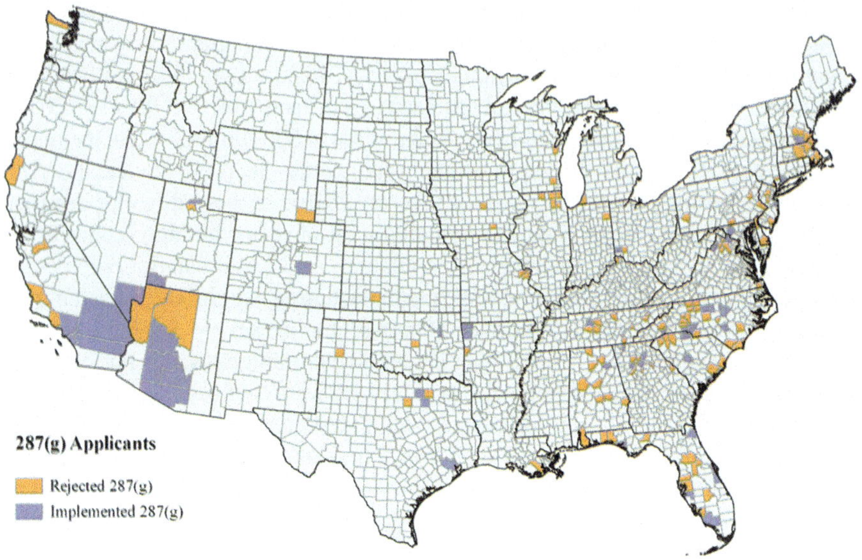

Fig. 14.1 County 287(g) Agreements, 2000–2010

undocumented populations and discouraging immigration. By 2009, more than 70 such 287(g) agreements had been signed, many of which were located in the New South where immigrants had arrived in rapid numbers during the preceding decade (for a thorough overview, see American Immigration Council 2017; Coleman 2009). Figure 14.1 shows a map of the geographic distribution of these agreements.

One of the primary motivations for elevated interior enforcement is the potential to reverse perceived labor market competition with native workers and improve economic well-being for citizens and other authorized residents. There are several theoretical reasons why local immigration enforcement, and the 287(g) program in particular, could impact employment. First, the direct removal of unauthorized migrants should reduce employment among unauthorized – largely Hispanic – workers. Beyond the direct impacts of enforcement on formal removal, programs like 287(g) are likely to incite fear and anxiety into immigrant communities that drive them to either relocate to less-hostile areas or to minimize situations where they may be susceptible to police contact (e.g., reduce commute times). Available evidence is partly supportive of these arguments, finding that 287 (g) implementation is associated with modest declines in noncitizen and recently-arriving immigrant populations (Kostandini et al. 2014; Capps et al. 2011; Watson 2014). We anticipate thus that 287(g) participation negatively impacts Hispanic employment, particularly in industries where unauthorized, low-skill workers are known to concentrate such as the agricultural, construction, and accommodation & food service sectors.

The impacts on other – likely-authorized workers – are less clear. Part of the motivation for local enforcement arrangements and other anti-immigrant policies was to ease native competition for work; thus, this logic would suggest that reductions in Hispanic employment would be offset by growth in non-Hispanic employment. In this scenario, the effect of 287(g) on total employment may be null, but one would anticipate contrasting effects for Hispanic and non-Hispanic workers. However, given limited evidence that low-skill unauthorized workers are perfect substitutes for native workers (Card 2005; Ottaviano and Peri 2012), it is plausible that Latino job losses are not filled by other workers. It is also possible that native or non-Hispanic workers are adversely affected by local immigration enforcement if demand for the services and goods their employment provides is reduced. As unauthorized residents (and their families) are not just workers, but also consumers, economic sectors providing service to immigrant communities – e.g., retail, housing, health and social services – may suffer from population losses associated with deportations.

These dynamics may also impact male and female employees differently. The vast majority of deportations are among male immigrants, with men composing 94% of interior removals between 2003 and 2015 (Moinester 2019). Accordingly, to the extent that changes in employment are driven by deportation, impacts of 287(g) on Hispanic employment should be stronger for men than women. Indirect negative effects on female Hispanic employment are likely, however, if households respond to hostile enforcement by leaving areas or if fears associated with commuting to work prompt women to avoid work. Conversely, 287(g) policies may actually increase Hispanic women's employment if the deportation of income-earning, male household members pushes previously unemployed women into work. For non-Hispanics, gender differentiation in 287(g) employment effects may exist if men and women differentially compete with immigrant workers or if shifts in labor demand differentially affects industries where men and women concentrate.

Existing research that examines the effect of local immigration enforcement on employment generally finds small to modest negative effects. Pham and Van (2010) compare economic indicators in counties with restrictive immigration laws to those without such policies, and find a modest decline (−4.0%) in total employment and a non-significant, but small decline (−2.6%) in wages in the counties with anti-immigration policies. They find that the job and payroll losses are concentrated in Food Services, where they estimate 5.3% employment declines and 2.9% wage declines. Assessing the impact of Secure Communities – a more-recent extension of the jail-based form of 287(g) – on economic conditions, East et al. (2018) find that local enforcement reduces the male employment rate by about 1%. East et al. (2018) find, however, that these employment effects are, as predicted, largest among low-skill non-citizen men (−5.2%) and smallest among citizen men (−0.5%). In work most closely related to this project, Bohn and Santillano (2017) look at the effect of 287(g) program on economic conditions by comparing employment and wage outcomes in counties implementing 287(g) with their non-implementing, contiguous neighbors. They find that little evidence that 287(g) reduced overall employment, but report that employment decreased in administrative services by about 7–10%, decreased modestly (but imprecisely) in construction and food services, but increased in manufacturing by about 4% after the implementation of 287(g).

This chapter builds off this past work in three key ways. First, our identification strategy does not rely on contiguous border counties, which may be problematic if there are spatial spillovers in the policy effects. These could exist if immigrants move to neighboring areas in response to 287(g), a dynamic that would bias employment losses negatively. They could also exist if immigrants in neighboring counties work in 287(g) counties and/or respond in ways similar to immigrants in neighboring 287(g) counties out of fear or confusion over the jurisdiction of enforcement policies, or to actual policing activities that extend across county borders. Rather, as described below, we leverage plausibly exogenous variation in the outcomes of counties applications to participate in the 287(g) program, that breaks the spatial correlations inherent to border-county analysis. Second, we are directly able to test how 287(g) affects the employment of racial groups differently. While, unlike East et al. (2018), we cannot distinguish between citizens and non-citizens, by comparing outcomes for Hispanic and non-Hispanic workers, we are able to test arguments about how certain subpopulations are impacted in different ways. Lastly, past work has focused either on estimating effects for male and female workers together, or on just assessing outcomes for male workers. Here, we look at both men and women, by race/ethnicity and industry, to evaluate gender differences in the effects of local enforcement policies.

Data and Methods

In order to assess the impact of 287(g) on employment patterns, we rely on applications from counties and municipalities to enter into agreements with the Department of Homeland Security (DHS) that grants local police officers authority to detain those suspected of immigration law violations. Specifically, we use data on applications from 158 counties, obtained via FOIA requests to DHS (see Pedroza 2019), of which 54 were approved and implemented the program, while the remainder were either denied by DHS or withdrawn by the applicant.[1] In addition to revealing the outcome of the application, the applications also contain information on the timing of the agreement and the specific enforcement agreement proposed (e.g., jail- or task-force-based). The geographic distribution of these agreements is shown in Fig. 14.1, with rejected applicant counties colored orange and implemented

[1]Specific information on how applications were evaluated is not available, but a DHS performance report indicates that concerns over whether county policing agencies had the resource capacity to carryout immigrant-related enforcement were relevant (see Department of Homeland Security 2010).

counties shaded purple. The map highlights both the variation in 287(g) applications across the country, as well as, the concentration of agreements in the Southern Atlantic region (particularly Florida, Georgia, the Carolinas, and Tennessee).

The firm-level outcome data come from the U.S. Equal Employment Opportunity Commission (EEOC). Federal statute and regulations require private establishments with more than 100 employees to file a Standard Form 100 (EEO-1) report each year, which includes each reporting establishment's basic information (e.g., name, address, industry code), as well as counts of the number of employees by race/ethnicity, gender, and broad occupational category.[2] Firms with 50–99 employees are also required to submit EEO-1 forms if they hold a federal contract during the focal year. The primary advantage to EEO data is that they allow us to obtain annual enumerations of employees, by race, gender, and industry, even in relatively small counties where several of the applications came from. The weakness is that these data do not capture entire county workforces, as small firms are not required to report employee counts. Comparisons to the Bureau of Labor Statistics employment reports indicate that EEO-1 data capture about half of the national employment, with coverage higher in manufacturing, transportation, and mining, but lower in construction, wholesale & retail trade, and services (see Stainback and Tomaskovic-Devey 2012; Robinson et al. 2005). Most importantly, comparisons between EEO-1 data on large establishments to Census data on county employment reveal that, while white workers are modestly underrepresented and nonwhite workers modestly overrepresented, the correlations between counts of employees, by race, are very high (above 0.85) (see Ferguson and Koning 2018). In order to avoid selection into reporting, we restrict our analytical sample to establishments with at least 100 employees each year and to establishments that have reported every year during the 2000–2012 period. In our preferred specification, we also restrict the analyses to years before 2013 to avoid potential confounding of the policy effects with the rollout of Secure Communities. The final sample thus consists of a panel of 6847 unique establishments in 158 counties that applied to 287g between 2000 and 2010.[3]

We consider a number of factors that may differentiate the relationship between 287(g) and employment. Given the different occupational and workplace experiences for male and female immigrant workers (see Hall et al. 2010, 2019), we explore how any impacts of 287(g) work operate differently by sex. Based on reports that agreements signed in the latter part of the 2000s more aggressively targeted unauthorized migrants (Capps et al. 2011; Pham 2018), we distinguish between early (2000–2007) and late (2008–2010) adopting counties. Similarly, we also test arguments that the implementation of immigration enforcement was more forceful in the

[2]These counts include leased, part-, and full-time workers, but exclude contract and temporary employees.

[3]The results are quantitatively and qualitatively similar when we remove some of the sample restrictions to (1) include all establishments, and not just those reported every year during the 2000–2012 period; and (2) expand the analysis period to 2000–2015, instead of 2000–2012.

Southern Atlantic by estimating models separately for Southern and non-Southern counties.[4]

To account for any pre-existing differences between implemented and rejected counties, we include a number of statistical controls, including annual county-level measures of total population (logged) and percent Hispanic from the Census Bureau's population estimates, unemployment rate and average weekly earnings from the Bureau of Labor Statistics, and reported total housing permits (logged) from the U.S. Census Bureau's Building Permits Survey. As our design incorporates county and year fixed effects, other county-level factors that are invariant over the study period, as well as year-specific common shocks that affect all counties, are absorbed by the model.

To assess the relationship between 287(g) and employment, we employ a difference-in-differences framework that allows us to determine how changes in employment in counties that implemented 287(g) differs from employment changes in counties that applied for, but did not implement 287(g). The model follows the basic form:

$$\log\left(y_{jict}\left(y_{jict}^{2}+1\right)^{\frac{1}{2}}\right)=\delta D_{ct}+\beta \mathbf{X}_{ct}+\alpha_{c}+\alpha_{t}+\epsilon_{jict} \tag{14.1}$$

where y_{jict} is employment for racial group j at establishment i in county c at year t. In order to account for non-employment within race-establishment-county-year cells, we apply an inverse hyberpolic sine (IHS) transformation that is substantively equivalent to a log tranformation.[5] D_{ct} is an indicator of whether the county had implemented 287(g) by year t; $\mathbf{X}_{ct}$ is a vector of time-varying county-level controls; α_c and α_t are county and year fixed effects, respectively. Our interest is in δ, which is the difference-in-differences (DD) estimator of the impact of 287(g) on employment. Standard errors are clustered at the county level, which takes into consideration the potential correlation of employment shocks between establishments within a county, including shocks that may be correlated across years.

Results

Estimates from our DD models for all workers (i.e., both male and female) are shown in Table 14.1. Each cell represents an estimate of δ in Eq. (14.1), including county-level time-varying controls, and year and county fixed effects. The rows represent

[4]The results by 287(g) adoption timing and geographic location are not included in the paper, but are available upon request.

[5]Results from models using $\log(y)$ or $\log(y + 1)$ transformations produce point estimates and yield interpretations equivalent to those from the IHS transformation presented here.

Table 14.1 Effect of 287g on employment by race and industry

	All workers	Hispanics	NH-Whites	NH-Nonwhites
	(1)	(2)	(3)	(4)
Total employment	.0055	−.0518 ***	−.0331 ***	−.0125
	(.0058)	(.0163)	(.0122)	(.0172)
Agriculture and mining	−.0946	−.2600 *	−.0337	−.1070
	(.1140)	(.1420)	(.1380)	(.1470)
Utilities	.0464	−.0285	.0228	.0866
	(.0547)	(.1100)	(.0563)	(.0706)
Construction	.0716	−.0264	.0795 **	.0479
	(.0447)	(.0990)	(.0374)	(.0788)
Manufacturing	.0152	−.0642 **	−.0128	−.0003
	(.0127)	(.0255)	(.0187)	(.0248)
Wholesale and retail trade	.0113	−.0247	−.0395 **	−.0145
	(.0100)	(.0215)	(.0161)	(.0186)
Transportation and warehousing	.0174	.0416	−.0631	.0232
	(.0422)	(.0588)	(.0542)	(.0583)
Information	−.0168	−.1030 *	−.0413	−.0513
	(.0428)	(.0576)	(.0428)	(.0572)
Finance and insurance	.0269	−.0033	−.0135	.0064
	(.0298)	(.0482)	(.0341)	(.0387)
Real estate and rental and leasing	.0783	.0800	.0355	.2150 **
	(.0846)	(.1100)	(.0880)	(.1020)
Professional, scientific, and technical services	.0169	−.0806	−.0118	−.0157
	(.0250)	(.0489)	(.0262)	(.0478)
Management of companies and enterprises	.0837	.4390	.1090	.1410
	(.2750)	(.4210)	(.2400)	(.3570)
Administrative and support and waste management	−.0172	−.0578	−.1060 **	−.0357
	(.0457)	(.0621)	(.0459)	(.0766)
Educational services	.0096	−.0502	−.0409	−.0012
	(.0890)	(.1420)	(.0713)	(.1900)
Health care and social assistance	−.0036	−.0543 *	−.0421	−.0404
	(.0135)	(.0306)	(.0257)	(.0303)
Arts. Entertainment, and recreation	.0487	.0963	−.0871	.1710 *
	(.0426)	(.0930)	(.0596)	(.0885)
Accommodation and food services	−.0076	−.0739 *	−.0117	−.0244
	(.0182)	(.0375)	(.0240)	(.0397)
Other services (except public administration)	.0558	.0115	.0057	.0611
	(.0456)	(.0752)	(.0513)	(.0824)

Notes: Each estimate is obtained separately by estimating Eq. (14.1) on the specified racial group and industry. All models include county-level controls as well as county and year fixed effects. Standard errors are clustered at the county level and are in parentheses. *** $p < 0.01$, ** $p < 0.05$, * $p < 0.1$

separate estimates across industries, while racial/ethnic groups are organized in the columns.

The estimate in the top left – total employment (IHS) for all workers – indicates that 287(g) has no meaningful impact on overall firm employment, with a small and statistically nonsignificant estimate. Working down column (1), the vast majority of the estimates are small and imprecise, with none being statisticfally significant at conventional levels.

Estimates for Hispanic workers are displayed in column (2) and we see evidence of negative employment effects. Specifically, total Hispanic employment in EEO-1 reporting firms is estimated to be about 5.2% lower in counties that implemented 287 (g), relative to those in rejected counties. These impacts, however, are higly concentrated in a handful of industries. Hispanic employment in agriculture and mining is estimated to be a full 26.0% lower in 287(g) counties. While of a smaller scale, we also find that Hispanic employment in manufacturing is 9.1% lower in 287 (g) counties, which contrasts with Bohn and Santillano's (2017) finding that total (all racial group) manufacturing employment was higher in 287(g) areas. We also find measureable negative impacts of 287(g) on Hispanic employment in the Information sector (−10.3%), Health Care & Social Assistance (−5.4%), and Accomodation & Food Services industries (−7.4%). These industries not only have reasonably large concentration of Hispanic workers, but – with the exception of the Information sector – are also industires where unauthorized workers are more prevalent.

Employment estimates for non-Hispanic white workers are shown in column (3) and reveal a modest negative effect of 287(g) of about 3.3%. This is tentative evidence that 287(g) does not fulfill the goal of some local areas to reduce immigrant labor and replace it with native workers. Our estimates indicates that this negative effect is driven largely by white employment declines in Wholesale and Retail Trade (−4.0%), along with substantial declines in Administrative & Support & Waste Services (−10.6%). Reductions in wholesale and retail sectors are expected if the loss of Latino workers reduces the demand for goods and services they consume; to the extent this occurs, white workers appear to be adversely impacted.

Lastly, the final column (4) reports non-Hispanic nonwhite workers, with mostly null effects. The two noteworthy estimates are the unexpected positive effects of 287 (g) on the Real Estate & Rental & Leasing and Arts, Entertainment, & Recreation industries.

In Figs. 14.2 and 14.3, we show event studies for Hispanic (Fig. 14.2) and non-Hispanic white (Fig. 14.3) employment. These are fit equivalentally to Eq. (14.1), but with event-time dummies. For each set of figures, we show total employment pre and post 287(g) in the upper left, and industry-specific trends in the remaining graphs. Overall, the figures show no pre-trends, suggesting that the parallel trends assumption of the difference-in-differences estimator holds. For Hispanic workers, Fig. 14.2 shows that, compared to 287(g) denied counties, the implementation of 287(g) has a lasting and negative impact on total Hispanic

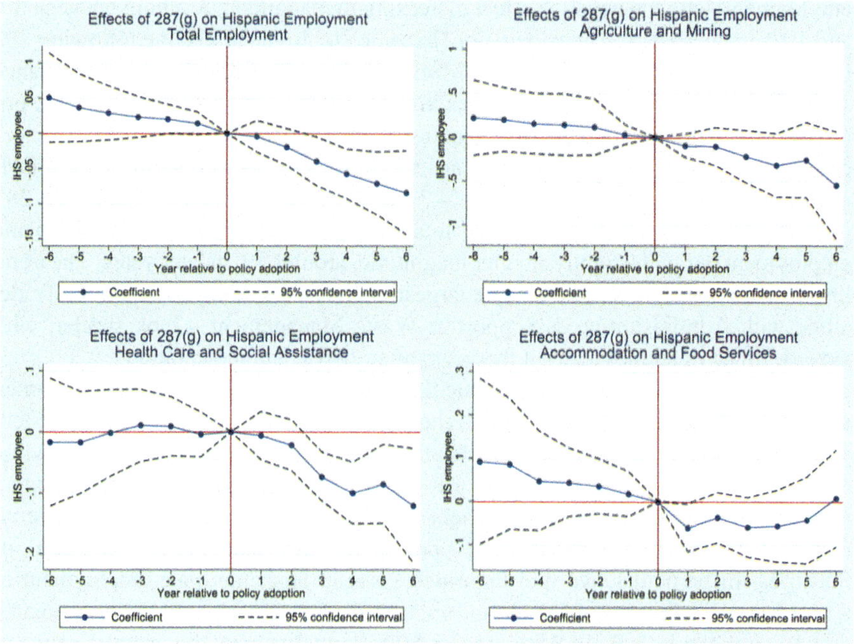

Fig. 14.2 Event Studies of 287(g) on Hispanic Employment, for Selected Industries

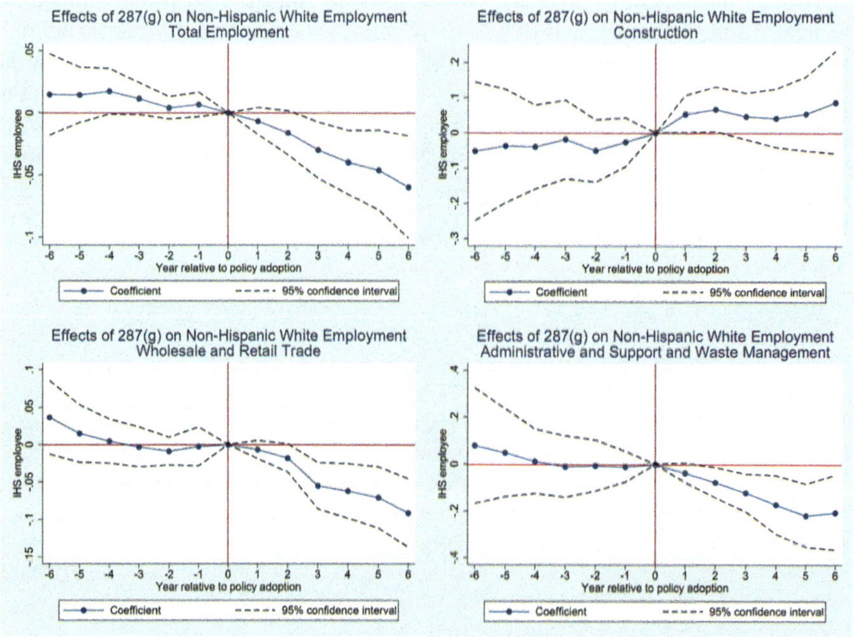

Fig. 14.3 Event Studies of 287(g) on Non-Hispanic White Employment, for Selected Industries

employment, with the negative effect increasing in magnitude, reaching between 5% and 10% in later years. The patterns of Hispanic employment decline following 287 (g) implementation in Agriculture & Mining and Health Care & Social Assistance sectors is similar, with gradual declines in the years following 287(g) adoption. For the Accomodation & Food Services sector, the employment losses appear to be more concentrated in the initial years, perhaps recovering by 6+ years following implementation. For white workers, Fig. 14.3 shows that the implementation of 287 (g) leads to a significant decrease in total employment, similar to total Hispanic employment but with slightly smaller magnitude, around 5% in later years. The trend lines for the two industries with the largest declines – Wholesale & Retail Trade, along with Administrative & Support & Waste Management – look similar, with slow declines in employment in the years after 287(g) implementation.

In Tables 14.2 and 14.3, we estimate the same set of models for men and women seperately. The DD estimates for men, shown in Table 14.2, are generally consistent with the models in Table 14.1. Specifically, the positive effects of 287(g) on white employment in the Construction sector is driven by male employment, registering a modestly large increase in employment (+7.9%). For Hispanic men, the overall employment losses are about 3.3%, but highly concentrated in a handful of inustiries. In particular, we find large decreases in Hispanic male employment in Agriculture & Mining (−29.0%); and modest declines in Manufacturing (−6.5%) and Accomodation & Food Services (−8.5%). For white men, the overall decline in employment is about 3.7%, which is particularly pronounced in Wholesale & Retail Trade, and Administrative Support. Lastly, for non-Hispanic nonwhite men, we find no evidence of significant employment effects from 287(g).

Results for female workers are shown in Table 14.3. For Hispanic women, we generally find smaller impacts of 287(g), which is expected given that deportation is largely concentrated among men. We do, however, find substantial declines in the Infomration sector and modest declines in Health Care & Social Assistance. For white women, our models indicates modest employment losses associated with 287 (g) with an overall effect of about −3.2%. These losses appear to be driven by reductions in employment among white women in Wholesale & Retail Trade and Health Care & Social Assistance (−3.0% and −4.4%, respecitively).

Table 14.2 Effect of 287g on male employment by race and industry

	All workers	Hispanics	NH-Whites	NH-Nonwhites
	(1)	(2)	(3)	(4)
Total employment	−.0005	−.0325 **	−.0368 ***	−.0076
	(.0063)	(.0130)	(.0124)	(.0165)
Agriculture and mining	−.1330	−.2900 **	−.0795	−.1620
	(.1240)	(.1380)	(.1460)	(.1960)
Utilities	.0824	.0290	.0605	.1290
	(.0561)	(.1060)	(.0590)	(.0811)
Construction	.0720	−.0246	.0786 *	.0307
	(.0526)	(.1060)	(.0427)	(.0876)
Manufacturing	.0148	−.0649 ***	−.0095	.0090
	(.0121)	(.0247)	(.0168)	(.0243)
Wholesale and retail trade	−.0028	−.0066	−.0554 ***	−.0190
	(.0105)	(.0209)	(.0191)	(.0177)
Transportation and warehousing	.0196	.0491	−.0529	.0392
	(.0462)	(.0580)	(.0553)	(.0624)
Information	−.0110	−.0884	−.0263	−.0383
	(.0441)	(.0612)	(.0446)	(.0601)
Finance and insurance	.0190	.0087	−.0175	−.0223
	(.0317)	(.0490)	(.0359)	(.0417)
Real estate and rental and leasing	.1060	.1550	.0573	.3260
	(.0848)	(.1240)	(.0822)	(.1400)
Professional, scientific, and technical services	.0308	−.0162	−.0019	.0165
	(.0273)	(.0459)	(.0304)	(.0496)
Management of companies and enterprises	.0045	.1740	.0405	.2040
	(.2910)	(.3970)	(.2550)	(.3950)
Administrative and support and waste management	−.0351	−.0259	−.1080 **	−.0609
	(.0455)	(.0679)	(.0463)	(.0747)
Educational services	−.0149	.0004	−.0427	−.0233
	(.1550)	(.1540)	(.1510)	(.1980)
Health care and social assistance	−.0033	−.0110	−.0312	−.0227
	(.0158)	(.0242)	(.0257)	(.0314)
Arts. Entertainment, and recreation	.0618	.1450	−.0711	.1340
	(.0430)	(.0961)	(.0577)	(.0935)
Accommodation and food services	−.0239	−.0850 **	−.0219	−.0180
	(.0187)	(.0412)	(.0274)	(.0456)
Other services (except public administration)	.0764	.0294	.0118	.0999
	(.0509)	(.0736)	(.0623)	(.0755)

Notes: Each estimate is obtained separately by estimating Eq. (14.1) on the specified racial group and industry. All models include county-level controls as well as county and year fixed effects. Standard errors are clustered at the county level and are in parentheses. *** $p < 0.01$, ** $p < 0.05$, * $p < 0.1$

Table 14.3 Effect of 287g on female employment by race and industry

	All workers	Hispanics	NH-Whites	NH-Nonwhites
	(1)	(2)	(3)	(4)
Total employment	.0133 *	−.0221	−.0315 **	−.0049
	(.0068)	(.0136)	(.0129)	(.0175)
Agriculture and mining	.0917	−.2160	.1650	.0065
	(.1010)	(.1640)	(.1200)	(.1260)
Utilities	−.0440	.0177	−.0753	.0432
	(.0666)	(.0966)	(.0584)	(.1000)
Construction	0.0763*	.1020	.0627	.0780
	(.0436)	(.0628)	(.0490)	(.0885)
Manufacturing	.0216	.0037	−.0256	.0131
	(.0182)	(.0263)	(.0271)	(.0266)
Wholesale and retail trade	.0239 **	.0027	−.0297 **	−.0009
	(.0120)	(.0199)	(.0139)	(.0228)
Transportation and warehousing	−.0151	−.0063	−.1240	.0126
	(.0580)	(.0549)	(.0748)	(.0714)
Information	−.0184	−.1620 ***	−.0349	−.0425
	(.0521)	(.0583)	(.0530)	(.0726)
Finance and insurance	.0238	−.0120	−.0174	.0030
	(.0306)	(.0480)	(.0342)	(.0421)
Real estate and rental and leasing	.0076	.0305	−.0089	.0005
	(.1160)	(.1380)	(.1130)	(.1490)
Professional, scientific, and technical services	.0005	−.0593	−.0263	−.0366
	(.0322)	(.0576)	(.0336)	(.0554)
Management of companies and enterprises	.1960	.6520	.1810	.0940
	(.2510)	(.3880)	(.2290)	(.3250)
Administrative and support and waste management	.0023	−.0273	−.1050	.0090
	(.0714)	(.0764)	(.0746)	(.0980)
Educational services	−.0077	−.1490	−.0370	−.1080
	(.0773)	(.1900)	(.0629)	(.2080)
Health care and social assistance	−.0038	−.0548 *	−.0442 *	−.0425
	(.0140)	(.0313)	(.0266)	(.0312)
Arts. Entertainment, and recreation	.0441	.0518	−.0886	.1800 *
	(.0506)	(.0805)	(.0673)	(.0939)
Accommodation and food services	.0014	−.0524	−.0105	−.0270
	(.0225)	(.0478)	(.0270)	(.0424)
Other services (except public administration)	.0084	−.0583	.0040	−.0181
	(.0496)	(.0757)	(.0465)	(.0965)

Notes: Each estimate is obtained separately by estimating Eq. (14.1) on the specified racial group and industry. All models include county-level controls as well as county and year fixed effects. Standard errors are clustered at the county level and are in parentheses. *** $p < 0.01$. ** $p < 0.05$, * $p < 0.1$

Conclusion

In this chapter, we examine the effect of 287(g) on employment in large firms, seperately for Hispanic and white workers. Using the data on county-level 287 (g) applications and establishment-level employee data from the Equal Employment Opportunity Commission, we employ a difference-in-differences identification strategy, comparing employment outcomes of establishments in 287(g) implemented counties to those in 287(g) denied counties, before and after the implementation of 287(g).

Overall, we find that 287(g) has significant and negative impacts on both Hispanic and white employment. The effects are concentrated in a few selected industries; for Hispanic workers, agriculture and mining, health care/social services, and accommodation/food services suffer the largest negative impacts, while for white workers, the adverse employment effects are mostly concentrated in wholesale and retail trade, as well as in administrative support/waste management and the health sectors. We also find that 287(g) differentially affects Hispanic and white workers by gender. For Hispanic workers, most of the effects are driven by male employment, consistent with the fact that deportation in the interior U.S. disproportionately impacts Hispanic men more than Hispanic women. For white workers, on the contrary, both male and female employment are negatively and statisticlly significantly affected by the implementation of 287(g).

This study provides new evidence of the impact of local immigration enforcement policies in the U.S., such as 287(g), that intend to restrict and reduce undocumented population in order to leave more employment opportunities to natives. Yet, the results show that the policy, like many other policies, brings unintended consequence to the group that it intends to help. Moreover, the current study provides insights to the heated debate on the effect of immigrants on natives. Immigirants, both documented and undocumented, are not only labor suppliers who might compete with natives in the labor market, but also consumers who demand goods and services. Therefore, when assessing immigration policies, it is cruicial to take into consideration both the supply side and the demand side effects of immigrants.

Finally, we recognize that the current study has several limitations. First, due to the nature of our outcome data, our results only speak to the employment impacts on large establishments, knowing that smaller establishments might have different employment trends and, thus, are affected by the policy differently. Second, our data do not allow us to separately assess the impact on citizens and non-citizens, which could limit the generalizability of the results in the debate on the effect of immigrants on natives. Future reseach can fill the gap by focusing on smaller-sized establishments and explore whether local immigration policies have different impacts on smaller establishments.

References

American Immigration Council. (2017). The 287(g) Program: An Overview.

Beniflah, J. D., Little, W. K., Simon, H. K., & Sturm, J. (2013). Effects of immigration enforcement legislation on Hispanic pediatric patient visits to the pediatric emergency department. *Clinical Pediatrics, 52*, 1122–1126.

Bohn, S., & Santillano, R. (2017). Local immigration enforcement and local economies. *Industrial Relations, 56*, 236–262.

Capps, R., Rosenblum, M., Rodriguez, C., & Chishti, M. (2011). *Delegation and divergence: A study of 287(g) state and local immigration enforcement*. Washington, DC: Migration Policy Institute.

Card, D. (2005). Is the new immigration really so bad? *Economic Journal, 115*, 300–323.

Coleman, M. (2009). What counts as the politics and practice of security, and where? Devolution and immigrant insecurity after 9/11. *Annals of the Association of American Geographers, 99*, 904–913.

Coleman, M. (2012). The "local" migration state: The site-specific devolution of immigration enforcement in the US south. *Law & Policy, 34*, 159–190.

Dee, T. S., & Murphy, M. (2019). Vanished classmates: The effects of local immigration enforcement on school enrollment. *American Educational Research Journal*. https://doi.org/10.3102/0002831219860816.

Department of Homeland Security. (2010). *The performance of 287(g) agreements*. Washington, DC: Office of Inspector General.

East, Chloe N., Philip Luck, Hani Mansour, and Andrea Velasquez. 2018. "The labor market effects of immigration enforcement." IZA discussion paper series no. 11486.

Ferguson, J.-P., & Koning, R. (2018). Firm turnover and the return of racial establishment segregation. *American Sociological Review, 83*, 445–474.

Gonzalez-Barrera, A., & Krogstad, J. M. (2014). *U.S. deportations of immigrants reach record high in 2013*. FactTank, Pew Research.

Hall, M., Greenman, E., & Farkas, G. (2010). Legal status and wage disparities for Mexican immigrants. *Social Forces, 89*, 491–513.

Hall, M., Greenman, E., & Yi, Y. (2019). Job mobility among unauthorized immigrant workers. *Social Forces, 97*, 999–1028.

Kostandini, G., Mykerezi, E., & Escalante, C. (2014). The impact of immigration enforcement on the US farming sector. *American Journal of Agricultural Economics, 96*, 172–192.

Massey, D. S., Durand, J., & Pren, K. A. (2016). Why border enforcement backfired. *American Journal of Sociology, 121*, 1557–1600.

Meissner, D., Kerwin, D. M., Chishti, M., & Bergeron, C. (2013). *Immigration enforcement in the United States: The rise of a formidable machinery*. Migration Policy Institute.

Moinester, M. (2019). A look to the interior: Trends in U.S. immigration removals by criminal conviction type, gender, and region of origin, fiscal years 2003-2015. *American Behavioral Scientist, 63*, 1276–1298.

National Academies of Sciences, Engineering, and Medicine. (2017). *The economic and fiscal consequences of immigration*. Washington, DC: The National Academies Press. https://doi.org/10.17226/23550.

O'Neil, K. (2010). Hazleton and beyond: Why communities try to restrict immigration. *Migration Information Source, 15*.

Ottaviano, G. I. P., & Peri, G. (2012). Rethinking the effect of immigration on wages. *Journal of the European Economic Association, 10*, 152–197.

Passel, J. S., & Cohn, D. V. (2019). *Mexicans decline to less than half the US unauthorized immigrant population for the first time*. Pew Research: FactTank.

Pedroza, J. M. (2019). Where immigration enforcement agreements stalled: The location of local 287(g) program applications and inquires (2005–2012). *SocAr, Xiv*(January 4). https://doi.org/10.31235/osf.io/2a8kz.

Pham, H. (2018). 287(g) Agreements in the Trump Era, 75 Wash. & Lee L. Rev. 1253.
Pham, H., & Van, P. H. (2010). The economic impact of local immigration regulation: An empirical analysis. *Cardozo Law Review, 32*, 485–518.
Potochnick, S., Chen, J. H., & Perreira, K. (2017). Local-level immigration enforcement and food insecurity risk among Hispanic immigrant families with children: National-level evidence. *Journal of Immigrant and Minority Health, 19*, 1042–1049.
Rhodes, S. D., Mann, L., Simán, F. M., Song, E., Alonzo, J., Downs, M., Lawlor, E., Martinez, O., Sun, C. J., O'Brien, M. C., Reboussin, B. A., & Hall, M. A. (2015). The impact of local immigration enforcement policies on the health of immigrant Hispanics/Latinos in the United States. *American Journal of Public Health, 105*, 329–337.
Robinson, C. L., Taylor, T., Tomaskovic-Devey, D., Zimmer, C., & Irvin, M. W., Jr. (2005). Studying race or ethnic and sex segregation at the establishment level: Methodological issues and substantive opportunities using EEO-1 reports. *Work and Occupations, 32*, 5–38.
Rugh, J. S., & Hall, M. (2016). Deporting the American dream: Immigration enforcement and Latino foreclosures. *Sociological Science, 3*, 1053–1076.
Stainback, K., & Tomaskovic-Devey, D. (2012). *Documenting desegregation: Racial and gender segregation in private sector employment since the civil rights act.* Russell Sage Foundation.
Steil, J. P., & Vasi, I. B. (2014). The new immigration contestation: Social movements and local immigration policy making in the United States, 2000–2011. *American Journal of Sociology, 119*, 1104–1155.
Stuesse, A., & Coleman, M. (2014). Automobility, immobility, altermobility: Surviving and resisting the intensification of immigrant policing. *City & Society, 26*, 51–72.
Varsanyi, M. (2010). *Taking local control: Immigration policy activism in US cities and states* (Vol. 5). Stanford: Stanford University Press.
Wang, J. S.-H., & Kaushal, N. (2018). Health and mental health effects of local immigration enforcement. *International Migration Review*. https://doi.org/10.1177/0197918318791978.
Watson, T. (2014). Inside the refrigerator: Immigration enforcement and chilling effects in Medicaid participation. *American Economic Journal: Economic Policy, 6*, 313–338.

Chapter 15
Experience of Domestic Violence by Young Women in India: Does the Nature of Occupation Play Any Role?

Akanksha Choudhary and Ashish Singh

Introduction

> Dhol Ganwar Shudra Pashu Nari Sakal Tadna Ke Adhikari. . . . .
>
> "Goswami Tulsidas" in "Sunderkand" in "Ramcharitmanas". (Awadhi version of "Ramayana" [7600 BCE] quoted from (Tapasyananda 1983).

The above verse which is from "Ramcharitmanas" (completed about 1633 CE; a re-telling of Sanskrit "Ramayana") means that: Dhol (a kind of musical instrument), Illiterates, Shudras (the lowest in the Hindu "Varna" system; untouchables etc.), Animals and Women – all deserve to be tortured.

Gender discrimination and the inferior status accorded to women compared to males in India is not a recent phenomenon and has been investigated comprehensively by both Indian as well as international scholars. It can be traced back to the ancient times, for example, to the times of "Manusmriti", one of the most widely studied ancient text among many "Dharmaśāstras" (Religious rule books) of Hinduism and which dates back to at least second century BC. "Manusmriti" mentions that "as a girl, a woman should obey and seek protection of her father, as a young woman her husband, and as a widow her son; and also woman should always worship her husband as a god" (verses, 2.67–2.69 and 5.148–5.155, Olivelle and Olivelle 2005). In one of the verses it also mentions that "a woman must never seek to live independently" (Olivelle and Olivelle 2005). Existence of phenomenon such as "Pati-Parmeshwar (husband is GOD)" not only promotes but also sustains the subordination of women to males (Jaising and Wolfe 1988).

A. Choudhary
School of Management and Entrepreneurship, Indian Institute of Technology Jodhpur, Jodhpur, India

A. Singh (✉)
SJM School of Management, Indian Institute of Technology Bombay, Mumbai, India

B. Jivetti, Md. N. Hoque (eds.), *Population Change and Public Policy*, Applied Demography Series 11, https://doi.org/10.1007/978-3-030-57069-9_15

That said a common consequence of the subordination of women to men is the extensive domestic violence which is faced by the married women on the hands of their husbands' in India (ICRW 2000). Here domestic violence refers to the injurious physical assault, coercive and controlling behavior causing psychological, sexual or physical harm, which often accompany or precede the use or threat of physical force (Barnish 2004). A lot of studies have empirically showed the existence and extensive nature of domestic violence in India (IIPS and ORC Macro 2007). According to ICRW (2000), about 52% of women in India undergo physical, mental or sexual abuse by their husbands at least once in their life.

Domestic violence is not just limited to India, but it is a worldwide phenomenon. Among the developed countries, about 30% of women have reported to have experienced domestic violence in the United States (Straus and Gelles 1986). Among the developing countries – about 42% women have reported of being beaten up by their husbands in Kenya (Raikes 1990). Also, in China which is often compared to India because of the nature of economic growth, about 57% of women experience domestic violence at least once in their married life (Xu 1997).

The negative consequences of domestic violence experienced by women have been well-documented in existing literature. Domestic violence not only hampers them physically and socially, but the countries in which women suffer from domestic violence also have to pay large socio-economic costs associated with it (WHO 2002; Carrillo 1992; Heise et al. 1994, Menon-Sen and Kumar 2001; Morrison and Orlando 1999; Stephenson et al. 2006; Straus et al. 1980). For example, a study in Canada showed that the socio-economic cost associated with domestic violence in the country in the year 1993 was more than $3.2 billion (Greaves 1995). Also, the socio-economic costs of domestic violence in Chile and Nicaragua were more than 2% of their respective GDPs in the year 1996 (Morrison and Orlando 1999). Considering the large amount of socio-economic cost which countries pay because of the existence of domestic violence it becomes an important issue which needs to be investigated in detail in a developing country like India.

One of the remedies often suggested in general by literature for domestic violence experienced by married women is – women empowerment through their employment (Panda and Agarwal 2005; Bhattacharyya et al., 2011). However, the extant literature on the subject shows the existence of two types of theories related to the employment and domestic violence experienced by women. One theory, also known as "male backlash theory" and was developed by Macmillan and Gartner (1999), mentions that the chances of experiencing domestic violence increase with the increase in employment of women. This theory explains that when the gap in wages of husband and wife reduces, husband tends to lose control in the house and to maintain the earlier status quo, he uses domestic violence as a tool. Also, in some cases self-sustainability of a female lashes the male ego of her husband which prompts him to go for domestic violence (Naved and Persson 2005). The other theory says that – as women get employed they are less likely to face domestic violence in their entire life span because their contribution in the family income reduces the financial stress and thus creating happy environment in family (Panda and Agarwal 2005; Gibson-Davis et al. 2005; Bhattacharyya et al. 2011).

Since the evidence relating employment status to domestic violence is conflicting and perhaps there is no study (to the best of our search) in the Indian context which has investigated the relationship between the nature/type of occupation (and not just employed/unemployed category) of women and domestic violence experienced by them, we in this paper have examined how the nature of occupation is related to domestic violence faced by women in India.[1] First, we have conducted the analysis at a pooled level for all the states covered in the survey and at the second level the analysis has been conducted for different states separately. The results obtained from the analysis show that the nature of occupation of women significantly affects their life as far as experience of domestic violence is concerned.

The Evidence So Far[1]

Domestic violence has always been considered as one of the evils of the societies across world; therefore, various researchers have put in efforts to reach to the bedrock of this phenomenon. However, only a limited amount of research in general (and in Indian context) has been conducted on studying the relationship between women's employment and domestic violence experienced by them in general. And a miniscule amount of research has been conducted to examine the nature of occupation and its impact on domestic violence.

Kim et al. (2007) proclaimed that economic empowerment leads to reduction of domestic violence among women in rural South Africa. On contrary Bowlus and Seitz (2006) asserted that women in Canada are exposed to more domestic violence when they are employed. Existing studies on India as well as other countries have mostly looked into effects of employment in a dichotomous form (employed vs. unemployed) on domestic violence. To the best of our search we could not find any study in the Indian context which has looked into how being in different occupations such as being in socially or economically lower occupations like agricultural and non-agricultural laborer vs. being in socially higher occupations like administrative or managerial jobs differently affects domestic violence faced by the ever married women.

Paul (2014, p.1) mentions that "Violence against women is a manifestation of historically unequal power relations between men and women, which have led to domination over and discrimination against women by men and to the prevention of the full advancement of women". Burns (2013) writing, "A woman stays in an abusing relationship because the fear of leaving is greater than the fear of staying; she will leave when the fear of staying is greater than the fear of leaving" is a hint towards that women who stay in abusing relationship mostly do so because of lack of financial independence.

[1]Details presented about existing relevant studies in the section on "The Evidence So Far".

A report on domestic violence submitted to Planning Commission, has some interesting findings (YES 2003). The causes of domestic violence as per the findings of the report are – parents' failure to pay full amount of dowry (11%), refusal to bring money from patriarchal family (13%), parents' failure to keep promises and commitments made at the time of marriage (8%), pre-marital love affairs suspected (9%), extra marital relations (10%), resistance to sexual abuse (4%), medically unfit for conception (7%), female-child delivered every time (7%), husband's unemployment (13%), husband problems at work place (19%) and husband's alcoholism (25%). Also, the study highlights that around 26% of the respondents resisted, 25% each fought back and had a feeling of helplessness. This also explains the feeling of helplessness for those women who fought back but could not improve their condition because those women were financially not independent (YES 2003).

Rao (1997), a qualitative study concludes that the chances of experiencing spousal violence can be reduced by increasing women labor force participation. This study has a limitation of scope both in terms of scale (it is restricted only to the Indian state of Karnataka) as well as generalization.[2] Further, the study is silent on nature of relationship between type of employment and domestic violence. That is what we have attempted in this paper using a sub-nationally representative data.

Data and Methods

Data

We have used data from the "Youth in India: Situations and needs" survey. The survey was conducted in the year 2006–07 by the International Institute for Population Sciences (IIPS), Mumbai (India) and the Population Council, New Delhi (India) under the stewardship of Ministry of Health and Family Welfare, Government of India (IIPS and Population Council, 2010).

The Youth in India: Situation and Needs study (also referred as the Youth Study), is the first-ever sub nationally representative study conducted to understand the conditions and needs of the young people in India. Its objectives were: to identify key transitions experienced by youth, including those pertaining to education, work force participation, sexual activity, marriage, health and civic participation; to provide state-level evidence on the magnitude and patterns of young people's practices, decision-making and attitudes; and, finally, to identify key factors underlying, attitudes and life choices (Choudhary and Singh 2017; IIPS and Population Council, 2010; Singh et al. 2014).

[2]Among other studies, the ecological framework proposed by Heise (1998) recognized the possible role of individual, familial, societal/community factors in placing women at the risk of violence within marriage.

The Youth Study focused on married and unmarried young women and unmarried young men aged 15–24 in both rural and urban settings. Married men in the age group 15–29 were also interviewed in the survey. The study was conducted in the Indian states of Andhra Pradesh, Bihar, Jharkhand, Maharashtra, Rajasthan and Tamil Nadu; these states together represent 39% of the country's population. These six states were purposively selected to represent the different geographic and socio-cultural regions within the country (IIPS and Population Council, 2010; Choudhary and Singh 2017; Singh et al. 2014).

The sample in rural areas was selected in two stages. Villages were selected at first stage using probability proportional to size (PPS) sampling scheme. Households were selected at the second stage using systematic sampling scheme. In urban areas, the sample was selected in three stages. At the first stage of selection, wards were selected systematically using PPS sampling Scheme. At the second stage, census enumeration blocks (CEB) containing approximately 150–200 households were selected using PPS sampling scheme. Households were selected at the third stage using systematic sampling scheme. A total of 50,848 married and unmarried young women and men were successfully interviewed. Response rates for individual interviews were in the range of 84–90%. Unmarried women registered the highest response rate (90%), (Choudhary and Singh 2017; IIPS and Population Council 2010; Singh et al. 2014).

For this study purpose, we have restricted our analysis to married women aged 15–24 years only. Appropriate sampling weights are used to derive the estimates at sub-national and state levels. The details of the sampling weight are given in the survey report (IIPS and Population Council, 2010).

Outcome of Interest: Measure of Gender Based Domestic Violence

The below mentioned criteria has been used to identify whether an ever married female has faced domestic violence: she is a victim of domestic violence if her husband has done any one of the following to her – slapped her; twisted her arm or pulled her hair; pushed her, shook her or threw something at her; punched her with his fist or with something that could hurt her; kicked her, dragged her or beat her up; tried to choke her or burn her on purpose; or threatened or attacked her with a knife, gun or any other weapon. This identification has been adopted from the survey report (IIPS and Population Council, 2010) and is also in line with Singh et al. (2013).

Main Independent and Control Variables

Nature of the occupation of the ever married women is the independent variable of interest. The "Youth Study" captures occupation of the women using the following categorization: not working, house wife/ unemployed (1); cultivator/farmer (2); agricultural laborer (3); non-agricultural laborer (4); administrative/executive /managerial and related (5); skilled workers (manual, machinery) (6); clerical (7); and other workers (8). The above categorization is closely related to social classes or status groups, while also taking into account the nature of the tasks the women perform. It is also broadly in line with the Indian National Classification of Occupations and the classification of occupations proposed in Motiram and Singh (2012) in the Indian context. Drawing from the published literature on determinants of domestic violence (Bedi et al. 2009; Caesar 1988; Heise 1998; Hotaling and Sugarman 1986; Jejeebhoy 1998; Sugarman and Hotaling 1989; Kalmuss 1984; Singh and Singh 2013; Straus and Gelles 1990) and other relevant factors in the Indian context, we have included the following control variables in the regression based analysis.[2]

(i) Individual level variables:

   (a) Education (categorical) – the survey captures education for the women using the following categorization: no formal schooling (1); 1–4 years of schooling (schooling less than primary) (2); 5–7 years of schooling (completed primary but less than middle school) (3); 8–9 years of schooling (completed middle school but less than secondary) (4); 10–11 years of schooling (completed secondary but less than higher secondary) (5); 12 and above years of schooling (higher secondary and above) (6). The above categorization is based on the Indian schooling system where primary, middle school, secondary, higher secondary (also called intermediate) and above are important milestones of educational attainment (Choudhary and Singh, 2017).
   (b) Duration of marriage in years – we also considered inclusion of age of the women in the analysis but the correlation between "age in years" and "duration of marriage in years" was substantially high (0.6) and significant at 0.1% level of significance; given that the women included in the analysis are 15–24 years old which is not a wide range and "duration of marriage in years" is a more relevant variable in the present context, we decided to include it and drop "age in years" from the regression based analysis; and.
   (c) Choice of marriage (categorical) – arranged marriage or love marriage.

(ii) Household/familial level variables:

   (a) Occupational status of the father of the female (categorical) – the survey captures the occupational status of fathers as: not working/ unemployed/ retired (1); cultivator (2); Agricultural laborer (3); non-agricultural laborer (4); administrative/ executive and managerial (5); skilled (manual, machinery) (6); clerical (7); and other workers (8); the reasons for the above categorization have already been explained while describing the occupation status of the women; in the Indian society, the social and economic status of a family is mostly linked to the occupation of the male household head[3], therefore in order to capture the status of women's natal family we have included occupational status of the father of the women in the analysis.

(b) Whether mother has experienced domestic violence (categorical – yes or no, the identification criteria for domestic violence being same as that used to capture the outcome variable) – we included this variable because there are a few studies which have found linkage between domestic violence experienced by daughters and domestic violence experienced by their mothers (Caesar 1988; Ellsberg et al. 1999; Hotaling and Sugarman 1986; Kalmuss 1984; Singh and Singh 2013; Straus and Gelles 1990; Sugarman and Hotaling 1989).
(c) Occupational status of the husband (categorical) – the survey captures the occupational status of husbands as: not working/ unemployed/retired (1); cultivator (2); Agricultural laborer (3); non-agricultural laborer (4); administrative/ executive and managerial (5); skilled (manual, machinery) (6); clerical (7); and other workers (8).
(d) Caste (categorical – scheduled castes and tribes [SC/ST], other backward castes [OBC] and Others) – the above categorization is a meaningful representation of the Indian social fabric along caste lines (Singh et al. 2013); scheduled tribes, scheduled castes and other backward castes are historically most disadvantaged and most discriminated against groups in India respectively; and.
(e) Religion – religion has been divided into three categories, namely "Hindu" (the majority religious group), "Muslim" (largest group among religious minorities) and "Others".

Before moving forward to community/societal level variables, it is important to discuss why we have not included variables like husband's education and wealth status of the household in the list of household level variables. That is because there is a very high degree of correlation (0.46; significant at 1% level of significance) between education of the women and education of their husbands. At the same time, the correlation between occupation of women and occupation of their husbands is very low (0.05). The same is true for education of women and occupation of their husbands. Therefore, we have included husband's occupation in the list but not husband's education. The reason for not including wealth status of the household is similar, the degree of correlation between education of women and wealth status of her household is again extremely high and significant, 0.55 and significant at 1% level of significance. It might be noted that in India majority of marriages take place between families (bride's and groom's) of similar stature (by wealth, occupation of household head etc.) and at the same time income in India is highly correlated with education and occupation. If we see the above two observations together, then we can see that not including husband's education and household wealth status in the analysis is not likely to cause any serious bias in the results because husbands' education is highly correlated with their wives' education which is included in the analysis and at the same time husbands' occupation is also included in the analysis; so the education of women and their husbands' occupation, together will be able to control for wealth of the household.

(iii) Community/societal level variables – rural/urban and state dummies for the pooled sample analysis.

Regression Strategy

We have used logistic regression to model the relationship between the occupational status of the ever-married young women and domestic violence experienced by them. Domestic violence faced by the women is the outcome of interest with nature of occupation as the main independent variable of interest along with some control variables as discussed in the previous sub-section. There is always a possibility of a high degree of correlation between a female's education and her occupation as well as her education and occupation with her father's occupational status. We checked for the above correlations and found insignificant levels of these correlations: (i) women's education and occupation – 0.11; (ii) women's education and fathers' occupation – 0.17 and; (iii) women's occupation and fathers' occupation – 0.01. We also performed tests for multicollinearity (Variance Inflation Factor [VIF] test as well as Condition Index test proposed by Belsley (1991)), but found no evidence of any substantial or significant multicollinearity.[4]

Results

Descriptive Statistics

Table 15.1 presents the composition of the sample by various individual, household and community level variables. The table presents the above for the All-India level as well as for the various states covered in the survey.

Sample Composition by Individual, Household and Community Level Factors: All India

Some stylized findings from Table 15.1 are as follows: (1) 35.64% of women have no formal schooling and only about 9% women have 12 or more years of formal schooling; (2) about 70% of women are either housewives or unemployed, whereas, less than 1% of the women are in administrative, managerial or related jobs and around 23% women are either agricultural or non-agricultural laborers; (3) nearly 94% of women have opted for arranged marriage and the remaining have gone for love marriage; (4) about 70% of women have completed 0–5 years in marriage; (5) about 42% of respondent's husbands are involved in agricultural or non-agricultural labor activities and around 15% are cultivators; (6) about 27% of respondent's mothers have experienced domestic violence; (7) about 30, 24 and 18% of fathers are cultivators, agricultural laborers and non-agricultural laborers, respectively, whereas only about 7% of fathers are either in administrative or managerial and related jobs; (8) about 27, 52 and 21% of the women are from SC/STs, OBCs

Table 15.1 Distribution of socioeconomic characteristics in the sample

	Percentage composition						
	All India	Andhra Pradesh	Bihar	Jharkhand	Maharashtra	Rajasthan	Tamil Nadu
Education							
Non-literate, literate with no formal schooling	35.64	29.81	60.86	47.24	13.72	50.02	8.43
1–4 years of schooling	8.27	10.90	8.01	12.14	6.00	7.84	6.33
5–7 years of schooling	19.42	25.69	13.36	16.02	21.47	17.50	20.42
8–9 years of schooling	16.54	11.89	6.85	15.32	27.77	13.02	29.26
10–11 years of schooling	11.08	13.96	7.52	5.61	15.60	5.08	16.90
12 and above years of schooling	9.04	7.75	3.39	3.68	15.44	6.54	18.65
Occupation							
Not working, house wife/student/unemployed	70.37	56.36	73.68	73.53	70.31	80.36	75.94
Cultivator	1.50	1.33	0.32	0.85	4.24	0.90	12.94
Agricultural laborer	19.71	30.62	21.43	18.25	18.88	8.19	3.43
Laborer (excluding agricultural)	3.02	3.65	1.22	3.59	2.36	4.98	1.27
Administrative/ executive and managerial	0.90	1.30	0.41	0.91	0.78	1.06	5.99
Skilled (manual, machinery)	4.10	6.59	2.71	2.68	2.46	4.22	0.43
Clerical	0.33	0.15	0.12	0.06	0.88	0.23	0.00
Other worker	0.07	0.00	0.11	0.12	0.09	0.08	0.00
Choice of marriage							
Arrange marriage	94.77	93.41	98.87	93.15	95.28	99.35	82.51
Love marriage	5.23	6.59	1.13	6.85	4.72	0.65	17.49
Years of marriage							
less than 1 year	10.32	9.69	10.38	7.47	12.11	8.18	12.98
1 years	12.90	11.14	10.06	11.85	14.33	14.58	17.63
2 years	14.03	12.67	11.66	11.80	16.46	12.99	19.67
3 years	12.68	12.13	11.69	12.99	14.25	12.12	13.40
4 years	11.54	12.13	9.58	10.77	12.36	12.91	11.36
5 years	10.85	11.70	11.10	11.63	10.56	10.78	8.67
6 years	9.47	10.23	11.07	10.03	7.36	10.45	7.17
7 years	6.66	7.35	8.17	7.07	4.70	7.05	5.07
8 years	5.30	6.30	6.36	6.36	3.88	6.22	1.98

(continued)

Table 15.1 (continued)

	Percentage composition						
	All India	Andhra Pradesh	Bihar	Jharkhand	Maharashtra	Rajasthan	Tamil Nadu
9 years	3.30	3.54	4.94	5.50	1.82	3.01	1.43
10 years	2.10	2.19	3.52	2.78	1.45	1.55	0.45
11 years	0.71	0.82	1.35	1.41	0.43	0.11	0.19
12 years	0.10	0.10	0.11	0.15	0.22	0.00	0.00
13 years	0.00	0.00	0.00	0.06	0.00	0.00	0.00
14 years	0.03	0.00	0.00	0.12	0.06	0.05	0.00
Husband's occupation							
Not working/ unemployed	2.07	0.24	2.49	3.75	1.93	4.53	0.45
Cultivator	15.29	15.11	13.53	17.78	13.99	24.04	5.15
Agricultural laborer	18.14	30.49	13.52	8.85	21.36	5.4	19.6
Laborer (excluding ag. laborer)	24.61	16.94	34.19	34.66	20.87	26.18	20.3
Administrative/ executive and managerial	10.21	7.58	8.33	8.2	13.31	12.93	11.66
Skilled (manual, machinery)	25.99	26.38	25.85	23.65	21.01	24.27	39.76
Clerical	2.63	2.51	1.6	1.27	5.31	1.87	2.26
Other worker	1.06	0.76	0.48	1.84	2.22	0.78	0.82
Mother experienced domestic violence							
Yes	27.21	39.61	22.40	24.55	23.11	18.24	34.61
No	72.79	60.39	77.60	75.45	76.89	81.76	65.39
Father's occupation							
Not working, house husband/ unemployed/ retired	0.59	0.57	0.52	0.70	0.72	0.33	0.83
Cultivator	30.78	30.49	29.17	29.74	30.43	45.31	12.65
Agricultural laborer	23.89	34.06	21.52	10.81	25.41	7.60	35.41
Laborer (excluding agricultural)	17.66	14.71	20.64	22.97	17.33	17.34	16.61
Administrative/ executive and managerial	7.71	4.89	8.06	8.87	8.01	9.53	9.86
Skilled (manual, machinery)	15.90	12.72	17.99	20.92	12.84	15.79	21.79
Clerical	2.77	2.11	1.77	3.49	3.99	3.65	2.48
Other worker	0.71	0.44	0.32	2.49	1.29	0.44	0.36
Caste							
SC/STs	26.82	28.14	23.62	34.25	22.94	31.20	26.20

(continued)

Table 15.1 (continued)

	Percentage composition						
	All India	Andhra Pradesh	Bihar	Jharkhand	Maharashtra	Rajasthan	Tamil Nadu
OBCs	52.42	51.08	66.26	57.62	27.82	50.76	72.10
Others	20.76	20.78	10.12	8.13	49.24	18.04	1.71
Religion							
Hindu	85.00	84.75	87.51	76.37	80.72	87.27	88.82
Muslim	9.59	6.19	12.48	13.45	9.53	10.81	7.70
Others	5.41	9.06	0.01	10.18	9.76	1.92	3.48
Area of residence							
Urban	22.23	22.82	7.67	16.12	36.40	20.37	41.14
Rural	77.77	77.18	92.33	83.88	63.60	79.63	58.86
States							
Andhra Pradesh	21.29	21.29					
Bihar	22.45		22.45				
Jharkhand	6.60			6.60			
Maharashtra	21.29				21.29		
Rajasthan	16.43					16.43	
Tamil Nadu	11.93						11.93
All India	100.00						

Source: Authors' computations based on the India Youth Study (2006–07). Sample size: 10483

and "Others" caste groups, respectively; (9) nearly 85% of women are Hindus, whereas only 9.6% women are Muslims; (10) almost 77% of women are staying in rural areas; (10) about 16, 22, 7, 21, 21 and 12% women belong to the states of Rajasthan, Bihar, Jharkhand, Maharashtra, Andhra Pradesh and Tamil Nadu, respectively.

Sample Composition by Individual, Household and Community Level Factors: States

Within India there are substantial inter-state differences among the various states and some of the key points (from Table 15.1) are as follows: (1) around 60% of women from Bihar have no formal schooling, whereas, percentage of women having no formal schooling in Tamil Nadu is as low as 8% and also around 18% women of Tamil Nadu have 12 or more years of schooling whereas states like Bihar and Jharkhand have only about 3% women who have 12 or more years of schooling; (2) about 22% of women from Bihar and Jharkhand are indulged in labor activities whereas in Tamil Nadu only 5% of women are involved in labor activities and also about 6% of women in Tamil Nadu are involved in administrative, managerial or related jobs whereas in Bihar, Jharkhand and Maharashtra not even 1% of women are in administrative, managerial or related jobs; (3) almost 99% of women in

Rajasthan go for arranged marriage whereas in Tamil Nadu it is about 82%; (4) about 40% of mothers' in Andhra Pradesh have experienced domestic violence, whereas in Rajasthan the proportion is about 18%; (5) almost 92% of women in Bihar are from rural areas whereas in Tamil Nadu the proportion of women from rural areas is only about 59%.

Experience of Domestic Violence by Individual, Household and Community Level Factors: All India

There are many prominent patterns which can be seen from Fig. 15.1 which shows the prevalence of domestic violence among the women by various socio-economic factors. Figure 15.1 includes information only for All-India level due to shortage of space. The same statistics for the various states are presented separately in Table 15.2. The key patterns from Fig. 15.1 are as follows: (1) the overall prevalence of domestic violence in India is 24.5%, it is substantially higher in rural areas compared to urban areas; (2) the prevalence of domestic violence experienced by women decreases with the increase in years of education, for example, about 31% of women having no formal schooling have experienced domestic violence whereas the same figure for women having 12 or more years of schooling is only 11%; (3) about 36–39% of women who are either agricultural or other laborer have experienced domestic violence, whereas for other occupational categories the figures are much lower; (4) an interesting finding from Fig. 15.1 is that the proportion of women having experienced domestic violence is 7% higher in the case of those women who have opted for love marriage as compared to those who have followed the arranged marriage route; (5) it can also be observed and is not surprising that the incidence of domestic violence increase with the increase in years of marriage; (6) women whose husbands are in either administrative/managerial or related jobs or not working are least exposed to domestic violence (about 15%); (7) interestingly, about 42% of women whose mothers were victims of domestic violence have also experienced domestic violence, clearly there is an intergenerational transmission of domestic violence in India; (8) as expected, the prevalence of domestic violence is substantially lower in the case of women whose fathers have been in administrative, executive/managerial and related occupations compared to all other occupational categories and finally; (9) the prevalence of domestic violence is substantially higher in the case of SC/ST women compared to women of other caste groups.

Experience of Domestic Violence by Individual, Household and Community Level Factors: States

The experience of domestic violence by women by various socio-economic factors for different states has been reported in Table 15.2. Some stylized findings from the table are as follows: (1) Surprisingly, the economically and demographically advanced states such as Tamil Nadu and Maharashtra are comparable to the

(a) Domestic violence experienced by ever married women by their education and occupation

(b) Domestic violence experienced by ever married women by choice and years of marriage

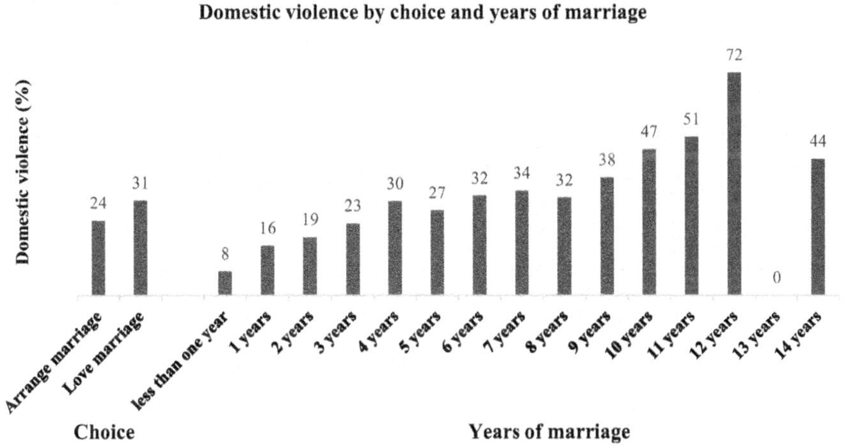

Fig. 15.1 Domestic violence experienced by ever married women by various individual, household and community level factors
Source: Based on authors' calculations from India Youth Study (2006–07)
(a) Domestic violence experienced by ever married women by their education and occupation
(b) Domestic violence experienced by ever married women by choice and years of marriage
(c) Domestic violence experienced by ever married women by husband's occupation
(d) Domestic violence experienced by ever married women by mother's experience of domestic violence and father's occupation
(e) Domestic violence experienced by ever married women by caste, religion, area and states

economically and demographically poorer states like Bihar and Jharkhand when it comes to prevalence of domestic violence; (2) Astonishingly, Tamil Nadu which is often considered one of the most developed states has highest proportion of women

(c) Domestic violence experienced by ever married women by husband's occupation

(d) Domestic violence experienced by ever married women by mother's experience of domestic violence and father's occupation

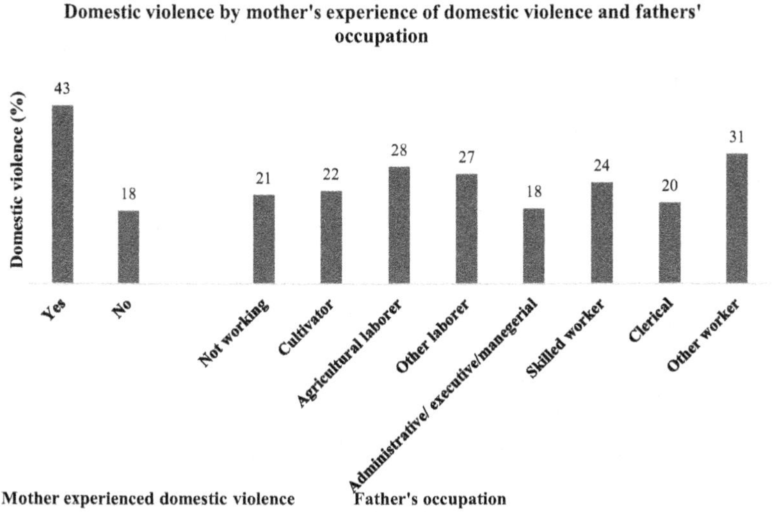

Fig. 15.1 (continued)

facing domestic violence in the schooling category of "12 or more years of schooling"; (3) women involved in administrative/managerial or related jobs experience lesser domestic violence than women from other occupational categories (except the category of "other workers") in almost all states; (4) choice of marriage does make a difference in determining the chance of domestic violence, on one hand there is Tamil Nadu where almost 34% of the women who have opted for love marriage suffer from domestic violence and on the other hand there is less developed state of Rajasthan where only 3% of women who have opted for love marriage have

(e) Domestic violence experienced by ever married women by caste, religion, area and states

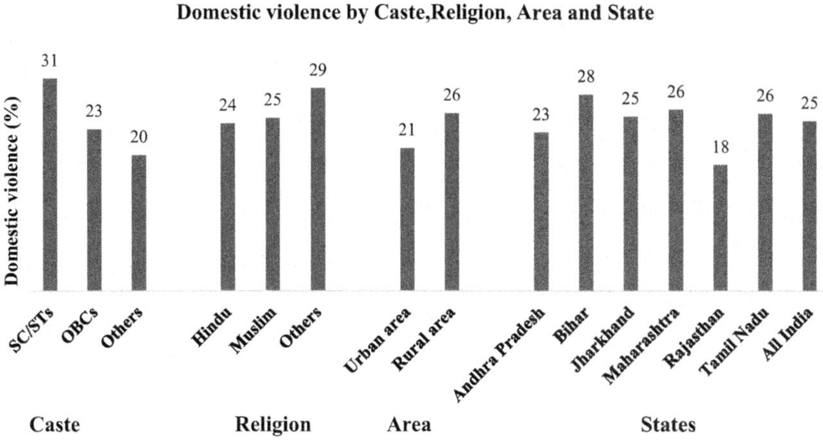

Fig. 15.1 (continued)

experienced domestic violence; (5) peculiar variations are observed while analyzing the domestic violence experienced by women by their husband's nature of occupation – on one hand about 44% of women whose husbands are not working experience domestic violence in Andhra Pradesh and on the other hand the proportion is as low as 6% for state of Rajasthan; (6) the correlation between women experiencing domestic violence and their mothers' having suffered domestic violence is highest in the case of Maharashtra, which is one of the economically and demographically advanced states in India; (7) in almost all states (except Rajasthan) the prevalence of domestic violence is lowest in the case of the father of the women being in administrative, executive, managerial and related occupations and last but not the least; (8) the prevalence of domestic violence is highest in the state of Bihar, among the SC/STs and among non-Hindu/Muslims, respectively.

Regression Results[3]

Regression Results: All-India

The main results from the multivariate analysis (presented in Table 15.3) are as follows: first, the odds of experiencing domestic violence significantly increase if a woman is either agricultural laborer (1.26 times) or non-agricultural laborer (1.32 times), compared to, if she is a housewife/not working or unemployed. The above

[3]Results of multicollinearity tests can be provided upon request. As a rule of thumb, a $VIF_i > 10$ indicates harmful collinearity (Kennedy 1998). For all the variables in our analysis the VIFs were much below 5, except in case of father's occupation being cultivator where it was 5.

Table 15.2 Prevalence of domestic violence faced by women: By socioeconomic characteristics

	Proportion of women experienced Domestic Violence						
	All India	Andhra Pradesh	Bihar	Jharkhand	Maharashtra	Rajasthan	Tamil Nadu
Education							
Non-literate, literate with no formal schooling	31.13	32.88	34.07	30.51	38.55	21.87	36.52
1–4 years of schooling	30.79	28.88	33.12	25.74	31.71	30.14	36.35
5–7 years of schooling	22.87	19.70	18.77	20.00	30.84	15.62	29.68
8–9 years of schooling	21.67	19.16	15.24	18.04	26.40	8.45	27.40
10–11 years of schooling	16.94	13.21	17.00	20.51	18.85	11.25	20.56
12 and above years of schooling	10.61	8.34	1.00	9.38	13.18	2.97	15.25
Occupation							
Not working, house wife/ student/ unemployed	20.82	18.18	22.61	22.77	22.82	16.16	23.14
Cultivator	28.15	49.52	32.74	6.05	22.57	27.16	
Agricultural laborer	35.97	27.09	48.72	31.40	36.71	27.95	43.52
Laborer (excluding agricultural)	38.99	43.10	37.39	46.62	40.71	35.85	30.26
Administrative/ executive and managerial	18.29	21.42	2.02	13.80	28.35	2.08	30.12
Skilled (manual, machinery)	22.31	25.74	21.49	22.59	29.87	13.29	19.33
Clerical	18.77	21.27	0.00	0.00	30.74		0.00
Other worker	11.57	0.00	0.00	57.16	0.00	25.00	
Choice of marriage							
Arrange marriage	24.16	21.97	28.32	24.88	26.11	18.00	24.11
Love marriage	30.64	34.36	22.35	26.56	26.34	3.54	33.68
Years of marriage							
less than 1 year	7.65	6.81	9.78	3.22	9.03	4.99	6.72
1 years	15.94	11.09	16.03	13.75	19.09	16.06	16.90
2 years	18.67	16.32	16.06	18.33	23.90	9.19	25.47
3 years	23.10	20.19	26.56	26.45	23.96	16.46	26.69
4 years	30.31	25.63	38.11	26.08	36.41	20.56	31.37
5 years	27.41	27.46	23.28	27.07	26.95	27.45	37.69
6 years	32.13	33.44	35.58	30.20	30.76	24.87	35.97

(continued)

Table 15.2 (continued)

	Proportion of women experienced Domestic Violence						
	All India	Andhra Pradesh	Bihar	Jharkhand	Maharashtra	Rajasthan	Tamil Nadu
7 years	33.78	33.06	36.95	36.91	32.40	20.76	50.14
8 years	31.52	28.58	41.18	31.51	33.99	20.06	31.74
9 years	37.96	34.57	40.45	34.03	62.58	28.33	17.57
10 years	47.04	38.29	60.15	39.54	56.01	6.32	100.00
11 years	51.18	42.64	46.63	65.81	65.99	79.59	32.86
12 years	71.84	0.00	100.00	16.56	100.00	0.00	0.00
13 years	0.00	0.00	0.00	0.00	0.00	0.00	0.00
14 years	44.11	0.00	0.00	0.00	100.00	0.00	0.00
Husband's occupation							
Not working/ unemployed	14.62	44.83	15.32	20.07	23.02	6.00	14.00
Cultivator	20.12	19.77	20.30	21.15	23.80	16.66	23.57
Agricultural laborer	30.2	22.30	38.51	26.72	33.75	32.99	34.90
Laborer (excluding agricultural)	28.38	29.05	31.88	28.79	29.35	23.10	22.01
Administrative/ executive and managerial	14.34	11.60	18.94	18.59	15.19	10.24	13.89
Skilled (manual, machinery)	25.4	25.00	27.21	25.99	27.44	17.90	27.36
Clerical	15.72	11.49	21.07	6.85	17.04	4.10	25.22
Other worker	18.22	27.45	21.24	20.68	18.24	7.58	7.81
Mother experienced domestic violence							
Yes	42.87	34.88	48.06	40.62	49.58	46.43	42.82
No	17.64	14.85	22.54	19.91	19.07	11.54	16.76
Father's occupation							
Not working, house husband/ unemployed/ retired	21.47	5.73	21.03	23.22	25.12	26.47	31.41
Cultivator	22.41	22.55	24.24	24.56	24.15	17.75	25.92
Agricultural laborer	28.22	22.30	35.86	26.45	30.41	24.42	28.56
Laborer (excluding agricultural)	26.54	24.82	33.37	27.05	25.82	19.28	23.46
Administrative/ executive and managerial	18.18	17.94	18.23	23.96	17.67	13.53	21.07
Skilled (manual, machinery)	24.43	25.29	25.43	24.24	26.14	18.43	25.27
Clerical	19.56	19.26	17.29	22.22	28.42	6.94	21.04
Other worker	31.34	28.64	35.55	19.49	37.92	27.08	34.53

(continued)

Table 15.2 (continued)

	Proportion of women experienced Domestic Violence						
	All India	Andhra Pradesh	Bihar	Jharkhand	Maharashtra	Rajasthan	Tamil Nadu
Caste							
SC/STs	30.62	31.76	30.91	26.28	33.93	25.00	35.02
OBCs	23.31	20.68	28.12	25.74	24.09	16.68	22.67
Others	19.60	15.80	22.90	14.32	23.63	9.06	15.32
Religion							
Hindu	24.15	21.74	28.33	25.49	25.23	17.72	26.43
Muslim	24.92	20.97	27.76	24.69	27.00	20.13	23.16
Others	29.24	33.78	0.00	21.69	32.65	13.67	15.10
Area of residence							
Urban area	20.61	19.01	22.04	18.98	25.06	11.05	20.66
Rural	25.61	23.90	28.77	26.15	26.73	19.66	29.36
States							
Andhra Pradesh	22.89	22.89					
Bihar	28.33		28.33				
Jharkhand	25.15			25.15			
Maharashtra	26.15				26.15		
Rajasthan	18.18					18.18	
Tamil Nadu	25.59						25.59
All India	24.51						

Source: Authors' computations based on the India Youth Study (2006–07). Sample size: 10483

occupations are often considered the lowest in the social hierarchy of occupations in India. Therefore, the above result is not very surprising because it is the women of the socially weakest occupational class who are having the maximum odds of facing domestic violence. Also, when women are involved in occupations which are physically demanding such as agricultural laborer or other laborer they fall short of energy in doing their domestic work at home as compared to their other counterparts who are not having as much physical exertion during their occupation. This might also be a reason for increased rate on domestic violence in such cases by their spouses. Second, increase in educational attainment significantly reduces the chances of experience of domestic violence which is quite intuitive and supports the general belief that education empowers women when it comes to marital relations.

Surprisingly, the chances of experiencing domestic violence are significantly higher (1.5 times) in the case of love marriage as compared to arranged marriage. This finding requires further investigation for deeper explanation, however, anecdotally it can be said that husbands of women who have gone for love marriage believe that the natal family and the women herself will not do much or will not resist

Table 15.3 Odds of experience of domestic violence by young women by their socioeconomic characteristics

	All India	Andhra Pradesh	Bihar	Jharkhand	Maharashtra	Rajasthan	Tamil Nadu
Nature of Occupation (reference: housewife/not-working/unemployed)							
Cultivator	1.04	4.6688***	1.6349	0.2279	0.7903	1.8832	
	(0.2511)	(2.6532)	(2.0707)	(0.2446)	(0.2749)	(1.2292)	
Agricultural laborer	1.2751***	1.1751	2.1011***	1.0833	1.3599	1.1458	1.4653*
	(0.0979)	(0.2154)	(0.4093)	(0.1813)	(0.2703)	(0.289)	(0.2874)
Laborer (excluding agricultural)	1.3189**	1.8245**	0.7951	1.5854*	1.3519	1.2274	1.1108
	(0.1683)	(0.5372)	(0.3663)	(0.44)	(0.4925)	(0.3382)	(0.3787)
Administrative/ executive and managerial	1.0293	1.1093	0.3335	0.8795	1.6061	0.3684	1.6664
	(0.2784)	(0.6459)	(0.399)	(0.5873)	(1.0611)	(0.391)	(0.9032)
Skilled (manual, machinery)	0.9466	1.1645	1.1692	0.3415	1.3499	0.7868	0.7289
	(0.1216)	(0.2791)	(0.4692)	(0.719)	(0.5178)	(0.2613)	(0.2144)
Clerical	0.903	4.0965			2.2005		
	(0.4341)	(4.8162)			(1.351)		
Other worker	1.8319			5.7981		2.6715	
	(1.3532)			(7.4654)		(3.646)	
Education	0.8671***	0.8216***	0.7448***	0.907**	0.9431	0.8434***	0.9533
	(0.0171)	(0.0391)	(0.038)	(0.0438)	(0.0452)	(0.0496)	(0.0473)
Love Marriage (reference: arranged marriage)	1.5036***	2.0781***	1.4197	1.6272**	1.1141	0.6899	1.4755**
	(0.1488)	(0.4559)	(0.7033)	(0.3631)	(0.3168)	(0.7504)	(0.2397)
Years of Marriage	1.1688***	1.194***	1.2047***	1.1741***	1.1841***	1.0944***	1.1991***
	(0.0112)	(0.0291)	(0.0261)	(0.0243)	(0.031)	(0.0293)	(0.0336)
Husband's nature of occupation (reference: not working/ unemployed)							
Cultivator	0.8607	0.101**	1.0158	0.6037	1.2491	1.4852	0.9614
	(0.1657)	(0.1083)	(0.4728)	(0.1985)	(0.6394)	(0.6533)	(1.0995)
Agricultural laborer	0.9436	0.1167**	1.0508	0.5261*	1.2745	2.6508*	1.4858
	(0.1834)	(0.1237)	(0.4872)	(0.1952)	(0.6418)	(1.3736)	(1.6629)

(continued)

Table 15.3 (continued)

	All India	Andhra Pradesh	Bihar	Jharkhand	Maharashtra	Rajasthan	Tamil Nadu
Laborer (excluding agricultural)	1.0124	0.1985	0.9377	0.7199	1.2431	1.9661	0.9806
	(0.1847)	(0.2092)	(0.4035)	(0.214)	(0.6151)	(0.8524)	(1.0926)
Administrative/ executive and managerial	0.8557	0.1455*	0.9516	0.6573	0.8594	1.7579	0.9388
	(0.1667)	(0.1569)	(0.4274)	(0.2244)	(0.4464)	(0.8205)	(1.0589)
Skilled (manual, machinery)	1.0799	0.2082	1.1666	0.6642	1.289	1.7847	1.3896
	(0.1973)	(0.2186)	(0.5013)	(0.2022)	(0.6374)	(0.7773)	(1.5432)
Clerical	0.6731	0.1172*	1.0115	0.2549*	0.6774	0.3005	1.533
	(0.1671)	(0.1319)	(0.6072)	(0.1782)	(0.3828)	(0.3311)	(1.8095)
Other worker	0.7785	0.2502	0.8412	0.6129	0.8224	0.2983	0.4649
	(0.2308)	(0.3098)	(0.8592)	(0.3007)	(0.5362)	(0.3506)	(0.7141)
Mother experienced domestic violence (reference: no)	3.2202***	2.7578***	3.0111***	3.0096***	3.7492***	4.4009***	3.1093***
	(0.1735)	(0.3584)	(0.417)	(0.3864)	(0.5161)	(0.6511)	(0.4005)
Father's nature of occupation (reference: not working/unemployed)							
Cultivator	0.9243	2.4352	3.4959	0.5286	1.2758	0.405	0.6917
	(0.2939)	(2.7041)	(3.9713)	(0.2939)	(0.9758)	(0.4676)	(0.4674)
Agricultural laborer	0.8496	2.0028	3.1356	0.4771	1.2446	0.4453	0.5221
	(0.2717)	(2.221)	(3.567)	(0.2764)	(0.9581)	(0.525)	(0.3457)
Laborer (excluding agricultural)	0.9729	2.2851	4.4692	0.561	1.2449	0.3953	0.5295
	(0.3098)	(2.5406)	(5.0704)	(0.313)	(0.9558)	(0.4597)	(0.3526)
Administrative/ executive and managerial	1.0835	2.7484	5.2933	0.5619	1.1031	0.5781	0.6737
	(0.3526)	(3.1328)	(6.063)	(0.3244)	(0.8669)	(0.6793)	(0.4615)
Skilled (manual, machinery)	0.9433	2.7929	4.2732	0.4833	1.1608	0.4525	0.5933
	(0.3008)	(3.1058)	(4.8594)	(0.2709)	(0.8952)	(0.5258)	(0.3936)
Clerical	1.1565	4.7872	3.6504	0.5541	2.574	0.3195	0.7077
	(0.3984)	(5.6118)	(4.3386)	(0.3479)	(2.0754)	(0.3918)	(0.5592)

Other worker	1.2833	4.6195	7.8723	0.4332	2.5021	0.92	0.9128
	(0.5052)	(6.266)	(12.0657)	(0.2865)	(2.2314)	(1.2756)	(1.0624)
Caste (reference: Others							
SC/STs	1.4121***	1.4813*	1.0343	1.6393*	1.0587	1.8982***	2.1611
	(0.1256)	(0.3298)	(0.2621)	(0.4237)	(0.1893)	(0.4459)	(1.3218)
OBC	1.1562*	1.0317	1.1316	1.6806**	0.9194	1.2871	1.5293
	(0.0914)	(0.202)	(0.2359)	(0.3939)	(0.1445)	(0.2795)	(0.9224)
Religion (reference: Hindu)							
Muslims	1.0537	0.8005	0.84	1.073	1.1264	1.2184	1.2579
	(0.0882)	(0.233)	(0.1544)	(0.192)	(0.2468)	(0.2581)	(0.3211)
Other religions	0.9398	1.2116		0.6971	1.204	1.6234	0.6076
	(0.1064)	(0.2718)		(0.1621)	(0.2732)	(0.99)	(0.2562)
Rural (reference: urban)	1.1066*	1.2637	0.8898	1.1384	0.861	1.3169	1.1689
	(0.0662)	(0.2072)	(0.1247)	(0.1532)	(0.1458)	(0.2238)	(0.1778)
States (reference: Rajasthan)							
Andhra Pradesh	1.1714						
	(0.1136)						
Bihar	1.7113***						
	(0.1545)						
Jharkhand	1.372***						
	(0.1223)						
Maharashtra	2.493***						
	(0.2495)						
Tamil Nadu	1.9659***						
	(0.2031)						

Source: Authors' computations based on the India Youth Study (2006–07)
Notes: ***$p < 0.01$, **$p < 0.05$, *$p < 0.1$; Sample size: 10483

domestic violence because they (the women themselves) can be morally held responsible for the marriage, that is, the marriage was of their choice (in a country where an overwhelming majority of marriages involve grooms chosen by the family members of women) rather than forced upon them by their natal families.

Also, the odds of experience of domestic violence increases significantly with the increase in years of marriage. Further, daughters whose mothers have experienced domestic violence are at a significantly much higher risk (3.2 times) of experiencing domestic violence compared to women whose mothers have not experienced domestic violence. This finding has been discussed in detail in the next section. Moreover, women from the OBC and SC/ST communities (historically disadvantaged and discriminated caste groups) are significantly more likely to experience domestic violence compared to women of other caste categories. Finally, women from rural areas and from the states of Bihar, Jharkhand, Maharashtra and Tamil Nadu have significantly higher chance of experiencing domestic violence compared to women from urban areas and from the state of Rajasthan, respectively.

Regression Results: States

Some of the key findings for the states are as follows: Women in Andhra Pradesh are at a significantly higher risk of experiencing domestic violence when they are either cultivator or agricultural laborer; also, the chances of experiencing domestic violence are significantly higher for those women in Bihar and Tamil Nadu who are working as agricultural laborers; further, women who are in non-agricultural labor in Jharkhand are also at a higher risk of experiencing domestic violence.

Strangely, in Maharashtra and Tamil Nadu which are among the economically and socially advanced states in India, the odds of experiencing domestic violence do not decrease with increase in education. In Jharkhand, Tamil Nadu and Andhra Pradesh, the chance of experiencing domestic violence is significantly higher in the case of love marriage compared to arranged marriage. In Andhra Pradesh, the women whose husbands are agriculture laborers are 0.12 times less likely to experience domestic violence compared to women whose husbands are either unemployed or not working, whereas, in Rajasthan the women whose husbands are agriculture laborers are 2.65 times more likely to experience domestic violence compared to women whose husbands are either unemployed or not working. It is difficult to explain such inter-state variations and further in-depth primary survey is required to gauge out the reasons for such variations.

In all the states, the odds of experiencing domestic violence increase substantially and significantly if the mother has also experienced domestic violence. Last but not the least, SC/ST women in Rajasthan and OBC women in Jharkhand are at a significantly higher risk of experiencing domestic violence compared to women belonging to upper caste categories, respectively.

We now present our main conclusions along with some discussion on the main findings, thus concluding the paper.

Conclusion and Discussions

The paper perhaps is the first attempt in examining how young women's nature of occupation affects the domestic violence experienced by them in India. The analysis is conducted both at the subnational level as well as separately for the states covered in the survey. The main results show that women involved in agricultural/non-agricultural labor activities for cash and kind have significantly higher chances of experiencing domestic violence compared to women who are either housewives/not working or unemployed. Also, women whose mothers have experienced domestic violence are at a substantially and significantly higher risk of experiencing domestic violence themselves. Further, women belonging to Scheduled Castes, Scheduled Tribes and Other Backward Castes have significantly higher odds of experiencing domestic violence compared to women belonging to "Other/Upper" caste category. Moreover, the odds of experience of domestic violence reduces significantly with the increase in years of education. Surprisingly, women who choose love marriage and women from the economically and demographically most advanced states (Maharashtra and Tamil Nadu) have significantly higher chances of facing domestic violence compared to the women who choose arranged marriage and women who are from relatively much poorer (economically and demographically) states such as Rajasthan, Bihar and Jharkhand, respectively.

Some other results which might be useful are that women in skilled and clerical jobs have lesser chances (though not significant) of going through domestic violence compared to women who are either housewives/not working or unemployed. Also, the nature of occupation of fathers of women does not make any significant difference when it comes to experience of domestic violence.

The results of the present paper are in line with the earlier studies, such as, Singh et al. (2013) which has studied intergenerational mobility in domestic violence in India and has found that daughters whose mothers have suffered from domestic violence are more likely to face domestic violence than those whose mothers have not. Also, Visaria (2008) has found that all those children who have seen spousal violence in their family are more likely to experience it.

Our results also show that the prevalence of domestic violence among young women in India is substantial and therefore it requires some discussion. One of the reasons for such high prevalence of domestic violence in India is probably the belief system of the young men and women in India when it comes to domestic violence. For example, 66% of never married young men whose mothers have faced domestic violence and 52% of never married young men whose mothers never faced domestic violence feel that it is alright to indulge in violence with wives. Similarly, 66% of never married young women whose mothers have faced domestic violence and 50% of never married young women whose mothers never faced domestic violence feel that it is alright to suffer from domestic violence by husbands (Singh and Singh 2013).

Another point worth discussing is that when it comes to domestic violence faced by women in India, economic and demographic prosperity of the region does not matter. This can be ascertained from the fact that the prevalence of domestic violence faced by young women in the economically and demographically most advanced states is either comparable to the prevalence in many poorer states or more than the prevalence in many poorer states. This leads to another issue worth mentioning, which is, it is important to study why the prevalence of domestic violence faced by women is so high even in the relatively more developed states of Tamil Nadu and Maharashtra and this forms the scope of future research in this area.

That said, the present study also suffers from a few limitations, such as, it is restricted to young women and it covers only six states of India. But, if we examine the results, we find that the prevalence of domestic violence faced by women increases with the years of marriage and hence if women of higher age groups get to be included in the sample, then the prevalence will increase further. Therefore, our estimates of prevalence of domestic violence faced by women may be taken as a kind of lower limit of the prevalence of domestic violence experienced by women in India. Similarly, though only six states were covered in the study, these states together represent 39% of India's population (IIPS and Population Council, 2010). Also, these six states were purposively selected to represent the different geographic and socio-cultural regions within the country and this study is sub-nationally representative (IIPS and Population Council, 2010; Singh et al. 2014). Therefore, our results can be safely generalized to a large part of India.

References

Barnish, M. (2004). *Domestic violence: A literature review*. London: HM Inspectorate of Probation.

Bedi, A. S., Chhachhi, A., & Bhattacharyya, M. (2009). Marital violence and women's employment and property status: Evidence from North Indian Villages. Discussion Paper No. 4361.

Belsley, D. A. (1991). A guide to using the collinearity diagnostics. *Computational Economics, 4* (1), 33–50.

Bhattacharyya, M., Bedi, A. S., & Chhachhi, A. (2011). Marital violence and women's employment and property status: Evidence from north Indian villages. *World Development, 39*(9), 1676–1689.

Bowlus, A. J., & Seitz, S. (2006). Domestic violence, employment, and divorce. *International Economic Review, 47*(4), 1113–1149.

Burns, R. J. (2013). 'The last straw' (2007). Available at: https://thelaststraw.wordpress.com/2007/08/15/quotes-about-domestic-violence/

Caesar, P. L. (1988). Exposure to violence in the families-of-origin among wife-abusers and maritally nonviolent men. *Violence and Victims, 3*(1), 49–63.

Carrillo, R. (1992). *Battered dreams: violence against women as an obstacle to development*. New York: United Nations Development Fund for Women.

Choudhary, A., & Singh, A. (2017). Are daughters like mothers: Evidence on intergenerational educational mobility among young females in India. *Social Indicators Research, 133*(2), 601–621.

Ellsberg, M. C., Pena, R., Herrera, A., Liljestrand, J., & Winkvist, A. (1999). Wife abuse among women of childbearing age in Nicaragua. *American Journal of Public Health, 89*(2), 241–244.

Gibson-Davis, C. M., Magnuson, K., Gennetian, L. A., & Duncan, G. J. (2005). Employment and the Risk of Domestic Abuse Among Low-Income Women. *Journal of Marriage and Family, 67* (5), 1149–1168.

Greaves, L. (1995). *Selected estimates of the costs of violence against women*. London: Centre for Research on Violence Against Women and Children.
Heise, L. L. (1998). Violence against women an integrated, ecological framework. *Violence Against Women, 4*(3), 262–290.
Heise, L. L., Raikes, A., Watts, C. H., & Zwi, A. B. (1994). Violence against women: a neglected public health issue in less developed countries. *Social Science and Medicine, 39*(9), 1165–1179.
Hotaling, G. T., & Sugarman, D. B. (1986). An analysis of risk markers in husband to wife violence: The current state of knowledge. *Violence and Victims, 1*(2), 101–124.
ICRW. (2000). *Domestic violence in India. A summary report of a multi-site household survey*. Washington, DC: International Center for Research on Women.
IIPS and Macro International. (2007). *National Family Health Survey (NFHS-3),2005–06, India, Vol. II*. Mumbai: International Institute for Population Sciences.
International Institute for Population Sciences (IIPS) and Population Council. (2010). *Youth in India: Situation and needs 2006–2007* (p. 396). Mumbai: IIPS.
Jaising, I., & Wolfe, A. (1988). The ignoble servility of Pati Parmeshwar: Towards equality for women'. *The Lawyer*. December, 6.
Jejeebhoy, S. J. (1998). Wife-beating in rural India: a husband's right? Evidence from survey data. *Economic and Political Weekly*, 855–862.
Kalmuss, D. (1984). The intergenerational transmission of marital aggression. *Journal of Marriage and the Family*, 11–19.
Kennedy, P. (1998). *A guide to econometrics*. Cambridge, MA: MIT Press.
Kim, J. C., Watts, C. H., Hargreaves, J. R., Ndhlovu, L. X., Phetla, G., Morison, L. A., & Pronyk, P. (2007). Understanding the impact of a microfinance-based intervention on women's empowerment and the reduction of intimate partner violence in South Africa. *American Journal of Public Health, 97*(10), 1794–1802.
Macmillan, R., & Gartner, R. (1999). When she brings home the bacon: Labor-force participation and the risk of spousal violence against women. *Journal of Marriage and the Family*, 947–958.
Menon-Sen, K., & Kumar, S. (2001). *Women in India: How free? How equal?* United Nations Resident Coordinator, New Delhi.
Morrison, A. R., & Orlando, M. B. (1999). *Social and economic costs of domestic violence: The cases of Chile and Nicaragua*. Morrison, AR y Biehl, ML, Too close to home: Domestic violence in the Americas, capítulo, 3.
Motiram, S., & Singh, A. (2012). How close does the apple fall to the tree? Some evidence from India on intergenerational occupational mobility. *Economic and Political Weekly, XLVII*(40), 56–65.
Naved, R. T., & Persson, L. Å. (2005). Factors associated with spousal physical violence against women in Bangladesh. *Studies in Family Planning, 36*(4), 289–300.
Olivelle, P., & Olivelle, S. (2005). *Manu's code of law*. Oxford University Press.
Panda, P., & Agarwal, B. (2005). Marital violence, human development and women's property status in India. *World Development, 33*(5), 823–850.
Paul, S. (2014). Women labor force participation and domestic violence: Evidence from India. *Journal of South Asian Development, 11*(2), 224–250.
Raikes, A. (1990). *Pregnancy, birthing, and family planning in Kenya: Changing patterns of behaviour*. CDR Research Report 15.
Rao, V. (1997). Wife-beating in rural South India: A qualitative and econometric analysis. *Social Science and Medicine, 44*(8), 1169–1180.
Singh, A., & Singh, A. (2013). Intergenerational transmission of gender based domestic violence in India: Some new evidence. Paper presented at the annual conference of the International Union for Scientific Study of Population (IUSSP)at Seoul, South Korea.
Singh, A., Gaurav, S., & Das, U. (2013). Household headship and academic skills of Indian children: A special focus on gender disparities. *European Journal of Population/Revue européenne de Démographie, 29*(4), 445–466.

Singh, A., Singh, A., Pallikadavath, S., & Ram, F. (2014). Gender differentials in inequality of educational opportunities: new evidence from an Indian youth study. *The European Journal of Development Research, 26*(5), 707–724.

Stephenson, R., Koenig, M. A., & Ahmed, S. (2006). Domestic violence and contraceptive adoption in Uttar Pradesh, India. *Studies in Family Planning, 37*(2), 75–86.

Straus, M. A., & Gelles, R. J. (1986). Societal change and change in family violence from 1975 to 1985 as revealed by two national surveys. *Journal of Marriage and the Family*, 465–479.

Straus, M. A., & Gelles, R. J. (1990). *Physical Violence in American Families*. New Brunswick: Transaction Press.

Straus, M. M. A., Gelles, R. J., & Steinmetz, S. K. (Eds.). (1980). *Behind closed doors: Violence in the American family*. Transaction Publishers.

Sugarman, D. B., & Hotaling, G. T. (1989). Violent men in intimate relationships: An analysis of risk markers. *Journal of Applied Social Psychology, 19*(12), 1034–1048.

Tapasyananda, S. (1983). Valmiki Ramayana: Sundara Kanda. Sri Ramakrishna Math, Madras.

Visaria, L. (2008). Violence against women in India: Is empowerment a protective factor? *Economic and Political Weekly*, 60–66.

WHO. (2002). *World report on violence and heath*. Geneva: WHO.

Xu, X. (1997). The prevalence and determination of wife abuse in urban China. *Journal of Comparative Family Studies*, 280–303.

Yugantar Education Society. (2003). *A study of nature, extent, incidence and impact of domestic violence against women in the states of Andhra Pradesh, Chhattisgarh, Madhya Pradesh, Gujarat and Maharashtra*. Submitted to Planning commission of India.

Part IV
Data, Methods, and Policy

Chapter 16
Ways to Evaluate Redistricting Plans

Maurreen Skowran

Redistricting is the redrawing of boundaries of political districts (Levitt 2010). These legislative districts determine who represents voters.

In the United States, representatives in the U.S. House, state legislators, and many local legislators in counties, municipalities, and other jurisdictions are elected from districts (Levitt 2010). These districts divide states, other jurisdictions, and the people who live there into geographical territories (Levitt 2010).

Besides legislative districts that compose states, counties and municipalities, there are also special districts, such as water and sewer districts and school districts (National League of Cities). These special districts may be divided, such as for electoral or attendance purposes (Morrison and Bryan 2019).

When discussing redistricting, it is important to keep in mind the Constitution during the founding of America. The founding fathers considered equal representation important for the U.S. House. So they established the census to get an accurate count of the population. Every 10 years, district lines are subject to change based upon the outcome of the census.

Besides the population count, other districting principles have become traditional. We will discuss them, and then a few proposed measures to determine whether gerrymandering exists in a given redistricting plan.

Redistricting criteria are often discussed in the abstract, but they deserve greater attention. During the redistricting process, map drawers should provide to decision makers and the public concrete specifics on how their plans compare against each other. More fully informing the redistricting bodies and citizens would allow for greater engagement and better decisions.

M. Skowran (✉)
Traffic Research Unit, Geospatial and Population Studies, University of New Mexico, Albuquerque, NM, USA
e-mail: mskowran@unm.edu

B. Jivetti, Md. N. Hoque (eds.), *Population Change and Public Policy*, Applied Demography Series 11, https://doi.org/10.1007/978-3-030-57069-9_16

Redistricting Criteria

Over time, several common redistricting criteria have evolved. These are codified in various levels, from case law, through statutes and state constitutions, to just guidelines.

The primary principles are two federal rules, concerning equal population and race and ethnicity. Congressional districts must be equal "as nearly as practicable" (Levitt n.d., Where Are the Lines Drawn section). State and local legislative districts have more flexibility. They need only be "substantially" equal, with an allowance of about 10% between the smallest and largest districts. (Levitt n.d., Where Are the Lines Drawn section).

Protecting the voice of racial and ethnic minorities is the impetus of the federal Voting Rights Act of 1965 (Levitt 2010), the details of which are beyond our discussion.

Some other rules are common but not uniform throughout the states. For example, most states require legislative districts to be reasonably compact, but states differ on the standards. Idaho requires only that districts avoid odd shapes, and Colorado bases its evalutation on total perimeter (Levitt n.d., sections on Idaho and Colorado). In another aspect, most states call for respecting political boundaries. But in different states, the levels of borders to be considered range from precincts to counties (Levitt n.d., Where Are the Lines Drawn section).

The most common state rules concern contiguity, followed by rules on political boundaries, compactness, communities of interests, and other issues (Levitt n.d., Where Are the Lines Drawn section). These rules are often put in vague terms, such as a requirement to follow them "where practicable," and without priorities set among them (Levitt n.d., Where Are the Lines Drawn section).

As Levitt (2010) advises, it's best to set clear priorities among different criteria, so everyone knows what to expect. Without ranked priorities, different plans can all be argued as the "best," because they come out as No. 1 according to one criteria or another. Clear priorities constrain map drawing, to make the redistricting less likely to give political advantage to anyone.

The first section of this article will discuss district compactness, including its limitations and benefits, state rules and other methods of evaluation. That is followed by a short section on benefits of redistricting criteria. The issue of a standard to determine excessive gerrymandering is introduced. Possible methods for standards are considered: efficiency gap, partisan symmetry, and sampling with Markov chain Monte Carlo.

Compactness

The word "gerrymander" originated in Massachusetts in 1812. The practice is named after the then-Massachusetts governor, Elbridge Gerry, and a salamander shape (Trickey 2017) (Fig. 16.1).

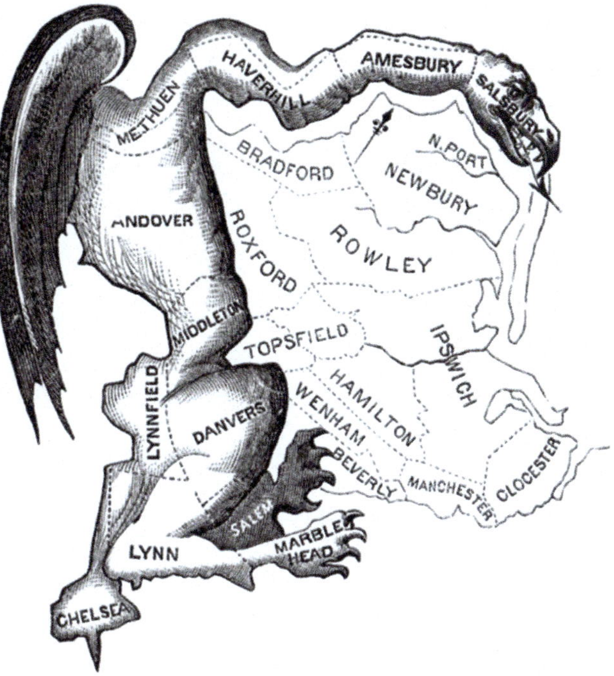

Fig. 16.1 Cartoon of district on which the word "gerrymander" is based (Source: Levitt n.d.)

The "eyeball test" is a very informal method of determining whether districts are compact. People unconsciously link odd shapes to nefarious purposes. But that does not mean the eyeball test is valid.

The maps for North Carolina and Maryland, below, fail the "eyeball test." (Figs. 16.2, 16.3).

A number of districts have even more-bizarre shapes. Above is a previous version of 7th Congressional District of Pennsylvania, nicknamed "Goofy kicking Donald Duck." It was drawn in 2011 but overturned by the Pennsylvania Supreme Court in January 2018 (Gabriel 2018).

A compact shape is no guarantee against gerrymandering. For example, a congressional redistricting plan in Pennsylvania scored high on five measures of compactness (Duchin 2018a, The Eyeball Test section). But analysis showed that it has the same extreme partisan skew as the plan it was intended to replace (Duchin 2018a, The Eyeball Test section).

And odd shapes might be justified, for reasons such as physical or political geography, or to unite a community of interest. (Duchin 2018a, The Eyeball Test section).

Nonetheless, the principle of compactness is among the traditional criteria for redistricting. Legislative districts are required to be reasonably compact by 37 states, and congressional districts are required to be compact by 18 states (Levitt n.d., Where Are the Lines Drawn section).

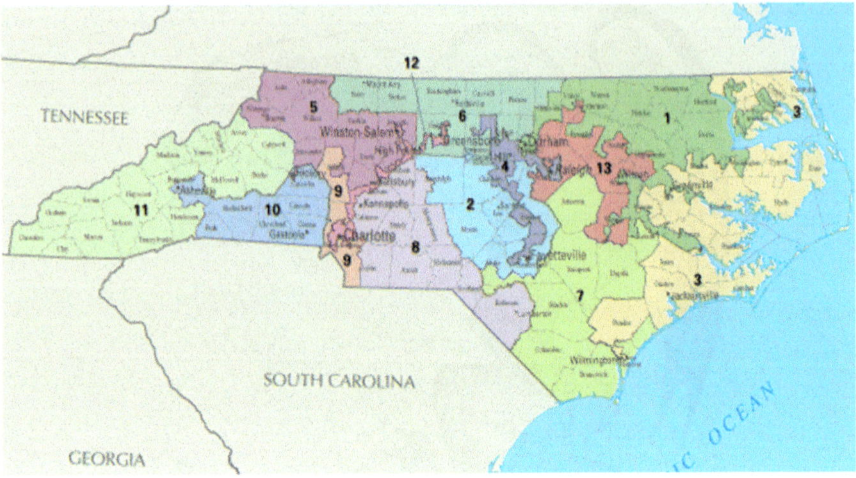

Fig. 16.2 North Carolina congressional districts, 113th Congress (Source: The National Atlas)

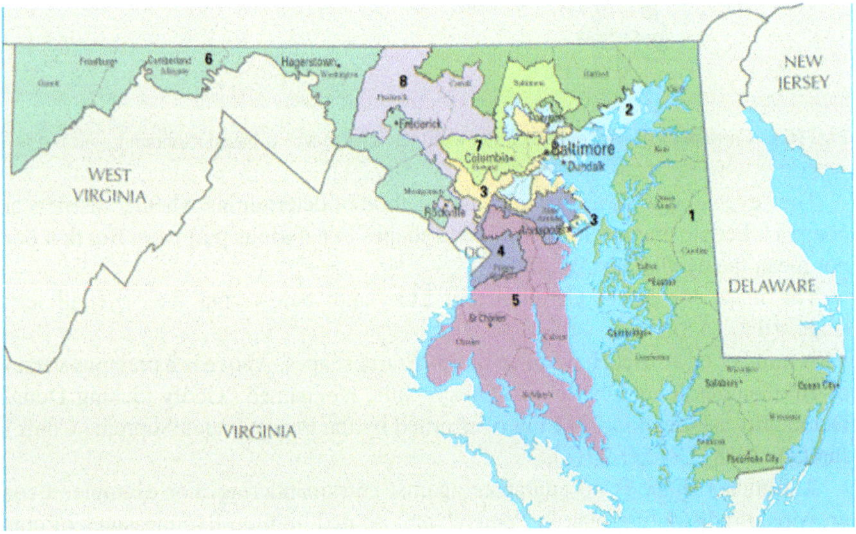

Fig. 16.3 Maryland congressional districts, 113th Congress (Source: The National Atlas)

But no redistricting principle is perfect. Rules on compactness are correlated with less influence by minorities (Barabas and Jerit 2004). The exception is large and segregated minority populations, when the compactness rules are linked to greater influence by minorities (Barabas and Jerit 2004).

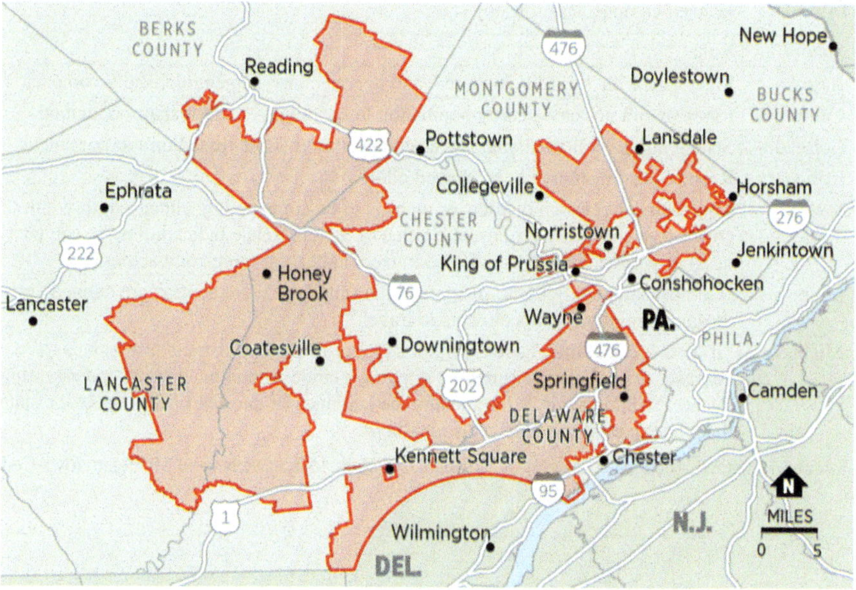

Fig. 16.4 Pennsylvania's 7th Congressional District of 2011 (Source: Every Vote Counts)

Still, compactness can have a number of advantages. These include:

- Making it easier for lawmakers, candidates and voters to travel across the district to communicate with each other.
- Making it easier for districts to represent communities of interest.
- Making it easier for voters to know which district they are in.

Most states have some requirement for compactness. But few spell out how that should be determined (Levitt n.d., Where Are the Lines Drawn section). And in many states, rules on compactness and several other criteria must be followed only as much as is practicable. New Mexico, for example, has only guidelines that call for districts to be "reasonably compact" (Levitt n.d., New Mexico section).

Table 16.1 shows some state rules on compactness.

As you can see from these maps from The National Atlas (n.d., section on Printable Maps, Congressional Districts, 113th Congress), these rules appear to have moderate success in keeping districts compact. (Figs. 16.5, 16.6, 16.7, 16.8, 16.9, 16.10 and 16.11).

Compactness can be measured in a number of ways. More than 30 formulas have been proposed (Levitt n.d.). One of the simplest is W/L, "where W and L are that of the circumscribing rectangle with minimum perimeter" (Niemi et al. 1990).

According to "Measuring Compactness: A Fair Redistricting Process," compactness measures can be divided into three broad categories:

Table 16.1 Selected compactness rules by state

State	Rule
Arizona	Must start with "grid-like pattern," and districts must be geographically compact.
California	Lines should not pass nearby population to encompass more-distant population.
Colorado	State legislative districts must be as compact as possible, based on total perimeter.
Idaho	Requires that districts not be shaped oddly.
Iowa	Compactness between districts or plans is to be compared by either length-width compactness (with lower absolute value of the difference indicating more compactness) or perimeter (with smaller perimeters indicating more compactness).
Maine	"Compact and contiguous" districts in the state legislative context are defined "as those that minimize impediments to travel."
Michigan	"If there are multiple districts within a city or township, districts (must) be as compact as possible, measured by drawing a circle around the district, and assessing the area within the circle (and within the landmass of the state)." the part of the state that is not part of the districts is measured.

Sources: Levitt n.d., sections on Arizona, California, Colorado, Idaho, Maine, and Michigan; Iowa Code

Fig. 16.5 Arizona congressional districts

Fig. 16.6 California congressional districts

Fig. 16.7 Los Angeles area, part of California's congressional districts

Fig. 16.8 Idaho congressional districts

1. Those that measure how far the district spreads out from its center, or dispersion.
2. Those that measure the smoothness, contortion or indentation of the boundaries.
3. Population measures. These are based on how the population is distributed within a district, and they will not be discussed further here. (Figs. 16.12 and 16.13).

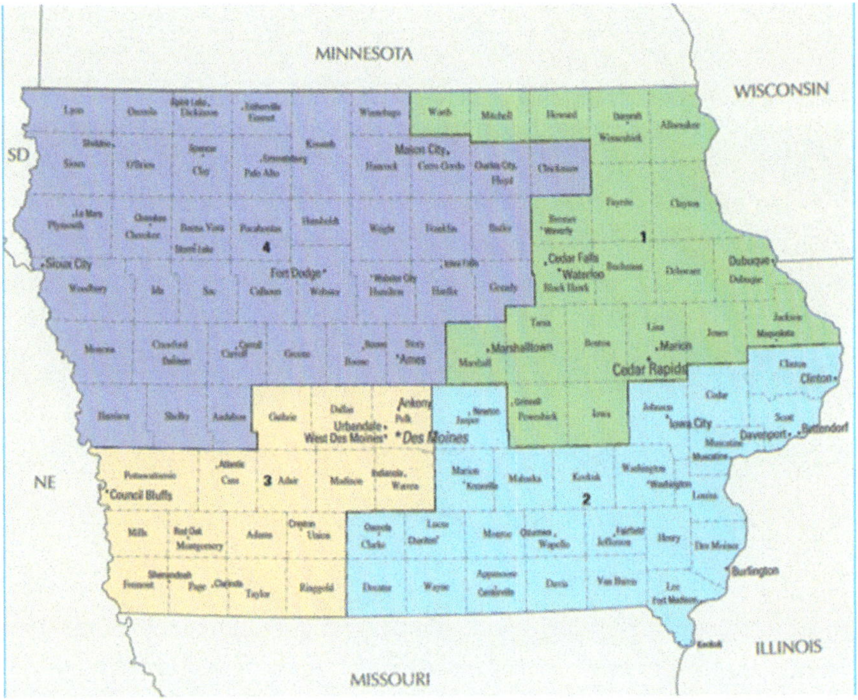

Fig. 16.9 Iowa congressional districts

Table 16.2 outlines four methods commonly discussed by experts to measure compactness: Roeck, Convex Hull, Polsby-Popper and Schwartzberg.

Niemi et al. (1990), found that the results of several types of compactness measures differed more when comparing individual districts than when comparing plans. This was found by comparing several compactness measures across different plans and jurisdictions. So, the specific measure chosen matters less than that the result is provided along with the maps under consideration.

Niemi's recommendations also include (1990): (A) Using measures only for comparison, not to eliminate any that don't meet a cutoff. (B) Using compactness as one of multiple criteria. (C) Using multiple compactness measures, of different types.

Political and Geographic Boundaries

Forty-two states call for consideration of political boundaries when drawing legislative districts., with 19 states calling for the same with congressional districts (Levitt n.d., Where Are the Lines Drawn section). Besides the principle against splitting political units, some states, such as New Mexico, encourage following physical geographic boundaries. (Levitt n.d.).

Fig. 16.10 Maine congressional districts

Fig. 16.11 Michigan congressional districts

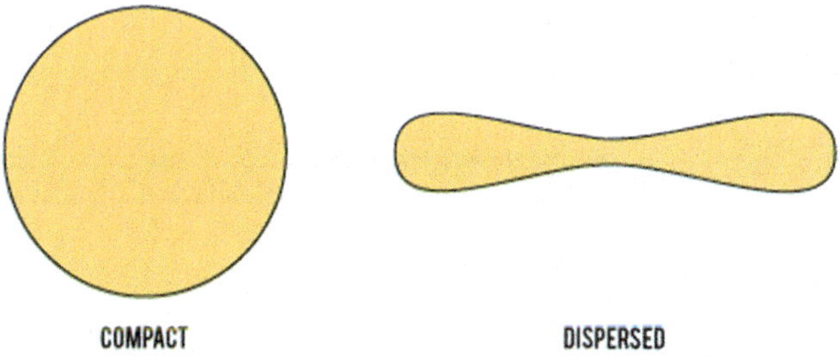

Fig. 16.12 Compactness vs. dispersion (Source: Azavea Redistricting Glossary n.d.)

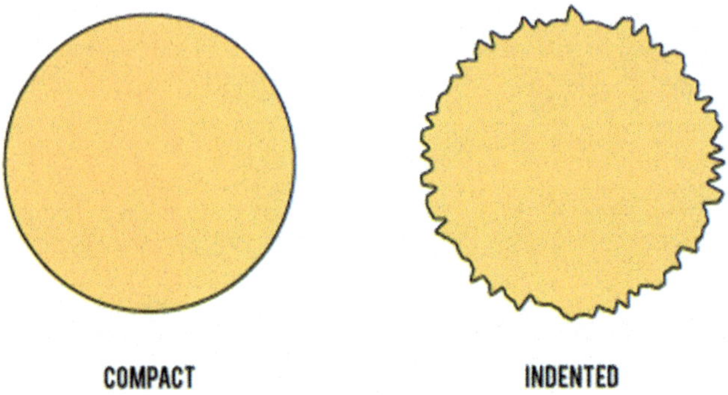

Fig. 16.13 Compactness vs. indentation (Source: Azavea Redistricting Glossary n.d.)

Table 16.2 Common methods to measure compactness

Figure	Method
	**Roeck method:** Compares district area to the area of the smallest circle that encompasses the district. Can penalize districts with islands. Measures dispersion.
	**Convex Hull method:** Compares district area to the area of the smallest "convex bounding polygon," or "convex hull," that encompasses the district. Can penalize districts with islands. Mainly measures dispersion.
	**Polsby-popper method:** Compares district area to the area of a circle with the same perimeter. Possibly the most common compactness measure. Measures indentation.
	**Schwartzberg method:** Compares district perimeter to the perimeter of a circle with the same area. Measures indentation.

Source: Redrawing the Map on Redistricting 2010 (pp. 8–10)

Fig. 16.14 Alabama congressional districts, 113th Congress (Source: The National Atlas, n.d.)

If the types of political boundaries are specified, they are most often for counties and cities, but they can be smaller units, such as precincts or townships (Levitt n.d., Where Are the Lines Drawn section).

This map of Alabama shows a good example of a state rule to follow county boundaries (Fig. 16.14).

But when considering splitting political units, states usually do not specify a preference between having the fewest jurisdictions that are split, or the fewest times that those jurisdictions are split (Levitt n.d., Where Are the Lines Drawn section). Exceptions are Maine and Michigan, requiring that political units be split as little as possible (Levitt n.d., sections on Maine and Michigan).

Counting the number of splits of political divisions is not the only way to evaluate according to this criteria. A method devised by Bowen, measuring "coterminosity," compares the boundaries of districts with the boundaries of political subdivisions (2014, pp. 856, 858). The "score is the proportion of a district's boundaries that are drawn over pre-existing subdivision boundary lines" (Bowen 2014, p. 868).

As with ranking priorities, it's best for legislators to be specific about political boundaries. For example, in redistricting legislation, they should indicate whether the preferences is for the fewest counties to be split, or the fewest times that those counties are split. With clear instruction, there is less to debate, and less opportunity for nefarious maneuvering.

Again, map drawers should show the decision makers and the public the result of an evaluation about political and geographic boundaries, such as indicating how many times counties were split in each plan, or the coterminity between each plan and other boundaries.

Benefits

Redistricting criteria have been linked to benefits.

Bowen (2014: 883) has tested the traditional districting principles of compactness and respect for political subdivisions. He found a correlation between compact and coterminous districts with respondents reporting that they were more ideological similar to their representative, more satisfied by contact with their representative, more likely to contact the representative, and more able to remember basic information about the representative .

Principles about communities of interest and political subdivisions have also been tied to lower rates of uncontested elections (Forgette et al. 2009:11).

Splitting postal codes area has also been linked to problems. For one thing, recognition of representatives decreases as splitting of postal code areas increase (Curiel and Steelman 2018, p. 23). Also, as the overlap between districts and postal code areas decrease, then contact of a representative by a resident decreases (Curiel and Steelman 2018:26–27).

"A Workable Standard" to Determine Extreme Partisan Gerrymandering

Although the U.S. Supreme Court has decided partisan gerrymandering is beyond the scope of federal courts, such gerrymandering may still be limited, such as by measures within any individual state.

Partisan bias, on both sides, is not always hidden. "I acknowledge freely that this would be a political gerrymander, which is not against the law," North Carolina Rep. David Lewis was quoted in a committee meeting transcript when redrawing district maps (Timm 2017, Purple State, Red Delegation section).

But courts have not found partisan intent enough to determine unconstitutional gerrymander.

The key to any gerrymandering case could be defining "a workable standard" to determine excessive political gerrymandering (Liptak 2017). In a 2004 Supreme Court case about political gerrymandering, *Vieth v. Jubelirer*, Justice Anthony Kennedy (now retired) wrote in his concurring opinion, "If workable standards do emerge to measure these burdens, however, courts should be prepared to order relief." Kennedy was the swing vote, writing that courts should not then review partisan gerrymandering, but possibly they should in the future.

A number of experts say that using a single measure is problematic, but we will discuss some standards that have been used in federal courts. They are the partisan symmetry standard, efficiency gap, and sampling (by the Markov chain Monte Carlo method) (Grofman and King 2007, 1; "What's at Stake in Common Cause v. Rucho," 2019; Charles et al. 2019).

It is important to use some measure, and to provide that to everyone, as an indication of the fairness of the process.

Partisan Symmetry

One measure of partisan gerrymandering is based on the idea of partisan symmetry. With this method, the percentage of vote given to each party is flipped, and then their performance is compared (Grofman and King 2007: 8). If each party is able to receive the same proportion of seats with the same proportion of the vote, then the symmetry standard is achieved. Grofman and King (2007: 6–7) say this method has achieved scholarly consensus and been used by courts and nonpartisan members of redistricting commissions.

This method does not equate to "proportional representation" (Grofman and King 2007:8). That is, the proportion of seats received and the proportion of votes received are not required to be equal (Grofman and King 2007:8).

Efficiency Gap

The efficiency gap is a new measure of partisan symmetry. A complaint in *Gill v. Whitford* argued that it should be used as a test of extreme pastisan gerrymandering ("What's at Stake in Common Cause v. Rucho," 2019).

The efficiency gap takes into account "wasted votes" for each party (Stephanopoulos and McGhee 2014, p. 831). Wasted votes are those for either

Table 16.3 Calculating the efficiency gap

District	Total votes, by party		Lost votes, by party		Surplus votes, by party		Wasted votes, by party		Difference	Efficiency gap
	A	B	A	B	A	B	A	B		
1	52	48	0	48	1	0	1	48	47	0.47
2	70	30	0	30	19	0	19	30	11	0.11

losing candidates or for a winning candidate, but more than needed to win (Stephanopoulos and McGhee 2014, p. 851). Districts with two candidates will all have the same number of "wasted votes" (half); they are just distributed differently.

The efficiency gap is "the difference between the parties' respective wasted votes, divided by the total number of votes cast" (Stephanopoulos and McGhee 2014, p. 831).

The court found that "efficiency gap analysis provides evidence that Defendants violated the governing constitutional standard: that a redistricting body must not adopt a districting plan that intentionally subordinates the interests of supporters of a disfavored party and entrenches a favored party in power" (Common Cause v. Rucho 279 F. Supp. 3d 587, M.D.N.C. 2018., p. 661).

"Efficiency gaps close to zero, which occur when the two parties waste approximately the same number of votes, reflect a districting plan that does not favor, invidiously or otherwise, one party or the other" (Common Cause v. Rucho. 318 F. Supp.3d 777, 799+, M.D.N.C. 2018, p. 886).

Stephanapolous and McGhee propose a standard of two seats for congressional plans and 8% for state house plans. Above that, the redistricting plans would be considered unconstitutional (Stephanopoulos and McGhee 2014, p. 837).

The efficiency gap approach does have the advantage of being a simple method, despite protestation from Chief Justice John Roberts that is was "sociological gobbledygook" (Matthews 2017).

But the efficiency gap has some problems. The main flaw is that "the efficiency gap punishes plans that have lots of competitive seats," according to political scientist Bruce Cain (Denton 2018).

Consider a couple of hypothetical districts: one very competitive, one lopsided.

Table 16.3 shows the results for a given hypothetical election, with 100 voters per district, and parties A and B.

District 1 is competitive, with a vote difference between the two parties of only 4 percentage points. Yet it has very high efficiency gap. District 2 has vote difference of 40 percentage points between the two parties, but its efficiency gap is low.

According to the court, "Efficiency gaps close to zero, which occur when the two parties waste approximately the same number of votes, reflect a districting plan that does not favor, invidiously or otherwise, one party or the other" (Common Cause v. Rucho. 318 F.Supp.3d 777, 799+, M.D.N.C. 2018, p. 84).

So, because our lopsided district has the lowest efficiency gap, it is seen to not favor a party. But our *competitive* district *is* expected to favor a party.

The creators of the efficiency gap acknowledge some limitations. These include uncontested elections and districts in which one party receives an extremely high share of the vote, such as at least 75% (Stephanopoulos and McGhee 2014:863).

The usefulness of the efficiency gap is also limited to larger entities, with more than six districts, according to Simon Jackman. (Common Cause v. Rucho, 2018).

Bernstein and Duchin (2017: 1002) found other problems with efficiency gap analysis, including that it "fetishizes three-to-one landslide districts". They see the efficiency gap as one tool that can be used as part of a broader analysis (Bernstein and Duchin 2017:1023).

Sampling with Markov Chain Monte Carlo (MCMC) Method

Another approach, the Markov Chain Monte Carlo method, can be used to measure not only partisan gerrymandering, but also certain other aspects, such as racial bias or competitiveness (Duchin 2018b: 4).

With this method, results of various valid plans for the same jurisdiction are compared against each other (Duchin 2018a). For example, comparing valid congressional districting plans for New Mexico controls for geography, demography and relevant laws. So the first step is to limit the possible plans by all the requirements of valid plans, such as keeping the pieces contiguous, limiting the population deviation, and keeping the same number of majority-minority districts, based on the laws in that jurisdiction (Duchin 2018b: 4).

Still, a multitude of plans could be made for any given jurisdiction. For example, if you wanted to get nine equal districts out of a nine-by-nine grid, you would have more than 700 trillion to choose from (Duchin 2018a). So it is at best not practical to compare *all* possible plans for a given jurisdiction. Instead, only a random sample of those plans are used. Just as in opinion polls, a large-enough sample represents all possible plans.

But the samples must be chosen randomly, by a particular method. As Duchin explains (2018b: 4): It is based on the smallest units of which districts can be composed. For example, New Mexico districts are composed of a number of precincts. Group those units into any valid plan for the jurisdiction, for the first member of the sample. Then, one unit is flipped from one district to another, making the second member of the sample. This flipping continues, making new plans, one unit at a time, through thousands to trillions of plans. Thus, *sampling* is used to define an *ensemble* of districting plans (Duchin 2018b: 4).

Using MCMC, a great many plans could be compared by, for example, how many Democrats or Republicans would have been elected with each plan. If the result of any given plan was an extreme outlier compared to the others in the sample or ensemble, that could be considered extreme partisan gerrymandering. Over time, a legal standard could be developed that would disallow extreme outliers (Duchin 2018b: 4).

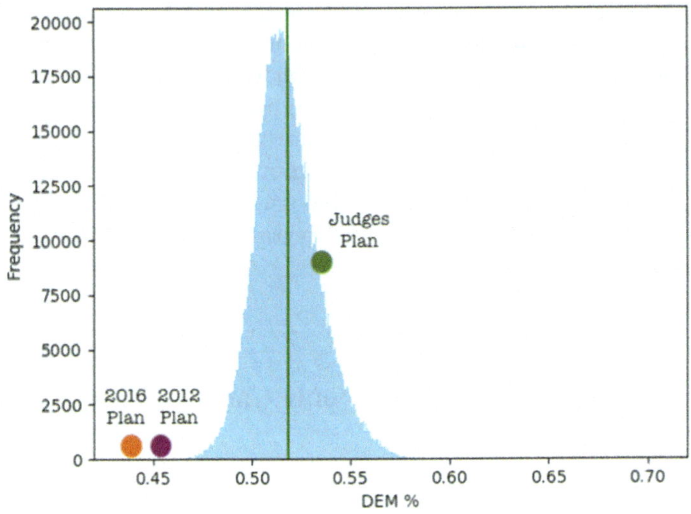

Fig. 16.15 North Carolina Legislature District 10 (Source: Metric Geometry and Gerrymandering Group)

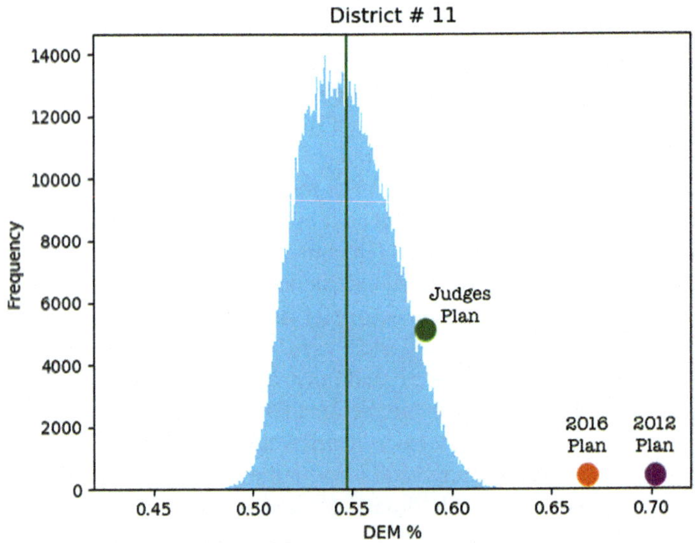

Fig. 16.16 North Carolina Legislature District 11 (Source: Metric Geometry and Gerrymandering Group.)

Consider the examples in Figs. 16.15 and 16.16 (Metric Geometry and Gerrymandering Group 2019). A million redistricting plans for North Carolina were sampled by MCMC, and those results are shown by the bell curves (Metric Geometry and Gerrymandering Group 2019). The red and purple circles represent plans enacted by the North Carolina legislature (Metric Geometry and Gerrymandering Group 2019). In these plans, there are very few Democrats in District 10, but a superabundance of them in District 11. These practices are known as "cracking" and "packing." Each plan is a far outlier.

Conclusion

We have outlined traditional districting principles, with a focus on measuring compactness.We have also discussed a few measures that can evaluate possible gerrymandering: the efficiency gap, partisan symmetry, and random sampling, such as with the Markov chain Monte Carlo method. Some other measures for possible gerrymandering can be found in Common Cause's Partisan Gerrymandering Writing Competition.

Although these criteria do not have universal standards, each state should set its own, including for the priorities among the criteria. Further, it is also important that the map drawers provide the decision makers and the public the results of evaluations for these criteria. That way, they can best make informed judgments.

Among the different methods to compare partisan gerrymandering, the Markov Chain Monte Carlo method is the most comprehensive. By comparing any given plan to many other possible plans, it gives the best indication of whether the plan is extreme or not.

Disclaimer This article is an original work by the named authors and is being submitted exclusively to the International Association of Applied Demography for publication consideration into an edited book on Applied Demography by Springer Publishing Company.

References

Barabas, J., & Jerit, J. (Winter 2004). Redistricting principles and racial representation. *State Politics & Policy Quarterly, 4*(4), 415–435. https://doi.org/10.1177/153244000400400404.

Bernstein, M., & Duchin, M. (2017). A formula goes to court: Partisan gerrymandering and the efficiency gap. *Notices of the American Mathematical Society, 64*(09), 1020–1024. https://doi.org/10.1090/noti1573.

Bowen, D. C. (2014). Boundaries, redistricting criteria, and representation in the U.S. house of representatives. *American Politics Research, 42*(5), 856–895. https://doi.org/10.1177/1532673x13519127.

Charles, G. E., Gupta, D., & Wilf-Townsend, D. (2019, March 8). *Amicus brief of mathematicians, law professors and students in support of appellees and affirmance*. Retrieved from https://mggg.org/SCOTUS-MathBrief.pdf

Common Cause. (n.d.). *Partisan gerrymandering writing competition*. Retrieved from https://www.commoncause.org/partisan-gerrymandering-writing-competition/

Common Cause v. Rucho, 279 F. Supp. 3d 587, M.D.N.C. (2018). Retrieved from https://casetext.com/case/cause-v-rucho-4

Common Cause v. Rucho, 318 F.Supp.3d 777, 799+, M.D.N.C. (2018). Retrieved from https://casetext.com/case/cause-v-rucho-5?

Curiel, J. A., & Steelman, T. (December 2018). Redistricting out representation: Democratic harms in splitting zip codes. *Election Law Journal: Rules, Politics, and Policy, 17*(4), 328–353. https://doi.org/10.1089/elj.2018.0528.

Denton, J. (2018, June 28). Breaking down the Supreme Court's inaction on gerrymandering. *Pacific Standard.* Retrieved from https://psmag.com/news/breaking-down-the-supreme-courts-inaction-on-gerrymandering

Duchin, M. (2018a). Geometry versus gerrymandering. *Scientific American.* Retrieved from https://www.scientificamerican.com/article/geometry-versus-gerrymandering/

Duchin, M. (2018b). Gerrymandering metrics: How to measure? What's the Baseline? *Arxiv.* Retrieved from https://arxiv.org/pdf/1801.02064.pdf

Every Vote Counts. (n.d.). *Partisan Gerrymandering.* Retrieved from https://www.evcnational.org/partisan-gerrymandering

Forgette, R., Garner, A., & Winkle, J. (2009, Summer). Do redistricting principles and practices affect U.S. state legislative electoral competition? *State Politics & Policy Quarterly, 9*(2), 151–175. https://doi.org/10.1177/153244000900900202.

Gabriel, T. (2018, January 26). In a comically drawn Pennsylvania District, the voters are not amused. *The New York Times.* Retrieved from https://www.nytimes.com/2018/01/26/us/pennsylvania-gerrymander-goofy-district.html

Glossary. (n.d.). Retrieved from https://redistricting.azavea.com/glossary/

Grofman, B., & King, G. (2007). The future of Partisan symmetry as a judicial test for Partisan gerrymandering after LULAC v. Perry. *Election Law Journal: Rules, Politics, and Policy, 6*(1), 2–35. https://doi.org/10.1089/elj.2006.6002.

Iowa Code § 42.4. Retrieved from https://www.legis.iowa.gov/docs/code/2016/42.4.pdf

Levitt, J. (2010). *A citizen's guide to redistricting.* Retrieved from https://www.brennancenter.org/sites/default/files/legacy/Democracy/CitizensGuidetoRedistricting_2010.pdf

Levitt, J. (n.d.). All about redistricting. Retrieved January 04, 2019, from http://redistricting.lls.edu/

Liptak, A. (2017, May 15). When does political gerrymandering cross a constitutional line? *The New York Times.* Retrieved from https://www.nytimes.com/2017/05/15/us/politics/when-does-political-gerrymandering-cross-a-constitutional-line.html

Matthews, D. (2017, October 12). *Chief Justice John Roberts is now feuding with the entire field of sociology.* Vox. Retrieved from https://www.vox.com/policy-and-politics/2017/10/12/16464188/john-roberts-sociological-gobbledygook-eduardo-bonilla-silva-gerrymandering

Measuring Compactness: A Fair Redistricting Process. (n.d..). Retrieved January 22, 2019, from http://www.redistrictingthenation.com/whatis-compactness.aspx

Metric Geometry and Gerrymandering Group. (2019, March 18). The mathematicians' brief in Rucho v. common cause. Retrieved from https://mggg.org/amicus/

Morrison, P. A., & Bryan, T. M. (2019). *Redistricting: A manual for analysts, practitioners, and citizens.* Springer.

National League of Cities. *Cities 101 – Types of local governments.* https://www.nlc.org/resource/cities-101-types-of-local-governments

Niemi, R.G.; Grofman, B., Carlucci, C., & Hofeller, T. (1990). Measuring compactness and the role of a compactness standard in a test for partisan and racial gerrymandering. *The Journal of Politics, 52*(4), 1155. Retrieved from https://www-jstor-org.libproxy.unm.edu/stable/2131686

Redrawing the Map on Redistricting 2010. (2010). Retrieved from https://cdn.azavea.com/com.redistrictingthenation/pdfs/Redistricting_The_Nation_White_Paper_2010.pdf

Rucho v. Common Cause (2019) No. 18–422, 588 U.S. ___ (2019). Retrieved from https://www.law.cornell.edu/supremecourt/text/18-422

Stephanopoulos, N O., & McGhee, E. M. (2014, October 1). Partisan gerrymandering and the efficiency gap. *University of Chicago Law Review 82*, 831–900. Retrieved from https://papers.ssrn.com/sol3/papers.cfm?abstract_id=2457468

The National Atlas of the United States of America. (n.d..). Printable maps. Retrieved June 24, 2019, from https://nationalmap.gov/small_scale/printable.html#congress

Timm, J. C. (September 21, 2017). *They're still drawing crazy-looking districts. Can't it be stopped?* Retrieved from https://www.nbcnews.com/politics/elections/they-re-still-drawing-crazy-looking-districts-can-t-it-n803051

Trickey, E. (2017, July 20). *Where did the term "Gerrymander" come from?* Retrieved from https://www.smithsonianmag.com/history/where-did-term-gerrymander-come-180964118/

Vieth v. Jubelirer, *541* U.S. *267* (2004). Retrieved from https://casetext.com/pdf-email?slug=vieth-v-jubelirer-4

What's at Stake in Common Cause v. Rucho. (2017, October 18). Retrieved from https://www.commoncause.org/democracy-wire/whats-at-stake-in-common-casue-v-rucho/

Chapter 17
The Number of Native Hawaiians and Part-Hawaiians in Hawai'i, 1778 to 1900: Demographic Estimates by Age

David A. Swanson

Introduction

Swanson (2019) provided an estimate of 683,200 for the population of Hawai'i in 1778, the year of first European contact and decennial estimates of the number of Native Hawaiians from 1780 to 1900.[1] This paper extends the earlier work by providing an estimate of the number of Part-Hawaiians in Hawai'i in 1780 and decennial estimates from 1780 to 1900.[2] For purposes of comparison, the estimates of Native Hawaiians are included and for both groups estimates by 5-year age groups are given. This historical information has never before been assembled for Part-Hawaiians and, of course, no historical comparisons have been made in terms of the historical demographic dynamics of the two populations.

As was the case with the historical estimates of Native Hawaiians developed by Swanson, this paper replies on demographic momentum. This momentum is captured by a form of "backcasting" that is consistent with the demographic components of change However, momentum is not sufficient to fill in all of the missing pieces. Thus, in this paper, backcasting is assisted by other tools, including interpolation and regression.

---

[1]Although there are arguments that the Spanish visited Hawai'i in the sixteenth century, most evidence suggests that the English were the first Europeans to do so, with the arrival of Cook in 1778 (Nordyke 1989: 15–18). This paper follows the argument that the English contact in 1778 was the first.

[2]In principle, the estimates of Part-Hawaiians and Native Hawaiians are based on US Census Bureau concepts, which, in turn, are based on the De Jure definition of a population and self-identification in terms of demographic and other characteristics.

D. A. Swanson (✉)
Edward A. Dickson Emeritus Professor, Department of Sociology, University of California Riverside, Riverside, CA, USA
e-mail: dswanson@ucr.edu

B. Jivetti, Md. N. Hoque (eds.), *Population Change and Public Policy*, Applied Demography Series 11, https://doi.org/10.1007/978-3-030-57069-9_17

Data and Methods

There are several points of departure for the development of historical estimates of Part-Hawaiians in Hawai'i. The first is a set of numbers for this population found in Table 3.1 for various years from 1853 to 1960 provided by Nordyke (1989: 178). The numbers prior to 1900 number represent census counts conducted by the Kingdom of Hawai'i before it was annexed by the U.S. and the numbers from 1900 to 1960 (in 10 year cycles) are taken from the U.S. Census. The Kingdom's census counts were for the years 1853, 1860, 1866, 1872, 1878, 1884, 1890, and 1896 (Nordyke 1989: 178).

The second point of departure is the set of decennial estimates developed by Swanson (2019) for the total number of Native Hawaiians in Hawai'i from 1778 to 1900. These date are shown in Table 17.1. Note the interrupted decline in the number of Native Hawaiians, from 683,200 in 1778 to 30,546.

The third point of departure is the U.S. Census Bureau's decennial data on Part-Hawaiians in Hawai'i from 1910 to 1930. From these data sources, the RCCRs along with extrapolation and a time-based regression method were used to develop preliminary estimates by age that were controlled to known numbers. The result in terms of the total number of Part-Hawaiians in Hawai'i from 1778 to 1900 is found in Table 17.2.

The numbers in Table 17.1 for 1778 and from 1780 to 1990 are taken from Swanson (2019). Because the backcasting method used to construct the Native Hawaiian population is discussed in detail by Swanson (2019), only a summary description is provided here. Using US Census data for 1910, 1920, and 1930, a standard forecasting technique known as Hamilton-Perry method (Smith et al. 2013: 176–179) was run in reverse. Briefly, the Hamilton-Perry Method forms a ratio of an age cohort in one census (e.g., the population aged 5–9 years in 1900) and the same

Table 17.1 Number of native Hawaiians in Hawai'i: 1780–1900

Year	Number
1778	683,200
1780	644,383
1790	480,978
1800	359,010
1810	267,971
1820	200,018
1830	149,297
1840	110,948
1850	80,574
1860	61,931
1870	48,579
1880	39,711
1890	33,457
1900	29,336

Descriptions of sources and methods are found in the text

Table 17.2 Number of part-Hawaiians in Hawai'i: 1780–1900

Year	Number
1778	0
1780	9
1790	17
1800	31
1810	58
1820	109
1830	203
1840	381
1850	714
1860	1337
1870	2785
1880	4459
1890	6186
1900	7857

Descriptions of sources and methods are found in the text

cohort in the successive census (e.g., the population aged 15–19 years in 1910). This is known as a cohort change ratio (CCR). The CCRs are then applied to a population by age (e.g., the 1910 population of Hawai‘i) to carry it forward in time as a forecast (e.g., to 1920). When the Hamilton-Perry Method is used for backcasting rather than forecasting, the cohort change ratios are run in reverse. A reverse cohort change ratio (RCCR) is the reciprocal of the corresponding CCR (Swanson and Tayman 2012: 340–353). That is, if the CCR for those aged 0–4 years in 1900 relative to those aged 10–14 years in 1910 is .858458, then the corresponding RCCR for those aged 10–14 in 1910 relative to those aged 0–4 in 1900 is 1/.858458 = 1.164878. With this method an adjustment was needed for the terminal open-ended age group (e.g., those aged 75 years and over) to avoid the fact that every 10 years (in the US Census context) this group would be 10 years younger and, as such, successively providing less information about the age structure of the population in question. (Swanson 2019). The RCCRs and the adjustments were initially applied to the 1910 census by age to generate a set of backcasted 1900 estimates by age for the Native Hawaiian population in Hawai‘i. The same RCCRs were then applied to the 1900 estimates by age to generate a set of backcasted 1890 estimates by age. This process was repeated until the 1770 population of Native Hawaiians by age was generated for Hawai‘i. The backcasting proceeded in decennial cycles from 1900 to 1770.[2] By interpolating between 1770 and 1780 an estimate for 1778 was produced.

In order to generate the population of Part-Hawaiians in Hawai‘i, RCCRs were also employed, but they were supplemented with other methods in some periods because of incomplete data. In addition, there was an important constraint on this population that was not a constraint on the population of Native Hawaiians: At the moment of first European contact there were no Part-Hawaiians, only Native Hawaiians. This is a benefit in terms of estimating the population of Part-Hawaiians from 1778 to the present and beyond because unlike the population of Native Hawaiians in 1778, there is no uncertainty about the number of Part-Hawaiians – it was zero. Thus, we know that the backcasting process has to conform to a population of zero by 1778.

Because it is convenient to start with the year 1780 in that it matches the decennial census cycle used by the US Census Bureau, both the Native and Part-Hawaiian estimates are set to 1780 as the point of departure. This means that there is some uncertainty about the number of Part-Hawaiians at this time. We know that at the moment of first contact there were zero Part-Hawaiians, but we do not know the exact number of Part-Hawaiians as of 1780, other than it was much, much smaller than the population of Native Hawaiians and that all of them had to be 2 years old or younger by 1780.

To begin the construction of the estimates of Part-Hawaiians from 1870 to 1850, a (reverse) rate (r) of population change was found for the period from 1860 to 1853 using data provided by Nordyke (1989: 178) on the number of Part-Hawaiians for these 2 years, 1337 and 862, respectively. The model in Excel terms is r = (POWER ((J12/J13),1/7) = 0.939159. This model was applied to the 1860 number (1337) to generate estimated numbers in 10-year cycles back to 1780. The numbers for 1860 to 1780 can be seen in Table 17.2 as can the number for 1778, which is zero.

Using the decennial estimates from 1860 to 1790 generated from the reverse extrapolation, a simple regression model ($n = 8$) was generated in order to develop estimates for 1870 and 1880 using year as the independent variable. Constructed in Excel, the model is $N = -3101.35 + 167.34*YR$ ($r^2 = .92$), where N = number of estimated Part-Hawaiians and YR = Year. For 1870, the model yielded an estimate of 2785 Part-Hawaiians ($2785 \approx -310{,}135 + (167.3371*1870)$) and for 1880, the model yielded an estimate of 4.459 ($4459 \approx -320{,}235 + (167.23371*1880)$. With these estimates for 1870 and 1880, an initial complete set of decennial estimates of Part-Hawaiians was available for Hawai‘i from 1780 to 1900.

Once the total population of Part-Hawaiians was constructed, Reverse Cohort Change Ratios (RCCRs) were constructed for the period 1920–1910 and used to generate initial estimates by age back to 1860 in decennial cycles. This step followed what was done by Swanson (2019) for the Native Hawaiian population. These estimates were then "controlled" to the independently estimated total populations for these years, respectively (Table 17.2). For the estimates by age from 1850 back to 1780, knowledge of the age of Part-Hawaiians was used. Of the nine estimated Part-Hawaiians in 1780, all were in age group 0–4 since none was older than 2 years. By 1790, none could have been older than 10–14, by 1800, none could have been older than 20–24, and so on to 1850 when none could have been older than 70–74. The survival rates described by Swanson (2019) were used to survive the population each decennial cycle by 10 years from 1780 forward to 1850 and the number of births was estimated by finding the difference between the survivors at any given time and the independently estimated total population. As an example of this process, in 1790 there were four survivors aged 10–14 remaining of the nine Part-Hawaiians in 1780 (who were aged 0–4). Assuming that births were evenly distributed over the 10 year period, of those born between 1780 and 1784, five were estimated to have reached age 5–9 by 1790 and of those born between 1785 and 1789, seven were estimated to have been alive (and aged 0–4) in 1790.

The Number of Part-Hawaiians, 1780 to 1900

Table 17.2 provides the total number of Part-Hawaiians in Hawai'i for 1778 and the decennial number from 1780 to 1900 in decennial cycles. Starting with an estimated population of zero in 1778, the number increases to nine in 1780 and to 7857 by 1900.

While the Native Hawaiian population is estimated to have declined from 644,383 in 1780 to 30,546 by 1900, a loss of 613,837, the Part-Hawaiian population is estimated to have increased from zero in 1778 to 7857 by 1900. The increase is monotonic and not dramatic. This is in contrast to the dramatic monotonic decline of the Native Hawaiian population during this same period, from 683,200 in 1778 to 29,336 by 1900. Largely, because of the decline in the Native Hawaiian population, the population of Part-Hawaiians is nearly one quarter its size by 1900.

The Number of Native and Part-Hawaiians by Age, 1780 to 1990

Table 17.3 provides the decennial estimates by 5-year age groups (0–4, 5–9,. . .,70–74, 75+) for the Part-Hawaiian population in Hawai'i from 1780 to 1900. In it, one can see the movement of the members of this population by cohort as they are survived into older age groups across time. Not until 1860 does this population have members from age 0–4 through 745 years and over.

Table 17.4 provides the same data for the Native Hawaiian population in Hawai'i. As one moves from 1780 to 1900, substantial declines are seen across all age groups. The declines to the population of child-bearing years is particularly important since this is the source of births. Given that the childbearing population is aged from 15 to 44, one sees that in 1780 there are 286,497 Native Hawaiians of childbearing age. By 1900, there are only 13,894 Native Hawaiians of childbearing age.

Table 17.5 provides the combined numbers by 5 year age groups for the Native and Part-Hawaiian populations in Hawai'i. In 1780 this group was composed of virtually all Native Hawaiians, with only nine estimated Part-Hawaiians that were no more than 2 years old out of a total of 644,391. By 1900, nearly one-quarter of this combined population is composed of Part-Hawaiians.

Table 17.3 Total part-Hawaiian population by age in Hawai'i, 1780 To 1990 (Excludes native Hawaiians)

Age	1780	1790	1800	1810	1820	1830	1840	1850	1860	1870	1880	1890	1900
0–4	9	7	12	18	29	49	83	145	109	230	393	577	789
5–9	0	5	7	13	22	37	65	116	98	215	425	616	815
10–14	0	4	5	10	19	33	59	107	80	249	386	556	717
15–19	0	0	4	7	14	25	47	86	66	226	367	611	780
20–24	0	0	3	4	11	21	40	75	96	180	413	538	682
25–29	0	0	0	3	6	14	28	53	98	134	337	460	672
30–34	0	0	0	2	4	10	21	42	97	179	247	477	546
35–39	0	0	0	0	3	6	14	29	107	212	214	453	544
40–44	0	0	0	0	2	4	10	23	94	186	252	293	497
45–49	0	0	0	0	0	3	6	16	111	220	322	274	509
50–54	0	0	0	0	0	2	3	10	84	167	244	278	284
55–59	0	0	0	0	0	0	2	5	82	163	238	293	219
60–64	0	0	0	0	0	0	2	3	69	137	201	247	247
65–69	0	0	0	0	0	0	0	2	61	122	178	219	237
70–74	0	0	0	0	0	0	0	1	34	68	99	122	132
75+	0	0	0	0	0	0	0	0	49	97	141	174	188
Total	**9**	**17**	**31**	**58**	**109**	**203**	**381**	**714**	**1337**	**2785**	**4459**	**6186**	**7857**

Table 17.4 Total native Hawaiian population by age in Hawai'i, 1780 to 1900 (Excludes part-Hawaiians)

Age	1780	1790	1800	1810	1820	1830	1840	1850	1860	1870	1880	1890	1900
0–4	38,229	28,535	21,299	15,898	11,866	8857	6122	4321	5065	4004	3502	3120	2945
5–9	39,476	29,466	21,994	16,416	12,254	9146	6827	3525	4522	3746	3782	3333	3043
10–14	43,967	32,818	24,496	18,284	13,648	10,187	7604	5255	3709	4348	3438	3007	2678
15–19	46,184	34,473	25,731	19,206	14,336	10,700	7987	5962	3078	3949	3271	3303	2911
20–24	49,864	37,219	27,781	20,736	15,478	11,553	8623	6437	4449	3140	3680	2910	2545
25–29	47,041	35,112	26,208	19,562	14,602	10,899	8135	6072	4532	2340	3002	2487	2511
30–34	46,825	34,951	26,088	19,472	14,535	10,849	8098	6044	4512	3118	2201	2580	2040
35–39	51,432	38,389	28,654	21,388	15,964	11,916	8894	6639	4955	3699	1910	2450	2030
40–44	45,151	33,701	25,155	18,776	14,015	10,461	7808	5828	4350	3247	2244	1584	1857
45–49	53,469	39,910	29,790	22,236	16,597	12,388	9247	6902	5152	3845	2870	1482	1901
50–54	40,464	30,203	22,544	16,827	12,560	9375	6998	5223	3899	2910	2172	1501	1060
55–59	39,524	29,501	22,020	16,436	12,268	9157	6835	5102	3808	2842	2122	1584	818
60–64	33,304	24,858	18,555	13,850	10,338	7716	5759	4299	3209	2395	1788	1334	922
65–69	29,530	22,042	16,452	12,280	9166	6842	5107	3812	2845	2124	1585	1183	883
70–74	16,449	12,278	9164	6840	5106	3811	2845	2123	1585	1183	883	659	492
75+	23,473	17,521	13,078	9762	7286	5439	4059	3030	2262	1688	1260	941	702
Total	**644,383**	**480,978**	**359,010**	**267,971**	**200,018**	**149,297**	**110,948**	**80,574**	**61,931**	**48,579**	**39,711**	**33,457**	**29,336**

Table 17.5 Total Hawaiian population by age in Hawai'i, 1780 To 1900 (Includes both native Hawaiians & part -Hawaiians)

Age	1780	1790	1800	1810	1820	1830	1840	1850	1860	1870	1880	1890	1900
0–4	38,238	28,542	21,310	15,916	11,896	8906	6205	4466	5174	4234	3896	3696	3734
5–9	39,476	29,471	22,001	16,429	12,275	9184	6892	3641	4620	3961	4206	3950	3858
10–14	43,967	32,822	24,501	18,295	13,666	10,220	7663	5362	3790	4597	3823	3563	3395
15–19	46,184	34,473	25,735	19,213	14,349	10,726	8034	6047	3145	4175	3639	3913	3690
20–24	49,864	37,219	27,784	20,741	15,488	11,574	8664	6512	4545	3320	4094	3448	3227
25–29	47,041	35,112	26,208	19,565	14,608	10,913	8163	6125	4630	2474	3339	2947	3183
30–34	46,825	34,951	26,088	19,475	14,538	10,859	8119	6087	4609	3297	2448	3057	2586
35–39	51,432	38,389	28,654	21,388	15,967	11,922	8908	6668	5062	3911	2124	2903	2573
40–44	45,151	33,701	25,155	18,776	14,017	10,465	7818	5851	4444	3433	2496	1877	2354
45–49	53,469	39,910	29,790	22,236	16,597	12,391	9253	6918	5263	4066	3192	1756	2410
50–54	40,464	30,203	22,544	16,827	12,560	9377	7001	5233	3983	3077	2416	1779	1343
55–59	39,524	29,501	22,020	16,436	12,268	9157	6837	5107	3890	3005	2360	1876	1037
60–64	33,304	24,858	18,555	13,850	10,338	7716	5761	4302	3278	2532	1988	1581	1169
65–69	29,530	22,042	16,452	12,280	9166	6842	5107	3813	2907	2245	1763	1402	1120
70–74	16,449	12,278	9164	6840	5106	3811	2845	2124	1619	1251	982	781	624
75+	23,473	17,521	13,078	9762	7286	5439	4059	3030	2310	1785	1402	1114	890
Total	**644,391**	**480,994**	**359,041**	**268,029**	**200,126**	**149,500**	**111,329**	**81,288**	**63,268**	**51,364**	**44,169**	**39,643**	**37,193**

Discussion

From 1778 to 1900, the Native Hawaiian population declined substantially. There was some outmigration of Native Hawaiians from Hawai'i, but the decline was overwhelmingly due to the fact that deaths consistently outnumbered births each year (Swanson 2019). For the Part-Hawaiian population, the story between 1778 and 1900 is just the opposite, births consistently outnumbered deaths. Why is this so? The answer is found in terms of births. First, in terms of demographic dynamics, one needs to understand that the only way a Part-Hawaiian came into the world until around 1800 or so was through a union between a Native Hawaiian and a non-Hawaiian because there were no Part-Hawaiians of reproductive age. Once the point was reached where there were Part-Hawaiians of reproductive age, a Part-Hawaiian could come into the world through a broader range of unions: (1) Part-Hawaiian and Part Hawaiian; (2) Part-Hawaiian and non-Hawaiian; (3) Part-Hawaiian and Native Hawaiian; and (4) Native Hawaiian and non-Hawaiian. From 1778 to today and into the future, the only way a Native-Hawaiian can come into the world is through a union between two Native Hawaiians in terms of demographic factors.

Thinking in these terms leads one to realize why today that Part-Hawaiians far outnumber Native Hawaiians. The US Census Bureau estimates, for example, that there are 194,660 Part-Hawaiians in Hawai'i as of 2010 and 78,389 Native Hawaiians (US Census Bureau, 2010). The estimated number of Native Hawaiians in 2010 clearly indicates that this population has made a comeback by more than doubling from 37,193 in 1900, but the growth in the Part-Hawaiian population is much more dramatic, growing from 7857 in 1900 to nearly 195,000 by 2010, which is almost 24 times the number estimated for 1900. In terms of demographic factors, the changes observed for both populations are largely due to births outnumbering deaths and the Part-Hawaiian population has a broader range of possibilities when it comes to births. One avenue of future work is to develop estimates of the two populations by age and sex. This will provide a clearer picture of the reproductive capabilities of each of them.

The growth in both the Native and Part-Hawaiian populations of Hawai'i between 1900 and 2010 is not only due to demographic factors (births, deaths, and migration), but also social factors. As a great deal of research shows, ethnicity and race are social constructs and fluid (see, e.g., Cornell and Hartmann 2007; Goldstein and Morning 2000; Nagel 1994, 1995; Omi and Winant 2015; Perez and Hirschman 2009; Yamashiro 2011). As such, a population defined on the basis of race or ethnicity is subject to change from factors that are not demographic. Such a change has been documented by Passel (1996), for example in regard to the American Indian population. Given Passel's findings and the "Hawaiian Renaissance" that started in the 1970s, it is likely that some of the recent changes in the numbers of both the Native and Part-Hawaiian populations, especially in Hawai'i, may be due to social and other non-demographic factors (see, e.g., Kana'iupuni and Liebler 2005; Kana'iupuni and Malone 2006; Ledward 2007; Stephan 1991). It is equally likely that changes impacting Hawai'i in the past led to changes in the definition of Native Hawaiian and Part-Hawaiian in terms of self-identification, as well as identification by others, and institutions.[3] Clearly, the documentation of change in the Native and Part-Hawaiian populations from 1900 to the current time (and perhaps into the future) should be undertaken so that as complete a record as possible exists back to the milestone year of 1778. This documentation should include an examination of both demographic and non-demographic factors as well as the development of estimates by age and sex.

Acknowledgments This paper benefited from comments and advice by a number of people, including Peter Burke, Sara Kehaulani Goo, Shawn Kana'iaupuni, Carolyn Liebler, Nolan Malone, Nancy Marker, David Takeuchi, Amy Ong Tsui, and Jane Yamashiro.

References

Berman, M. (1988). *All that is solid melts into air: The experience of modernity*. New York: Penguin Press (Re-issue Edition).

Cornell, S., & Hartmann, D. (2007). *Ethnicity and race: Making identities in a changing world*. Thousand Oaks: Pine Forge Press.

Goldstein, J., & Morning, A. (2000). The multiple-race population of the United States: Issues and estimates. *Proceedings of the National Academy of Sciences of the United States of America, 97* (11), 6230–6235.

Kana'iupuni, S., & Liebler, C. (2005). Pondering poi dog: Place and racial identification of multiracial native Hawaiians. *Ethnic and Racial Studies, 28*(4), 687–721.

Kana'iupuni, S., & Malone, N. (2006). This land is my land: The role of place in native Hawaiian identity. In J. Frazier & E. Tetty-Fio (Eds.), *Race, ethnicity and place in a changing America* (pp. 291–305). New York: Global Press.

Ledward, B. (2007). On being Hawaiian enough. *Hūlili, 4*(1), 107–143.

Nagel, J. (1994). Constructing ethnicity: Creating and recreating ethnic identity and culture. *Social Problems, 41*(1), 152–176.

Nagel, J. (1995). American Indian ethnic renewal: Politics and resurgence of identity. *American Sociological Review, 60*(6), 947–965.

Nordyke, E. (1989). *The peopling of Hawai'i* (2nd ed.). Honolulu: University of Hawai'i Press.

Omi, M., & Winant, H. (2015). *Racial formation in the United States* (3nd ed.). New York: Routledge.

[3]Because the initial population of Part-Hawaiians arose after Cook's visit, it may be the case, for example, that within Hawai'i they were initially identified as Native Hawaiian by the Hawaiian people, but over time, with the arrival of more foreigners, especially American Missionaries, these same people may have been later identified as Part-Hawaiian. While the "identification" issues are open questions that may or may not be answered, it is certain that as modern capitalism lapped at Hawai'i's shores and then overwhelmed the islands, Marx's observation that "all that is solid melts into air," (see, e. g., Berman 1988) came into play and ideas about ethnicity and race in Hawai'i were subject to the same rapidity and levels of change as were ideas concerning property, labor, the production and definition of wealth, social class, and socio-economic status.

Passel, J. (1996). The growing American Indian population, 1960-1990: Beyond demography. In G. Sandedur, R. Rindfuss, & B. Cohen (Eds.), *Changing numbers, changing needs: American Indian demography and public health* (pp. 79–102). Washington, DC: National Academies Press.
Perez, A., & Hirschman, C. (2009). The changing racial and ethnic composition of the US population: Emerging American identities. *Population and Development Review, 35*(1), 1–51.
Smith, S., Tayman, J., & Swanson, D. (2013). *A Practitioner's guide to state and local population projections*. Dordrecht: Springer.
Stephan, C. W. (1991). Ethnic identity among mixed-heritage people in Hawaii. *Symbolic Interaction, 14*(3), 261–277.
Swanson, D., & Tayman, J. (2012). *Subnational population estimates*. Dordrecht: Springer.
Swanson, D. A. (2019). A new estimate of the Hawaiian population for 1778, the year of first European contact. *Hulili, 11*(2), 203–222.
U.S. Census Bureau. (2010). *Hawaii. Sex by age, native Hawaiian only and native Hawaiian only in combination with one or more other races*. American Factfinder Extract from the 2006–2010 American Community Survey.
Yamashiro, J. (2011). Racialized national identity construction in the ancestral homeland: Japanese American migrants in Japan. *Ethnic and Racial Studies, 34*(9), 1502–1521.

Chapter 18
Housing Development and Enrollment Trajectories in K-12 School Districts

Charles Rynerson

Background

For existing members of a community, large scale residential developments are a vivid manifestation of growth, often inspiring pride among civic boosters but more commonly viewed as threats to quality of life. Increased traffic, loss of open space, impacts on property values, erosion and runoff, water and sewer capacity, and a myriad of other ills are cited by neighbors opposed to new development (Lewis and Neiman 2002). School overcrowding and the cost of new school facilities are among the most common reasons for residents to resist growth (Nguyen 2009; Machell et al. 2009).

School district administrators are also wary of new development, perhaps due to past experiences with insufficient capacity, recognition of the costs and challenges of adding capacity, or pressure from school communities including principals and existing families who fear school overcrowding. In response, school demographers are asked to estimate the enrollment impacts of proposed developments in order to support school district facility planning and to address community concerns (Yarzebinski 1992).

Estimates of the average number of public school students per housing unit may be derived for large geographic areas from the U.S. Census Bureau Public Use Microdata Sample (Wong et al. 2017). More refined estimates may be derived from case studies of comparable developments if geolocated student address data and

---

An original work by Charles Rynerson submitted exclusively to the International Association of Applied Demography for publication consideration into an edited book on Applied Demography by Springer Publishing Company.

C. Rynerson (✉)
Population Research Center, Portland State University, Portland, OR, USA
e-mail: rynerson@pdx.edu

B. Jivetti, Md. N. Hoque (eds.), *Population Change and Public Policy*, Applied Demography Series 11, https://doi.org/10.1007/978-3-030-57069-9_18

local residential parcel data are available. These estimates are called Residential Demographic Multipliers, Pupil Yield Factors, or Student Generation Rates (SGRs).

When the number and type of future housing units are known, estimates based on SGRs may be the best tool available to predict the future number of students residing in new developments. In addition to helping school districts plan for growth, providing evidence from comparable existing developments may counter inflated expectations of two or more school-age children in every home. The Population Research Center (PRC) at Portland State University has calculated SGRs for several districts in suburban Portland, Oregon, finding as few as 0.45 and as many as 0.84 K-12 residents per new detached single family home district-wide. In most districts, rental apartments, condominiums, and attached homes are typically home to even fewer K-12 students, on average.

Shortcomings of Student Generation Rate Calculations

While helpful to address concerns, generic SGRs may fail to capture the nuances of dynamic change in neighborhoods. In our experience working with school districts, PRC has identified several shortcomings of SGR-based estimates that may lead to faulty assumptions about future enrollments:

1. There may be wide variance in SGRs depending on specific characteristics of housing units. Simply knowing the number and type of future units may be inadequate to produce reliable estimates of future school enrollment.
2. Applying current SGRs to proposed new housing may not capture future shifts in household composition and birth rates.
3. Enrollment changes in existing homes may augment, diminish, or negate the growth from new housing.
4. SGRs are not static; a single set of SGRs measured at a point in time won't reflect the way that families and individuals age in place.

Variance Depending on Characteristics of Housing Zoning and density regulations imposed by jurisdictions typically regulate the number of allowable homes, but not characteristics such as number of bedrooms or affordability level. Projections of future enrollment based merely on residential capacity may be premature. Even after subdivisions are platted, detailed information about the characteristics of homes may not be available. Our findings that SGRs for three- and four-bedroom homes are higher than for one- and two-bedroom homes (Rynerson and Siu 2009) are not surprising. SGRs for single family homes may also vary depending on lot size (Rynerson et al. 2015), sale price (Rynerson and Siu 2006), proximity to schools, and neighborhood amenities such as playgrounds. For rental apartments the range of SGRs is even greater. Rates measured in 2016 in new multiple family housing in Portland ranged from 0.01 K-12 students per home in market rate studio and one bedroom apartments to 1.12 K-12 students per home in income-restricted developments in which most of the units had three or more bedrooms (Rynerson and Yang 2017).

Shifts in Birth Rates and Household Composition Because of the current downturn in births, future SGRs may be lower than those measured in recent years even in comparable developments. The number of babies born in the U.S. in 2017 was the lowest in 30 years, about 11 percent lower than the peak in 2007. In 2017 the number of births to residents in 49 of 50 states had fallen since 2007, with declines of more than 20 percent in Arizona and New Mexico (National Center for Health Statistics 2018). Jaffe (1969) cites changes in birth rates as a reason to combine the SGR-based projections with cohort-survival projections.

In just 10 years between 2007 and 2017, the number of U.S. households with one or more children under 18 fell by nearly 1.7 million, while the number without children under 18 grew by nearly 9.4 million. This shift resulted in the share with children under 18 falling from 34.4 percent to 30.8 percent of all households.[1] In the same 10 year period the share of U.S. households with a householder age 60 or older increased from 28 percent to 35 percent, accounting for all of the net growth in households and incentivizing development of age-restricted communities.[2]

Enrollment Changes in Existing Homes When residential development is on the horizon, it is expected that an increase in the number of homes will result in an increase in the number of students. While that is a likely result, very few developments are large enough to encompass an entire school catchment area. An incremental increase to existing housing stock is more typical. Stakeholders may expect to simply add the SGR-based estimates of students in new housing to existing enrollment figures. Equally important, and drawing less attention, is that the number of students in existing homes may be increasing or decreasing. When student population in existing homes declines, flat or declining enrollment in catchment areas containing extensive homebuilding activity may come as a surprise.

Aging in Place Enrollment losses at individual schools may mean the loss of teachers, administrators, and support staff, and even school closures. When enrollments decline we often hear anecdotes from hopeful residents (e.g. the neighborhood is turning over to younger families, and the sidewalks are full of strollers, etc.). Most homes do eventually turn over, but neighborhoods dominated by owner-occupied single family homes evolve slowly. Families don't necessarily move after their children finish school. Taylor et al. (1986) finds that the average number of elementary-age children in single family homes in London, Ontario peaked 7 years after homes were built and declined systematically before stabilizing at year 25. Smith (2017) experiments with curvilinear and oscillating models to estimate enrollments at individual grade levels based on age of housing. In our work, we consistently measure the highest elementary SGRs in homes less than 10 years old, and the highest high school SGRs in homes 10–20 years

[1] U.S. Census Bureau, 2007 and 2017 American Community Survey 1-year estimates, Table B11011.
[2] U.S. Census Bureau, 2007 and 2017 American Community Survey 1-year estimates, Table B25007.

old. Homes older than 20 years typically have the lowest SGRs at all school levels. We use 20 years as a threshold for "older" homes because we have found that once most of the original families have grown beyond school age, SGRs are relatively stable. Gradual turnover of older homes results in relatively little difference in the number of school age children between homes that are 20, 30, or even 50 years old.

Student Generation Rate Case Study

Most of the work done by PRC and other researchers to measure SGRs has been derived from observations of student populations at a single point in time. Here we take a different approach, using annual student address data for a 13-year period from 2006–07 to 2018–19 to measure the trajectory of student population in several contiguous subdivisions adjacent to an elementary school in suburban Portland.

The subdivisions were platted and built-out quickly during the mid-2000s housing boom; 177 homes were built in 2005 and 2006 on lots ranging in size from 5000 to 7000 square feet. Most of the homes have four or five bedrooms and three or four bathrooms and between 2600 and 3500 square feet of floor area. They are larger than the average new suburban Portland area home, having sold for about 50 percent more than the metro area median when new. However, they are similar in size and relative cost to future homes planned in new subdivisions within the same school catchment area. Given their large size and proximity to schools and high-paying jobs, they are likely to remain desirable to families with children.

Student address files are compiled for October 1st each year, when schools throughout Oregon are required to perform an annual student headcount. Information is limited to address, grade level, and school attending. Spatial taxlot layers and attribute data are from county tax assessor files, standardized and maintained by Metro, the Portland area regional planning agency (Metro 2018).

Figures 18.1a and 18.1b illustrate the distribution of public school students by grade level in the homes when they were newly built in fall 2006, and 12 years later in fall 2018. When the homes were new, their original occupants included many young families with children in primary grades. Kindergarten, 1st, 2nd, and 3rd grade students comprised half of the K-12th grade total in 2006. By 2018 there were one third fewer K-3rd grade students than in 2006, but twice as many middle school (6th-8th grade) students and three times as many high school (9th–12th grade) students. These results, expressed as SGRs in Table 18.1, demonstrate the risk of calculating SGRs at a single point in time.

The average rate of 0.32 high school students per home in 2018–19 is the highest of the 13 year period. While the 2006–07 elementary rate and the 2018–19 middle school rate are relatively high, neither represent the peak rate. Figure 18.2 shows the trajectory of enrollment, notably the rise and fall of elementary enrollment and subsequent increases in secondary enrollment. Peak enrollments include 92 elementary students (0.52 per home) in 2009–10, 44 middle school students (0.25 per home) in 2013–14, and 57 high school students (0.32 per home) in 2016–17 and again in

Fig. 18.1a Students per grade level, 2006–07

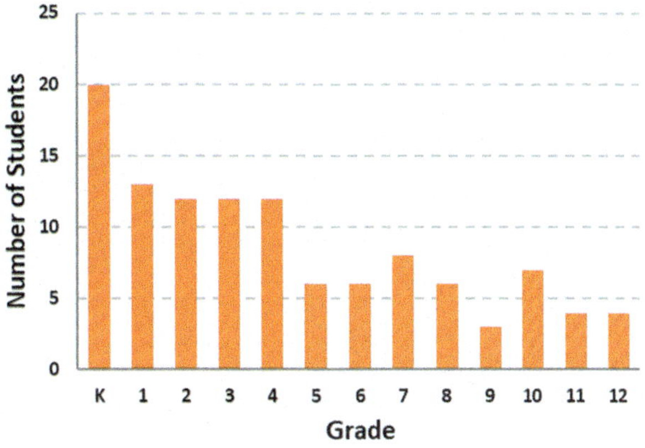

Fig. 18.1b Students per grade level, 2018–19

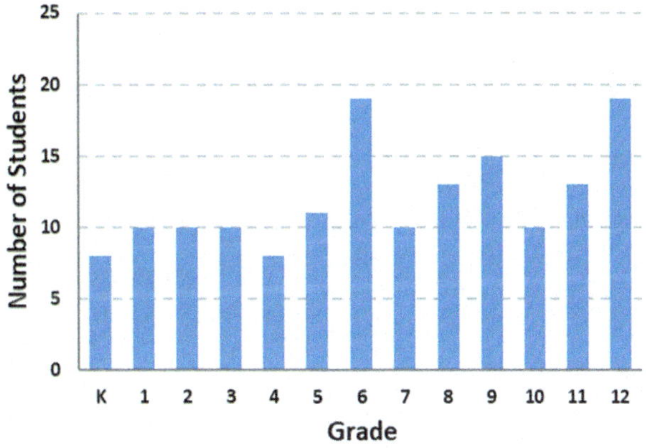

Table 18.1 Students per housing unit

School year	K-5	6-8	9-12	K-12
2006–07	0.42	0.11	0.10	0.64
2018–19	0.32	0.24	0.32	0.88

Public school students in 177 single family homes built 2005–2006

2018–19. Elementary students declined steadily as a share of all K-12 students, from a high of 66 percent in 2006–07 to a low of 37 percent in 2018–19.

Secondary (6th–12th grade) students outnumbered elementary (K-5th grade) students in the 2018–19 grade distribution. However, the grades were more evenly distributed in 2018–19 than in 2006–07. Even when compared with its 2009–10 peak, the decline in elementary enrollment has been more subtle than the growth in secondary enrollment. We explain this by stratifying the homes by length of occupancy of current residents, using the taxlot attribute field indicating the most recent sale date. As of 2018 about half of the 177 homes had not been resold since their initial purchase, based on a sale date in 2005 or 2006; nearly all of those were

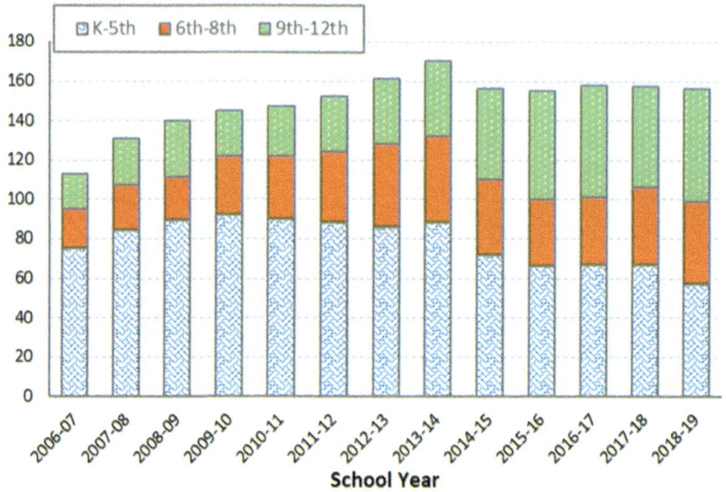

Fig. 18.2 Students per school level, annually 2006–07 to 2018–19

Table 18.2 Students per housing unit, 2018–19, by date of most recent sale

Most recent sale	K-5	6-8	9-12	K-12
2005–2006 (90 homes)	0.18	0.28	0.41	0.87
2007–2013 (39 homes)	0.46	0.21	0.26	0.92
2014–2018 (48 homes)	0.48	0.19	0.21	0.88

Public school students in 177 single family homes built 2005–2006

still occupied by their original owners. Among 90 homes with a sale date in 2005 or 2006, 83 had owner addresses that matched the situs address, and were assumed to be owner-occupied by the original purchasers. The number of high school students residing in the 90 homes that had not turned over to new owners increased from seven in 2006–07 to 37 in 2018–19 while the number of elementary students fell from a peak of 58 in 2009–10 and 2010–11 to a new low of just 16 students in 2018–19. The 48 homes sold since 2014 accounted for 23 elementary students. Table 18.2 shows that as of 2018–19 the overall K-12th grade SGRs were similar regardless of sale date, but homes with a recent sale included younger families, on average, than those still occupied by the original owners. Homes sold between 2007 and 2013 had similar elementary SGRs to those sold between 2014–2018, in spite of the sale occurring five to 11 years before the 2018–19 school year. Presumably those 2007–2013 homebuyers moved in with preschool age children, or gave birth after purchasing their homes.

Conclusion

The potential for student growth from large new residential developments understandably causes concern for school administrators. However, a typical pattern of suburban development on the periphery of metropolitan areas is that new homes are built in proximity to those built during previous waves of development. The number of students residing in those existing homes may be increasing, decreasing, or stable, depending on the age of the homes, the school level, and the rate of housing turnover.

A naïve approach to using SGRs in long range enrollment projections is to add future students to existing enrollment without acknowledging the trajectory of enrollment change in existing homes, which typically outnumber new homes in any catchment area. We were fortunate to have 13 consecutive years of student data for this case study, demonstrating the growth and decline at specific grade levels that occurs as homes age. We also found that turnover may stem losses in elementary grades. Higher rates of turnover may cause enrollment to stabilize faster than if all of the original owners remain in their homes.

Longitudinal student data may not always be available, but spatial parcel data is commonly available from jurisdictions throughout the U.S. at little or no cost, and may be used to estimate SGRs by age and type of housing unit for a single year of student data. These rates should be applied dynamically over time to both existing and new homes for more reliable long range enrollment projections. If larger demographic trends such as shifts in household composition and birth rates are a concern, housing-based projections may be controlled to cohort-based projections at a district or sub-district level.

References

Jaffe, A. J. (1969). *Handbook of statistical procedures for long-range projections of public school enrollment* (Vol. 24017). Washington, DC: US Office of Education, Office of Program Planning and Evaluation.

Lewis, P. G., & Neiman, M. (2002). *Cities under pressure: Local growth controls and residential development policy*. San Francisco: Public Policy Institute of California.

Machell, E., Reinhalter, T., & Chapple, K. (2009). Building support for transit-oriented development: Do community-engagement toolkits work?

Metro, Regional Land Information System, August 2018 release.

National Center for Health Statistics (NCHS). (2018). Division of Vital Statistics, Natality public-use data 2007–2017, on CDC WONDER Online Database, September 2018. Retrieved from http://wonder.cdc.gov/natality-current.html

Nguyen, M. T. (2009). Why do communities mobilize against growth: Growth pressures, community status, metropolitan hierarchy, or strategic interaction? *Journal of Urban Affairs, 31*(1), 25–43.

Rynerson, C., & Siu, V. (2006). *Three rivers school district population and enrollment forecasts, 2007–08 to 2016–17*. https://pdxscholar.library.pdx.edu/enrollmentforecasts/2/

Rynerson, C., & Siu, V. (2009). *Molalla river school district population and enrollment forecasts, 2009–10 to 2020–21*. https://pdxscholar.library.pdx.edu/enrollmentforecasts/45/

Rynerson, C., & Yang, X. (2017). *Portland public schools enrollment forecasts, 2017–18 to 2031–32: Based on October 2016 enrollments*. https://pdxscholar.library.pdx.edu/enrollmentforecasts/118/

Rynerson, C., Chun, N., & Stewart, S. (2015). *Tigard-Tualatin school district enrollment forecast update, 2016–17 to 2025–26*. https://pdxscholar.library.pdx.edu/enrollmentforecasts/96/

Smith, G. H. (2017). New methods for projecting enrollments within urban school districts. PhD (Doctor of Philosophy) thesis, University of Iowa, 2017.

Taylor, H. W., Code, W. R., & Goodchild, M. F. (1986). A housing stock model for school population forecasting and planning. *The Professional Geographer, 38*(3), 279–286.

Wong, S., Miles, D., Connor, G., Queenan, B., & Shott, A. (2017). Residential demographic multipliers: Using public use microdata sample records to estimate housing development impacts. *Cityscape: A Journal of Policy Development and Research, 19*(3), 415–427.

Yarzebinski, J. A. (1992). Handling the 'Not in my backyard' syndrome: A role for the economic developer. *Economic Development Review, 10(3), 35.*

Chapter 19
State Longitudinal Data Systems and Public Policy Research

George C. Hough Jr. and Vivien W. Chen

Introduction

In an effort to develop policies related to education and careers, numerous states have developed state longitudinal data systems (SLDS) for the purpose of better evidence-based decision making. These SLDS offer promise for supporting policymaking and research beyond meeting federal and state data reporting requirements, yet may not answer many of the most pressing research questions nor create meaningful collaboration between researchers and practitioners (Conaway et al. 2015).

With SLDS, the collection and organization of data across time provides a more dynamic approach to analysis than was previously provided by the cross-sectional data reported from the school district to the state education agency to the federal government. In addition, much of the data were contained in data silos representing various programs. Therefore, data had to be linked across these silos to get a complete story of a student's participation. Finally, SLDS developed linkages to perform cross-sector research and evaluation. K-12 data were linked to early learning, postsecondary education and eventually workforce participation. The Washington state Education Research and Data Center (ERDC) developed their

---

G. C. Hough Jr. (✉)
Education Research & Data Center, Office of Financial Management, Olympia, WA, USA
e-mail: George.Hough@ofm.wa.gov

V. W. Chen
Education Research & Data Center, Forecasting & Research Division, Helen Sommers Building, Olympia, WA, USA
e-mail: Vivien.Chen@ofm.wa.gov

B. Jivetti, Md. N. Hoque (eds.), *Population Change and Public Policy*, Applied Demography Series 11, https://doi.org/10.1007/978-3-030-57069-9_19

SLDS into a P20W[1] data warehouse, with the vision "To promote a seamless, coordinated preschool-to-career (P20W) experience for all learners by providing objective analysis and information" (ERDC, About Us 2015a).

After describing a brief history of the role of data governance in the evolution of the Washington state SLDS, this paper will provide critical analysis of the strengths and shortcomings of using this SLDS database for research. The following sections will then proceed to outline the six key uses of longitudinal data as developed by Dougherty et al. (2007) and provide examples of ongoing research by ERDC represented in these six key areas. The paper will conclude with remarks for future research using SLDS, and in particular the Washington state P20W.

The Role of Data Governance

The *No Child Left Behind* policy can be credited with changing the way educators across the nation looked at school performance (Snow-Renner and Torrance 2002), and the 1994 version of the Elementary and Secondary Education Act (ESEA) led to the beginning development of SLDS (Improving America's Schools Act 1994; Hough and Beard 2017). This paper advances this previous research by highlighting the importance of data governance in generating public policy research with data partners and public researchers.

The Washington State Legislature Placed ERDC in a Unique Role

In 2007, ERDC was created as Washington's P20 Workforce (P20W) office, based upon Governor Gregoire's study of education called Washington Learns. The Governor and others were frustrated by the limitations of the data systems within each education sector because there were no linkages across the early learning, K-12, higher education and workforce data silos. Revised Code of Washington 43.41.400, aka RCW 43.41.400 (2017) gave ERDC the authority needed to coordinate the data governance process. It is also noteworthy that ERDC is located in the Governor's budget agency, the Office of Financial Management/Forecasting and Research Division.

In 2009, the legislature expanded ERDC's duties that were incorporated into the ERDC's mission and agenda. These additional responsibilities mostly addressed the need for ERDC to direct the data governance process (ERDC, History 2015c): Lead

[1]P20W is a system of data that follows individuals from preschool (P=Pre-kindergarten), K-12, and postsecondary through post-graduate education (20), along with workforce data (W). If it incorporates additional outcomes data, it is often referred to as a P20W+ system (+ = e.g., corrections).

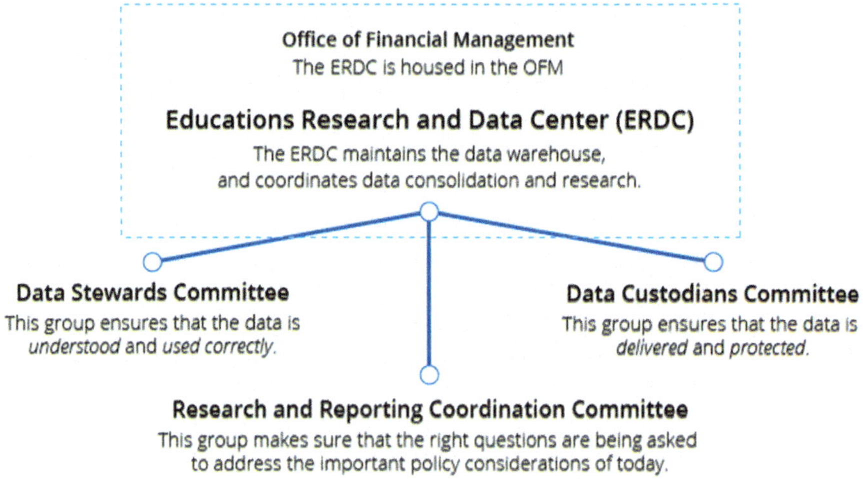

Fig. 19.1 Data governance structure of ERDC and P20W warehouse

P20W data governance; and Monitor and evaluate the education data collection systems of the organizations and agencies that contribute to ERDC, and ensure that they're flexible and that the data they provide is relevant. To complete the tasks listed in the duties above, ERDC outlined a data governance structure.

Data Governance Structure

ERDC has implemented a robust data governance structure that includes all data contributors, representatives from local education agencies and data users. Figure 19.1 displays this data governance structure. ERDC is housed in the Office of Financial Management (OFM), and maintains the data warehouse and coordinates data consolidation and research. ERDC works together with three informal committees composed of ERDC staff and outside members (ERDC, Data Governance 2015b).

Research and Reporting Coordination Committee

Staff from the ERDC as well as representatives of various agencies and stakeholder organizations make sure that the right questions are being asked to address the important policy considerations of today (see https://erdc.wa.gov/about-us/research-priorities). Based on their understanding of the policy questions and needs of their organizations, they make recommendations regarding ERDC's research focus and priorities, and coordinate with the other committees to make sure the ERDC has the data it needs to enact those priorities. The people on this committee are also policy matter experts, and offer advice to data requestors on their research questions and priorities, and review research before being publicly released. This committee updates policies and procedures related to data sharing (ERDC, Data Governance 2015b).

Data Stewards Committee

Data stewards include staff and researchers from data contributors who are directly familiar with the data contributed by their agency, as well as researchers at the ERDC. This group ensures that data is understood and used correctly (ERDC, Data Governance 2015b).

Data Custodians Committee

Data custodians are technical experts from contributing organizations and the ERDC. These custodians ensure that data is delivered to whomever needs it, and that it is secure and protected at all times. Technical experts from contributing agencies coordinate on how the data is exchanged between their agency and the data warehouse, and how the data are stored and protected (ERDC, Data Governance 2015b).

Centralized Data Collection

As mentioned above, ERDC maintains the P20-W data warehouse and coordinates data consolidation and research. "In a centralized data system, all participating source systems copy their data to a single, centrally located data repository where they are organized, integrated, and stored using a common data standard. As depicted in Fig. 19.2, data in a P20W+ centralized SLDS are periodically matched, integrated, and loaded into a central repository. Users query the system and can access the data that they have been authorized to view and use" (Institute of Education Sciences, September 2016).

The degree to which this centralized P20W+ data warehouse works is largely a function of the collaboration between the ERDC data partners (see https://erdc.wa.gov/data-resources/our-partners). Informed by the data contributors, ERDC enacted formal memoranda of understanding explicitly outlining procedures related to data security, privacy, and use. These governance practices, coupled with ERDC's success in providing useful information and research to its partners, have made state agencies more willing to share data with the center" (Institute of Education Sciences, July 2015).

Figure 19.2 displays the basic structure of the Washington State P20W+ data system. Data from early learning, K-12, postsecondary data, workforce data, as well

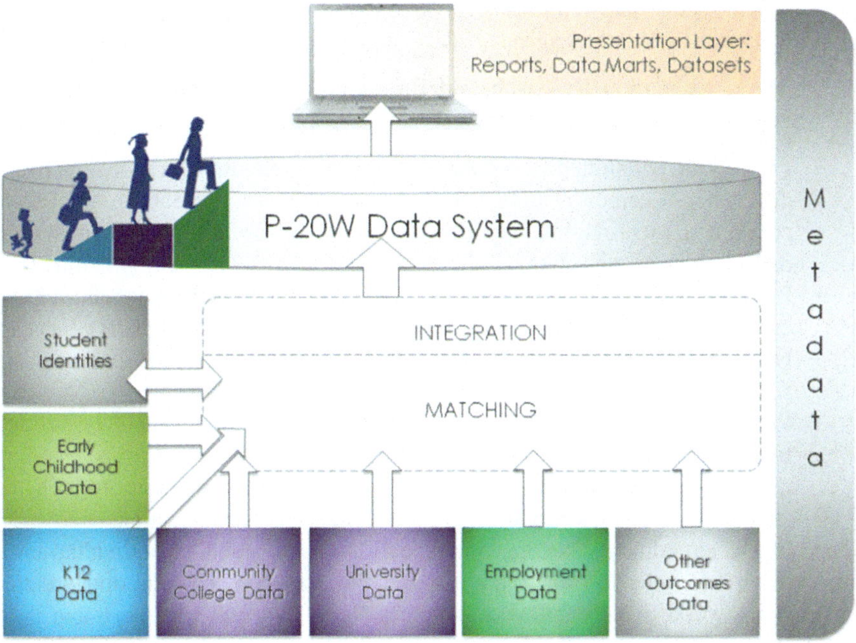

Fig. 19.2 Basic structure of centralized data system (Institute of Education Sciences, March 2013)

as other sources are matched utilizing the various identity information common to each data feed (name, date of birth, social security number, etc.). The system is then integrated with a common P20ID, with unique identifiers removed, to form the P-20 W+ data system. Data marts and datasets with summarized data are then constructed for research cohorts from which reports and data dashboards are constructed. Metadata are captured for each of the processes.

Memorandum of Understanding

A vital feature of data governance then is the manner in which the data are shared, stored, and protected. The ERDC Memorandum of Understanding (MOU) covers not only relationships between ERDC and their data partner agencies who contribute data, but also with external data requestors. The MOUs are guided by the following principles (ERDC, Memorandum of Understanding 2015e).

Principle 1: Education Research and Data Center provides cross-sector, linked data to all data consumers in a consistent, transparent way

- ERDC provides cross-sector, linked data and analysis. Cross-sector is defined as including at least two education or workforce sectors.

- Single-sector data requests will be referred to the appropriate agency. If requested by a partner agency, ERDC may assist with a single-sector analysis.

Principle 2: Education Research and Data Center maintains the P20W data warehouse

- Partners are committed to populating the data warehouse with more and better data as it becomes available to load into the data warehouse. ERDC is committed to providing opportunities for on-going consultation regarding data quality and new data.
- ERDC is established as an authorized representative of the state educational agencies for purposes of researching and analyzing data to support education budgeting and policymaking functions of the Legislature, the Governor, and state educational agencies.
- Partner data contributors agree to share student and workforce data through ERDC to the extent feasible. The data contributors share data enabling research and evaluations to improve their respective programs through the analysis of student, educator, fiscal and workforce data.
- ERDC coordinates the matching and de-identification of individual-level records in order to protect the confidentiality and privacy of individual students, workers and employers.

Principle 3: Protecting the privacy of individuals is a critical priority

- Three levels of data can be extracted from the data warehouse:
- Highly Restricted-Use Data (Level 1) – Data that include information about the identity of individuals and employers is confidential. Level 1 is the term for data that contains individually identifiable information. Level 1 data require specific procedures to protect confidentiality per Family Educational Rights and Privacy Act (FERPA) regulations and other state and federal requirements. These types of data are rarely shared and shall only be used for unit record matching purposes.
- Access to Level 1 data requires a formal request that must be approved by a Data Request Review Board or Institutional Review Board, and any other data contributors.
- Restricted-Use Data (Level 2) – Data at the unit record level but containing no individually identifiable information. Level 2 data are all information in those files and all information derived from those data including data resulting from merges, matches or other uses of the data. Merging Level 1 data records with Level 2 data records results in a merged file that is Level 1. Merely deleting identity fields from a Level 1 files does not necessarily create a Level 2 file. Disaggregation of Level 2 records, even without explicit identification fields, may result in a record where the identity of the subject could be reasonably inferred.
- Access to Level 2 data requires a data sharing agreement with ERDC (data contributors do not have to sign the agreement).

- Public-Use Data (Level 3) – Data aggregated from Level 2 data that contain no unit record data, is for public use, and acceptable for publication purposes. Access to Level 3 data is unrestricted.

Principle 4: Partner agency data contributors (at the state and local levels) are experts at understanding and explaining the data

- ERDC will refer single-sector data requests to the appropriate agency.
- ERDC will confer with all data contributors before responding to a request for cross-sector information.
- ERDC will notify pertinent state agencies when data have been shared.
- Recipients of ERDC data or reports will be provided an agency contact to communicate with regarding use of the data.
- Recipients of data will provide an opportunity for ERDC and data contributors to review and comment on research design, drafts of findings and conclusions, and final reports.

Principle 5: Common understanding and use of data increases its value

- With any public display of data, data recipients must include the methodology and assumptions used on a data set.

Having outlined the data governance process and the type of available data and privacy protections surrounding these public data, the following section will proceed to discuss some limitations of using SLDS data for conducting research. Following this discussion, outline the six key uses of longitudinal data as developed by Dougherty et al. (2007) and provide examples of ongoing research by ERDC represented in these six key areas. The research will also highlight the particular level of data utilized from the Washington state P20W+.

Limitations of SLDS/P20W Data for Research

Despite the ability of SLDS to make data readily available to the state, school districts and public data users, most state SLDS are not set up for research purposes. As Conaway et al. (2015: 18S) state, "from a state perspective, research is a by-product of the SLDS, not its raison d'être". ERDC, however, has secured external federal funding as well as additional state funds based upon the promise of conducting research utilizing the SLDS/P20W. Furthermore, federal grant programs now require that research is facilitated.

Data Accessibility

The Washington state P20W+ has been developed over almost a decade. One of the major issues for researchers and data partners is accessing all the data. The

Washington state P20W+ database is represented by a third normal form ("Data Normalization", n.d.). The hierarchy reflects a PRO Model representing a Person, performing a Role within an Organization (e.g., a student enrolled at a university during a given quarter/semester).

In order to make the data more accessible for most researchers, but less granular for Big Data enthusiasts, ERDC staff denormalized the P20W+ database into data marts referred to as a star schemas with a fact table as its core and dimension tables surrounding it representing the star's points ("Star Schema", n.d.). The creation of data stars representing cohorts of interest also helped to alleviate the ad hoc nature formerly employed to meet researcher's data demands. That is, the data would now be standardized via a query process to develop the data marts, rather than created individually by ERDC research staff. The Data Marts would now represent the following cohorts: Early Learning, 9th Graders, High School Graduates, Post-Secondary outcomes, and Justice. The goal is to process the raw data into facts or metrics, which hold the measurable, quantitative data about enrollment, achievement, workforce participation and attach these facts to personal and organizational characteristics.

Despite being manageable, these data marts have limitations of their own. College quarters/semesters are aggregated to represent annual totals, along with credit hours, achievements and GPAs. Sequences and events are minimized to annual totals also. So, the data are more readily available and manageable, but some level of detail has been lost. Annual averages may appeal to legislators, but do they best represent the original data?

Incomplete or Limited Data

For most of the Washington state P20W+, we only have Washington state public data. That is, the P20W+ represents a partial population register. "Partial population registers are set up for specific administrative programs and cover only those persons directly affected by the particular program or belonging to a particular group" (Bryan 2004: 32). The Pre-K information largely represents those in the Early Childhood Education and Assistance Program (ECEAP) in Pre-K managed state certified centers, as well as some students in the Pre-K program reported through the Office of Superintendent of Public Instruction (OSPI). The K-12 data are only for public school students (OSPI). For post-secondary data, we can collect information on almost all students enrolled in public and private colleges and institutions, but once again, we only have detailed data on course-taking, GPA, and other details from Washington state public 2-year and 4-year post-secondary data, the State Board for Community and Technical Colleges (SBCTC) and the Public Centralized Higher Education Enrollment System (PCHEES). Likewise, the P20W+ system only has workforce data for those in the labor force in Washington state, although Washington is one of the only states that includes wage hours. It would be ideal to have this information from all states, but especially from our neighboring states.

Data Reliability

All the data contained in the P20W+ have been profiled and validated extensively by the data stewards and custodians. Again, as these data are collected for administrative programs, entered by office staff in many cases, the data have internal quality checks, but may need additional validation when integrated into the P20W+ system. Additionally, many newly required data collection systems take time to be integrated and evaluated for data quality. Data systems are central to the P20W+ system, and need additional validation from ERDC staff and consultants to build and maintain the current P20W+ data system.

Despite issues of accessibility, incomplete or limited data and data reliability, there are still plenty of elements in the Washington state P20W+ data system to use for the purposes of policy research. In the following section, we will highlight a number of examples where ERDC and other researchers have utilized the P20W+ data system to answer critical questions posed by the lead education agency (OSPI), other data partners and researchers.

Six Key Uses of Longitudinal Data

As mentioned in the introduction, another key contribution will be to outline the six key uses of longitudinal data as developed by Dougherty et al. (2007) and provide examples of ongoing research by ERDC represented in these six key areas. Many of the examples provided in the Dougherty et al. (2007) article are sole sector analyses. ERDC research focuses mainly on cross-sector longitudinal studies, however not all of the research maximizes ERDC's capability of cross-sector data and longitudinal analyses. In addition, many of the Dougherty et al. (2007) work focuses on individual student progress whereas the ERDC studies focus on cohort analyses.

Progress Monitoring

Progress monitoring involves using longitudinal data to track students' progress. The same applies when monitoring schools or school districts, or even at the community college and university level. Longitudinal data are needed at the student level to monitor if the organizations are helping to improve academic and workforce growth. Many of the ERDC dashboards are representations of progress monitoring at an aggregated level:

- *High school feedback reports* follow up high school graduate cohorts 2004–05 to current postsecondary enrollment, progress and degree completion, and earnings at the state, school district and high school level over time (see https://erdc.wa.gov/data-dashboards/high-school-graduate-outcomes).

- The *Public 4-year dashboard* provides an overview of student enrollment, progress, and degree completion across Washington state public 4-year institutions over time for students enrolled 2007–08 to current (see https://erdc.wa.gov/data-dashboards/public-four-year-dashboard#annual-enrollment and https://erdc.wa.gov/data-dashboards/time-degree-visualization).
- The *Earnings for graduates dashboard* portrays higher education institution graduates' earnings outcomes by institutions, degree types, and degree major over time for years after graduation (see https://erdc.wa.gov/data-dashboards/earnings-for-graduates).

Diagnosis and Prescription

Longitudinal information on individual students' academic histories can be valuable in making diagnoses about how to remedy a condition or even promote progress for a group of students.

- State Need Grant recipients' educational progress and degree completion. This is a study comparing undergraduates' educational progression and degree completion, between those received State Need Grant and those who did not. It follows college freshmen up to 8 years to investigate the variation in degree completion, by comparing the outcomes across race/ethnicity, gender, and sector (Chen 2017).
- Homelessness and foster care study This is a study in response to Section 223(1)(bb) of ESSB 6032 proviso, which evaluates the outcomes of several grade cohorts, including kindergartens, 3rd graders, and 9th graders. The study follow up students for 6 years to evaluate the impact of homelessness and foster care on students' academic outcomes and college going. As a first-step of this type of study required to be completed in 3 months, the analysis is rather descriptive than predictive. However, the over-time patterns help identify key factors or potential mechanisms explaining the impact of these disadvantaged status on a student's learning outcomes (Chen, Pyle & Aldrich 2019).
- Washington Eighth Graders' Educational and Employment Trajectories. This study tracked 2005 eighth graders' educational and workforce trajectories over 10 years. It identifies the key stages of dropping out, enrollments and completions in high school and college, and also lightly touch students' incarceration status. It portrays a student's learning life course in a way to help future research explore significant mechanisms and develop potential interventions to improve high school graduation, college going, degree completion, and further improve employment in Washington state. This study is also used to facilitate the Governor's Results' Washington plan on identifying the *Opportunity Youth* and develop plans to improve this group of disadvantaged youth to further improve their lifelong outcomes (ERDC 2014).

Internal Benchmarking (Better Practice Within School/District)

This key, as we understand it, is to use longitudinal data to identify the areas of greatest success within a student's own school or school system in search of better practices. As far as I know, this type of research is missing from using our data. Two reasons explaining the lack of research in this area: a) ERDC research focuses on cross-sector analysis. Normally this type of research examines student's immediate outcomes before and after treatment within sector, and b) Specific practices/intervention within schools or districts are normally not collected by a statewide administrative data system. To do this type of analysis requires extra effort to collect data outside of SLDS.

- A study by the Washington state Center for Court Research, Administrative Office of the Courts (Gertseva and McCurley 2019) best represents this approach. In this paper entitled, Education Outcome Characteristics of Students Admitted to Juvenile Detention, the authors described the characteristics of students admitted to juvenile detention in 8th or 9th grade and examined whether being detained for any reason has adverse effects on education outcomes in adolescence and early adulthood. The authors used the Washington state P20W+ to conduct cross-sector linking of K12, higher education, and Administrative Office of the Courts records to conduct their analyses.

External Benchmarking (Extend Better Practices to Outside of School/District)

In this example, the search for better practices should extend outside of one's own school or district. In a study by Pyle and Chen (2019), the authors examined Substance Abuse Prevention and Intervention in Schools Programs (known as SAPISP or SAP) to answer the following research questions: (1) What are the characteristics of schools that provide SAP intervention programs and those that do not? And (2) Is there any change in school-level substance abuse and academic outcome measures among SAP and non-SAP schools, over time?

- The conclusions by Pyle and Chen (2019) are summarized as follows: (1) The findings of consistently higher substance abuse discipline incident rates among SAP-CPWI schools compared to the No SAP schools ("average" schools) is as expected, since SAP-CPWI schools were selected for the program based on the determination of high risk for substance abuse among students. (2) Among the SAP-CPWI schools, the substance abuse discipline incident rate trends measured between 2015 and 2018 did not show any significant increase or decrease. However, for the No SAP schools, there were significant increases in tobacco and alcohol incident rates.

- In another study (Reykdal 2019) found that Dual credit programs allow students to earn high school and college credit at the same time. There are two main types of dual credit programs: exam-based dual credit and concurrent enrollment dual credit. Research is clear: participation in dual credit courses increases high school and postsecondary success for students.

Dual credit, when used intentionally, is a tool for addressing inequity in Washington's school system. Since students take dual credit courses while in high school, the courses allow young people to buy down the cost of a college education. Growing equitable access to dual credit programs is one way to stem exploding student loan debt and better prepare our young people for an economically stable future.

In addition, while enrollment continues to grow in Washington, there is significant inequity in participation rates in dual credit courses. Students of color, students experiencing poverty, and students with disabilities participate in dual credit courses at a lower rate than their peers. Lawmakers believe expanding dual credit and reducing college costs to All students will reduce current inequities.

Predictive Analysis

Predictive analysis refers to the use of longitudinal data to articulate the historical relationship between earlier and later outcomes for groups of students followed over time.

- ERDC conducted a STEM pipeline study (Chen 2019c) that examined the effect of STEM learning through a student's schooling years, from 3rd to 12th grade and into college. Specifically, this study applied mediation analysis to compare the sizes of the direct effect (net effect) of high school STEM coursework, against indirect effects of mediators measured by 3rd to 12th grade achievement outcomes in math and science, on predicting the probability of earning college STEM course credits. It extensively improved our understanding of STEM pipeline through a student's learning life course.

Program and Policy Evaluation

The sixth use of longitudinal data "encompasses the evaluation of *programs* by comparing results for students receiving particular services compared to those who don't; the evaluation of *policies* by the study of the relationship between their implementation over time and longitudinal indicators of student success; and evaluation of *schools and teachers* using "value-added" analysis that controls for prior student achievement and growth" (Dougherty et al. 2007).

- The analyses, An evaluation of the General Educational Development certificate on earnings for Washington high school dropouts (Chen 2015), applied propensity score matching to analyze high school dropouts' earnings gaps between those who received a GED and those who did not. With a longitudinal spectrum of data, this study was able to investigate the heterogeneous effects of GED on long-term earnings by carefully taking into account individuals' prior dropout characteristics and skills. Such design could provide insights for improving dropout prevention program.
- In a study by Benson (2019), event history analysis was applied to understand the impact of need-based financial aid on students' degree completion 6 years after their first-time college enrollment. The findings of this paper indicated: (1) Students with better financial aid packages complete at higher rates and at a faster pace than students with lesser financial aid packages; (2) Needy students who receive greater amounts of financial aid are more likely to attend full-time; or, alternatively, needy students who attend college full-time are more likely to receive greater amounts of financial aid; (3) Student who did not receive financial aid in the second year were less likely to complete; (4) The amount of unmet need for students still receiving aid increases every year and the problem is worse for the students with lesser aid packages; and (5) Students with the lowest financial aid package drop out at substantially higher rates, but even a bit more financial aid goes a long way toward decreasing dropout rates.

To summarize, ERDC has produced significant research in using the state longitudinal data system (P20W+) to analyze public policy research. Before we conclude, there are a few avenues we think should be explored in future development of the P20W+ data system, as well as some limitations we think need to be addressed in order for the Washington state longitudinal data to be fully utilized from a methodological point of view.

Conclusions and Future Refinements

Reviewing the above sections, it is apparent that ERDC has organized and developed an advanced cross-sector P20W+ data system. It is also apparent that this database has led to several significant data profiles and research findings relevant for state policymakers. All this work has contributed to improve the capacity of using Washington state data for evidence-base policymaking. For example, ERDC staff are currently developing a research handbook to provide researchers a better understanding about the data and applications. Hopefully, through this publication, similar research with longitudinal data can become more plentiful and generalizable for other state SLDS. ERDC freely shares much of its P20W development knowledge through the Institute of Education Sciences (IES) at the National Center for Education Statistics (NCES).

Finally, more scholarly research could be accomplished if a few data issues were resolved in many of the P20W+ systems developed by the states. Below are the issues we feel need to be addressed.

Future Refinements

More data and data sources need to be incorporated into the data warehouse, both for expanding research opportunities as well as enhancing identity matching. Birth records, corrections data, early learning clients and the All-Payer Claim Database (APCD) for healthcare claims should be incorporated more fully into the current P20W+ system. This would not only allow the P20W+ system to begin at birth, but would also permit better identity matching for current and future research. Other additions to the data warehouse that will strengthen ERDC and affiliates capacity for research include grade history data (STEM), discipline and absence data (equity), and more detailed racial and ethnic data for K-12 students. Once entered into the larger P20W+ system, metrics will need to be developed to produce analytical files with metrics as data.

Currently, there is a minor issue with over-matching, or the comingling of identities. That is, more than one person is assigned a unique P20 ID (e.g., Jan Smith and Jane Smith with identical birthdays). Identities are easy to merge, but hard or near impossible to unmerge. Luckily, using Token IDs can assist as each Token ID is guaranteed to be associated with only one person (Sabel 2013). Hopefully, with more information on individuals within each system, the more likely the identity matching will provide a one-to-one correspondence between individuals and their P20 ID.

Calendar Time

Other pressing issues to address concern the notions of the timing of events and calendar time and dates. When evaluating and profiling various datasets, we have discovered that the date recording of events is unrelated to the actual event – oftentimes the date recorded refers to the beginning of the school year or the date the data were entered. These errors in data entry greatly affect the timing and sequencing of events. Some of the data entry errors we can correct (e.g., dropouts dated at the beginning of the school year can be backdated to the end of the school year), however some we cannot correct or just plain miss.

Another issue of great concern regarding the causality of events has to do with organizing an annual calendar of time. Figure 19.3. Displays the Core Data Systems of the P20W and their Annual Calendars for the 2016–17 period. As displayed, the data system would be hard pressed to create annual data that covers the same time period for all the major sectors. Early learning and K-12 largely operate on a September year(t) to August year(t + 1) calendar. The 4-year university (PCHEES)

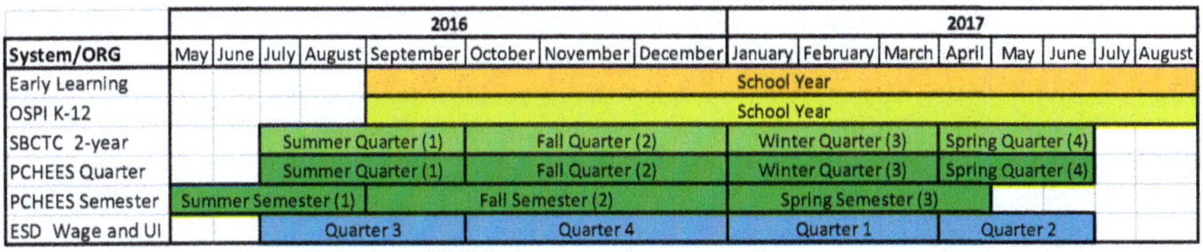

Fig. 19.3 Core data systems of the Washington state P20W and their approximate annual calendars, 2016–17

semesters largely operate on a trimester year, with 4 month periods running May to August, September to December, and January to April. The only data sources that seem to align are the quarterly data for 2-year college SBCTC and 4-year PCHEES terms, along with the ESD wage and unemployment insurance (UI) quarterly data.

The ultimate goal in organizing time is to develop biographies of individuals into structured histories of groups that allow for the analysis of sequences and timing of events. This would eventually allow for the full methods of event history to be employed. It would also allow the creation of cohort trajectories and studies will begin to provide a more complex understanding of the process of the education and career life course.

In conclusion, the Washington state P20W+ system provides longitudinal data that are useful for research in addressing policy questions in a number of areas. And moving forward, strengthening the quality of the data will allow for even further developments in research.

Acknowledgements The authors are grateful to Marc Baldwin and Jim Schmidt for the encouragement and support for all ERDC projects. We are also grateful to our ERDC colleagues for comments and suggestions, including Tim Norris, Carol Jenner, Karen Pyle, John Sabel, Huan Zhao, Jeff Thayne, Andrew Weller, Thomas Aldrich, Gary Benson, Lynn Cole, Teresa Greene, Kent Meneghin, and Micah Sanders.

References

Benson, G. (2019). *Impact of need-based financial aid on college completion: An event history analysis*. Olympia: Education Research and Data Center. Retrieved May 15, 2019 from https://erdc.wa.gov/publications/student-outcomes/impact-need-based-financial-aid-college-completion-event-history.

Bryan, T. (2004). Chapter 2: Basic sources of statistics. In J. Siegel & D. Swanson (Eds.), *The methods and material of demography* (2nd ed.). San Diego: Elsevier Academic Press.

Chen, V. W. (2015). *Assessment of the general educational development on earnings for Washington high school dropouts. Olympia: Education Research and Data Center.* Retrieved October 28, 2019 from https://erdc.wa.gov/publications/economic-returns/assessment-general-educational-development-certificate-earnings.

Chen, V. W. (2017). *A longitudinal profile of state need grant recipients' educational progress and degree completion* (Second engrossed substitute house bill 2376). Olympia: Education Research and Data Center. Retrieved from https://erdc.wa.gov/publications/student-outcomes/longitudinal-profile-state-need-grant-recipients-educational-progress#tableau.

Chen, V. W. (2019a). *Mathematics coursetaking pathway to College STEM for Washington State High School Students*. Olympia: Education Research and Data Center.
Chen, V. W. (2019b). *An overview of high school science coursetaking to College STEM in Washington State. Olympia: Education Research and Data Center.*
Chen, V. W. (2019c). *The impact of high school math and science coursework on Postsecondary STEM. Olympia: Education Research and Data Center*. Retrieved October 24, 2019 from https://erdc.wa.gov/publications/student-outcomes/impact-high-school-math-and-science-coursework-postsecondary-stem.
Chen, V., Pyle, K., & Aldrich, T. (2019). *Educational outcomes for children and youth experiencing homelessness*. In response to the 2018 Budget Bill, ESSB 6032 Section 223; contracted by the Washington State Department of Children, Youth, and Families. Retrieved October 24, 2019 from https://erdc.wa.gov/publications/student-outcomes/education-outcomes-children-and-youth-experiencing-homelessness.
Conaway, C., Keesler, V., & Schwartz, N. (2015). What research do state education agencies really need? The promise and limitations of state longitudinal data systems. *Educational Evaluation and Policy Analysis, 31*(1S), 16s–28s.
Database Normalization (n.d.). Retrieved from https://en.wikipedia.org/wiki/Database_normalization.
Dougherty, C., Meller, L., & Smith, N. (2007). Six key uses of longitudinal data. *Data Quality Campaign: Using Data to Improve Student Achievement, 2*(4), 1.
Education Research and Data Center (ERDC). (2014). *Washington 8th grader's socioeconomic trajectories and outcomes. Olympia: Education Research and Data Center.* Retrieved October 24, 2019 from https://erdc.wa.gov/publications/student-outcomes/washington-eighth-graders-educational-and-employment-trajectories.
Education Research and Data Center (ERDC), Office of Financial Management. (2015a). About us. Site accessed July 5, 2019. https://erdc.wa.gov/about-us-0
Education Research and Data Center (ERDC), Office of Financial Management. (2015b). Data governance. Site accessed July 5, 2019. https://erdc.wa.gov/about-us/data-governance
Education Research and Data Center (ERDC), Office of Financial Management. (2015c). Data history. Site accessed August 9, 2019. https://erdc.wa.gov/about-us/history-erdc
Education Research and Data Center (ERDC), Office of Financial Management. (2015d). Privacy considerations. Site accessed August 9, 2019. https://erdc.wa.gov/data-resources/privacy-considerations
Education Research and Data Center (ERDC), Office of Financial Management. (2015e). Memorandum of Understanding. Site accessed August 15, 2019. https://erdc.wa.gov/research-partners/our-partners/memorandum-understanding
Gertseva, A., & McCurley, C. (2019). *Education outcome characteristics of students admitted to juvenile detention*. Olympia: Center for Court Research, Administrative Office of the Courts.
Hough, G. C., Jr., & Beard, M. M. (2017). Chapter 11: State longitudinal data systems: Applications to applied demography. In D. A. Swanson (Ed.), *The frontiers of applied demography* (Applied demography series 9). Cham: Springer International Publishing.
Improving America's Schools Act. (1994). Retrieved December 9, 2015, from http://www2.ed.gov/legislation/ESEA/index.html
Institute of Education Sciences, National Center for Education Statistics. (2013, March). *SLDS issue brief: Building a centralized P-20W data Warehouse*. Retrieved August 7, 2019 from https://nces.ed.gov/programs/slds/pdf/centralized_warehouse.pdf.
Institute of Education Sciences, National Center for Education Statistics. (2015, July). *SLDS issue brief: Determining where to house P20W+ state longitudinal data systems*. Retrieved August 15, 2019 from https://slds.grads360.org/#communities/pdc/documents/8635.
Institute of Education Sciences, National Center for Education Statistics (Updated September 2016). *SLDS issue brief: Centralized vs. federated : state approaches to P20W+ data systems*. Retrieved August 7, 2019 from https://slds.grads360.org/services/PDCService.svc/GetPDCDocumentFile?fileId=23767.

Pyle, K. & Chen, V. W. (2019). *The characteristics of schools with and without substance abuse prevention and intervention in schools programs*. Office of Financial Management Education & Research Data Center. Retrieved November 27, 2019, from: https://erdc.wa.gov/publications/student-outcomes/characteristics-schools-and-without-substance-abuse-prevention-and

Revised Code of Washington 43.41.400. (2017). Office of Financial Management Education Data Center. Retrieved December 10, 2015, from http://apps.leg.wa.gov/rcw/default.aspx?cite¼43.41.400

Reykdal, C. (2019). *Report to the legislature: Covering the costs of dual credit for students and families*. Olympia: Office of Superintendent of Public Instruction. Retrieved November 27, 2019 from https://www.k12.wa.us/sites/default/files/public/communications/2019-11-Covering-Costs-of-Dual-Credit.pdf.

Sabel, J. (2013). *Identity matching at ERDC*. Education Research and Data Center (ERDC). Retrieved December 3, 2019 from https://erdc.wa.gov/publications-and-reports/erdcs-identity-matching-process

Snow-Renner, R., & Torrence, M. (2002). *No child left behind policy brief: State information systems*. Retrieved December 9, 2015, from https://eric.ed.gov/?q=snow-renner&id=ED468187

Star Schema (n.d.). Retrieved from https://en.wikipedia.org/wiki/Star_schema.

Chapter 20
Patterns of Geographic Variation of Mortality by Causes of Death for Small Areas in Brazil, 2010

Bernardo Lanza Queiroz, Flávio Freire, Everton E. Campos de Lima, Marcos Gonzaga, and Emerson Augusto Baptista

Introduction

In the last decades, Brazil has experienced an accelerated decline in infant, child and adult mortality. The median gain in life expectancy at birth between 1950 and 2010 was about 15 years, a much faster process of improvement in years of life than the observed in developed countries. In 1950, life expectancy at birth in Latin America, for example, was 51 years, reaching 67 years in 1990 and more than 70 years in 2010. Interesting fact, the mortality rates observed around 2000s were like the observed rates in the US and Canada many decades before (Palloni and Pinto 2011), thus showing that there is still room for improvement in terms of life expectancy in the region as well as in Brazil. In addition, there is a wide variation in levels of mortality and life expectancy among and within regions in the country.

We thank Brazilian Science Foundation (CNPq) for financial support for the Project (421183/2018-7). We would also like to thank Julia Calazans, at Cedeplar, for her assistance with Datasus microdata.

B. L. Queiroz
Universidade Federal de Minas Gerais, Belo Horizonte, Brazil
e-mail: lanza@cedeplar.ufmg.br

F. Freire · M. Gonzaga
Universidade Federal do Rio Grande do Norte, Natal, Brazil
e-mail: fhfreire@ccet.ufrn.br

E. E. C. d. Lima
Unicamp, Campinas, Brazil
e-mail: evertone@unicamp.br

E. A. Baptista (✉)
Shanghai University (Asian Demographic Research Institute), Shanghai, People's Republic of China

B. Jivetti, Md. N. Hoque (eds.), *Population Change and Public Policy*, Applied Demography Series 11, https://doi.org/10.1007/978-3-030-57069-9_20

In recent years, there are signs of convergence in infant mortality among the regions of Brazil, as infectious diseases have reduced their number of deaths, but there are still great differentials in life expectancy at birth and adult mortality in all Brazilian regions (Castro and Simões 2010; França et al. 2017). Recent studies (França et al. 2017; Borges 2017) analyzed the states of the federation (UFs) or large regions and they show how changes in the pattern of causes of death impact the evolution of life expectancy at birth and the wider mortality differentials observed in the country. However, the Brazilian regions are also marked by large inequalities and the understanding of this dynamic in smaller areas is fundamental for demographic and public health studies.

Studies of small areas, however, are usually compromised by data quality problems issues and the rarity of events occurrence. Recent researches analyze small areas mortality by causes of death (Baptista et al. 2018; Baptista and Queiroz 2019a, b; Pinheiro and Queiroz 2020) separately, but there is no analysis of the set of causes and compatibility of the results. Ross et al. (2018) estimate the burden of HIV and Tuberculosis for municipalities in Brazil under the Global Burden of Diseases project (França et al. 2017). They use a Bayesian spatially mixed effects regression model (Dwyer-Lindgren et al. 2016) and find large inequality across states and regions of the country. We propose in this paper, a methodology that combines a more detailed demographic method (relation model) with spatial analysis.

The main objective of this chapter is to present estimates of mortality by five major death causes—neoplasms, cardiovascular diseases, external causes, respiratory diseases and infectious diseases—in small areas of Brazil in 2010, based on data from the Mortality Information System (SIM in Portuguese) of Datasus, and analyze spatial patterns of causes of death. We focus in these specific causes because they represent the majority of deaths observed in the country in that year and show an interesting variation over every age at the life cycle. Consequently, we investigate deaths that are usually concentrated at younger, prime-age adults and older ages. A second objective is to analyze the spatial distribution of death causes across Brazilian small areas since some causes may be concentrated in a number of areas presenting a significant spatial correlation within or between states. Increases in external causes of deaths especially among males, for example, explains the high concentration of male adult mortality in some areas of the country (Souza and Lima 2007; Waiselfisz 2011; França et al. 2017). In addition, there are evidences of a concentration of males mortality from cardiovascular diseases in the less developed northeastern parts of the country (Baptista et al. 2018).

There is clear evidence that quality of mortality data in Brazil and its regions is improving over time, and a large part of the country shows almost complete coverage of death counts. Improvement in data quality is also observed in the decline of deaths registered as ill-defined causes of deaths that are less than 6% in 2017. The improvement is mostly explained by public investments in collection health data. As we show in this chapter, we proposed a simple and directed method to produce small-area mortality estimates by causes of deaths. We also produced age-standardized mortality estimates at the local level for a series of causes to investigate the spatial pattern of mortality. We find that sex differentials in mortality remained high over the period of analysis due to the increase in external causes of deaths especially among males. This increase also explains the high concentration of male adult mortality in some areas of the country. We find a concentration of mortality from cardiovascular diseases in the less developed Northeastern parts of the country, especially for males. Finally, we find that in 2010 there is a convergence process for causes of deaths across regions in the country and our proposed method to estimate complete age profiles for small areas by causes of death seem to be very robust and could be used in other countries with limited data.

Literature Review

Followed by a rapid mortality transition from high to low levels, Brazil also experienced a rapid epidemiological transition (Prata 1992; Borges 2017). This process, however, did not follow the same pattern as verified in most industrialized countries and other Latin American countries, such as Chile, Cuba, and Costa Rica (Prata 1992; Schramm et al. 2004; França et al. 2017; Borges 2017). Empirical evidence shows that there is overlap in transitional health states, with the persistence of some tropical infectious diseases (e.g. dengue, cholera, malaria, etc.) in parallel with the increase mortality rates by chronic and degenerative diseases. The conclusion is that there is no linear pathway across all stages of the epidemiological transition in Brazil, leaving it in the counter-transition state (Schramm et al. 2004; França et al. 2017; Borges 2017). In addition, the two morbidity-mortality groups are still also at high levels, characterizing themselves as a long-term transition process. At the same time, there are contrasting epidemiological situations in different regions of the country, also creating a scenario of epidemiological polarization (Schramm et al. 2004; França et al. 2017and Borges 2017).

Estimation of complete schedule of sex and age-specific mortality rates for small areas in Brazil remains a challenge for demographers and population heathy studiers for several reasons. First of all, there is a universal problem related to small exposition and high sampling variability in recorded deaths (Riggan et al. 1991; Bernardinelli and Montomoli 1992; Pletcher 1999; Gonzaga and Schmertmann 2016). Second, in many Brazilian regions death records is incomplete still in recent years (Queiroz et al. 2017). Finally, the quality of deaths registration by causes can vary across time and space (Lima and Queiroz 2014; da Cunha et al. 2017).

Related to the first and second problems, recent methods have been proposed to deal with the instability in observed deaths/exposition rates (De Beer 2012; Gonzaga and Schmertmann 2016). For Brazilian small areas Schmertmann and Gonzaga (2018) propose a Bayesian method to estimate data quality and mortality rates in 2010. The methodology allows to properly evaluate the data for small areas and obtain more robust estimates. In the last three decades, the Brazilian government has made significant investments to improve the coverage of death records and quality of death counts registration (Paes 2005). Although, the vital registration system is showing clear signs of improvement in recent years, the quality of death counts coverage and information on causes of death varies widely by region. França et al.

(2008) argues that, although Brazil has shown some signs of improving death registration coverage, there is still huge regional heterogeneity and limited quality of the death causes' registers. The percentage of ill-defined causes increased from 15% in 2000 to 18% in 2010, and the distribution of ill-defined death is also presenting substantial variation on the regional scale (Lima and Queiroz 2014), ranging from 11% in the South to 30% in the Northeastern part of the country. Lima and Queiroz (2014) find that the classification of causes of deaths is considerably improving in the Northeastern and Northern regions. They also suggested that improvements are related to governmental and administrative investments in data quality and have weak relation to economic indicators. da Cunha et al. (2017) find a clear improvement in the registration of causes of death in the last decades in Brasil. Similar evidence was observed by other authors (Ishitani et al. 2017).

The spatial pattern of mortality by cause, specifically for small areas, is still less well known. Most of the studies on the causes of death have focused on the whole country, large geographic regions (Southeast, Northeast, etc.), or very specific municipalities in Brazil (Nowbar et al. 2014; Rasella et al. 2014; Guimarães et al. 2015; Borges 2017; França et al. 2017). There is an absence and need for studies on the variation of causes of death mortality in time and space, or studies that aim to estimate age-specific mortality rates by causes of deaths for in less populated smaller areas.

In recent years, there has been increasing interest in identifying regional differences in mortality, and causes of death in different parts of the world. Kulkarani et al. (2011) argue about the importance of regional mortality and health studies as guiding points for public health policy planning and health system coverage. In their analysis, based on recent data from the United States, the authors show that there is an enormous variability in life expectancy between sites, with some of them showing mortality rates much higher than those observed in other developed countries. Similar research by Kibele et al. (2015), in Germany, shows that the factors that explain the variation of mortality in the past are still affecting the regional mortality variation present in the country.

Vierboom et al. (2019) investigate the increasing regional inequality in mortality in the US. They show that smoking related mortality explain most of the increase in the regional inequality in the last two decades. Barbieri et al. (2015a, b) analyzes regional trends in adult mortality in the US from 1959 to 2004. The study shows that the decline in mortality in the late 1960s was unexpected, but also widespread and driven by declining mortality from cardiovascular disease. In the same direction, Roth et al. (2017) estimated age-standardized mortality rates for cardiovascular diseases among US counties between 1980 and 2014 and they found extensive spatial disparity. They observed large differences among municipalities in mortality rates. Fenelon (2013) investigates the divergence in adult mortality rates for the American states between the 1960s and 2000s. He observes different levels of mortality variation that lead to an increase in the differential between North and South states. The results indicate that part of the divergence is explained by smoking-related mortality. Mokdad et al. (2017) investigate mortality from neoplasms in American counties and note that despite the decline in cancer mortality in the United States, there is substantial regional variation in overall mortality decline patterns and mortality from specific types of cancer. Dwyer-Lindgren et al. (2017) analyze the life expectancy differential in the American counties between 1980 and 2014. The results indicate persistent and large differentials related to issues of health access, health behavior and social and economic differentials.

We argue that small areas in Brazil are, like the United States and Germany, characterized by differences in mortality for various reasons, and there is also a growing demand for adequate health policies given the current aging process of the Brazilian population. The importance increases to develop research that seeks to understand the evolution of mortality, its determinants and variations over the last years.

In the case of Brazil, Queiroz et al. (2018) analyze the differential of adult mortality between 1980 and 2010 in Brazil and observe, as for the US, how social and economic factors influence trends in time and space. Baptista et al. (2018) analyze the recent evolution of mortality due to cardiovascular diseases and decompose the effects of changes in levels of mortality rates and age structure of the population. They note an increased concentration of high mortality rates from this disease in the Northeast. Along the same lines, Pinheiro and Queiroz (2020) estimate mortality rates for motorcycle accidents in small areas and observe a concentration in the frontier areas of the country, especially in the countryside of less developed states. In common, the two studies on Brazil combine statistical and demographic methods in the analyses.

This chapter is relevant since public health policies and well-being must be spatially designed and focused to serve an aging population subject to varying risk of mortality from different causes.

Methodology and Data

Death Counts and Population Data

We use death record from the Ministry of Health database, DATASUS (http://www2.datasus.gov.br). The database provides information on deaths, causes of deaths, by age and sex at the municipality level. The data are available since 1979, but we use information only from 2010, the last census year. Mortality data is organized using codes from the ICD Revision 10th from 1996 on). Data cleaning and compilation is done at the municipal, provincial and state level, and an electronic data file is transferred to the national office every 3 months. Population by age and sex, at the local level, comes from the Brazilian Census 2010.

The original data is available at the municipality level. The main limitation in using city level data in Brazil is that the number and composition of cities change over time. In 1980, there were 3974 municipalities and in 2010 there were 5565. To avoid problems using this information, we aggregated municipalities by comparable small areas, using the IBGE definition of comparable microregions. The microregions serve only for statistical purpose; therefore, they do not represent a political or administrative entity. The main advantage of working with these geographical areas is that they have not changed their boundaries over the period of analysis and they are areas with regional and socioeconomic similarities. Microregions are stable and comparable over the period of analysis. Although we are analysing data only for 2010 census year, results at microrregion level could be usefull for further comparisons.

Ministry of Health data, Datasus, is publicly available. Historically, states of the North and Northeast observed lower coverage and worse quality of death declarations in relation to the states of the Southeast and South (Queiroz et al. 2017; Lima and Queiroz 2014; Agostinho and Queiroz 2010). However, since the 2000s, there has been an impressive improvement in both conditions (Queiroz et al. 2017; Lima and Queiroz 2014; Agostinho and Queiroz 2010). There are still relevant questions about the quality of age declaration for older ages and how it might affect mortality estimates at older ages (Nepomuceno and Turra 2020; Pinheiro and Queiroz 2019; Reis and Turra 2017; Gomes and Turra 2009). The improved quality of mortality information in the North / Northeast has radically altered the trends, mainly due to noncommunicable diseases. Several articles argue and show that the quality of the information has improved since 1996 (Guimarães et al. 2015; Borges 2017; França et al. 2017; Ross 2018). It allows an adequate comparison between the regions.

Mortality Estimates and Topals Relation Model

The first step of the analysis is to obtain the corrected value of mortality across regions in the country. We use estimates produced by Queiroz et al. (2017, 2018) for small-areas and states in this paper. They use a combination of smoothing techniques with death distribution methods to evaluate and adjust mortality death counts. First, they follow Gonzaga and Schmertmann (2016) and use a method that combines Poisson regression with the TOPALS relational model (De Beer 2012). The method can be applied for any small areas and allows to build complete schedules of sex and age-specific mortality rates via mathematical adjustments to a specified standard schedule without assuming a fixed age pattern of mortality rates for small areas (Gonzaga and Schmertmann 2016). Second, they use Death Distribution Methods (Hill et al. 2009; Queiroz et al. 2017, among others) to evaluate coverage of deaths counts in relation to population counts (Hill et al. 2009). We present the methods briefly, because there is an extensive literature detailing its formalization (Bennett Neil and Horiuchi 1981; Hill 1987; Hill et al. 2009; Dorrington 2013a, b). These methods compare the age distribution of deaths with the age distribution of the population and provide the age pattern of mortality for a set period. There are three main methods: general growth balance (GGB), proposed by Hill (1987), synthetic extinct generations (SEG) proposed by Bennett Neil and Horiuchi (1981), and the synthetic extinct generations adjusted (SEG-adj) proposed by Hill et al. (2009). The deaths of distribution methods have strong assumptions: closed population, the degree of coverage of deaths is constant by age, the degree of population count has constant coverage by age and ages of living and deaths declared without errors.

The estimates of completeness of death counts coverage and adjustment factors are obtained by all regions in the study. However, the adjustment is only one for all ages and total deaths. We use the same adjustment factor for each cause of death. We combine a variety of analysis to produce estimates of mortality by sex, age and causes of deaths for small areas in Brazil.

Mortality estimates below 15 years of age were indirectly calculated using surviving children information from Census 2010. The main idea of the method, proposed by Brass (1975), is to estimate the probability of death in childhood from the dead children ratio according woman's age.

Later, we produce age-specific mortality rates by causes of death for each region in the country following Gonzaga and Schmertmann (2016). The technique is based on a relational model that estimates and smooths the specific mortality rates for any municipality in Brazil (Gonzaga and Schmertmann (2016)). Although it requires a standard that describes the behavior of age rates, the proposed technique is not sensitive to the choice of this standard, making the model very flexible for application in any area whose observed rates vary greatly in age. In our case, we set a different standard (age specific mortality by cause of death) for each of the cause we study. We use the Brazilian 2010 age specific mortality rates by cause of death as a standard to obtain our estimates.

The proposed method, called TOPALS regression, is a Poisson regression model based on De Beer's (2012) proposal for smoothing and projecting death probabilities. Gonzaga and Schmertmann (2016) proposed a method that estimates the logarithm of mortality rates by sex and age (0, 1, 2, ..., 99+) for all Brazilian municipalities based on the following relational model:

$$\underset{101x1}{\lambda(\alpha)} = \underset{101x1}{\lambda^*} + \underset{101x7}{\mathbf{B}}\ \underset{7x1}{\alpha} \tag{20.1}$$

where λ is a vector of the logarithm of mortality rates in a small area of interest; λ^* is a vector of the logarithm of the standard mortality rates; B is an array of constants where each column is a B-Spline linear function and α is a vector of parameters to be estimated representing deviations from the mortality curve to be estimated in relation to the standard curve. The ages 0, 1, 10, 20, 40, 70, 100 were arbitrarily defined as points of knots of the Spline function. These ages were defined as cut points characteristic of the function of mortality in view of the typical behavior of the function between each of these ages.

In the relational model in (20.1), the linear function B-Spline (B) is responsible for smoothing the logarithm of the mortality rates between the points of knots, considering the α parameters estimated based on the standard function of mortality. Thus, for any set of mortality rates (D_x/N_x) and population exposed to risk (N_x), where D and N represent, respectively, the number of deaths and the population exposed at each age x (for x = 0,1, 2, ..., 99), it is assumed that deaths are random variables that follow a Poisson distribution, such that the logarithm of its likelihood function is defined as:

$$\log L(\alpha) = \sum_{x} [D_x \lambda_x(\alpha) - N_x \exp(\lambda_x(\alpha))] - \sum_{k=0}^{5} (\alpha_{k+1} - \alpha_k)^2 \tag{20.2}$$

The term given by the sum to the left of item (20.2) aims to penalize very divergent estimates of α thus avoiding an implausible mortality pattern for the area of interest. The assumption translated by the addition of this penalty term into the logarithm of the likelihood function is that very small areas tend to have a structure of mortality rates by age different from the standard, but not too far from what should be a reasonable mortality pattern for human populations and, in particular, to the area of interest (Gonzaga and Schmertmann 2016).

Indirect Standardization

In the first step of the analysis, we use age-standardized mortality rates. Standardization can take place in two ways: direct (direct standardization) or indirect (indirect standardization). In the case of direct standardization, it is necessary to have the total of events, distributed by age groups, and the age distribution of the study population. In this chapter, the standard age structure will be that of Brazil, year 2010. In this way, it will be possible to compare crude mortality rates by each cause obtained by microregion and to get the differential level of the variable studied here, since they will only reflect the differences in specific rates of the variable in the analyzed populations. Therefore, using the direct standardization technique, it is as if the crude rates of the various populations of the Brazilian microregions had the same age composition, but each one maintaining its own specific rates, that is, it standardizes the population structure of the population and leaves the mortality by different causes of deaths observed in each locality.

Empirical Bayesian Models

For our second set of estimates, we propose the application of the empirical Bayesian estimator, whose initial idea developed by Marshall (1991) was the use of a contraction estimator for under-5 mortality rates in census tracts in Aukland, New Zealand. The central objective of the method is to smooth the random fluctuation of mortality rates by approaching an observed rate in each area smaller than an average global rate or even at an average rate of the neighboring sectors, considering the population size of the area question.

However, the crude rates observed in small municipalities are not indicated for this type of analysis, since they can be influenced by random fluctuation and complications of calculations due to small numbers. Thus, we incorporate location into the empirical Bayesian estimator by defining neighborhood structures for each area. The definition of the neighbors of each observation will serve as the basis for the definition of a priori parameters. That is, the rates of the small areas, mainly, will converge towards the average of the rates of their neighbors. In relation to the global approaches, the estimate that considers the local average of the neighbors will present a spatial smoothness closer to the reality of the events of interest.

In this chapter, we chose the James-Stein empirical Bayesian estimators, proposed by Marshall. In this model, the estimators are defined according to Eqs. 20.3 and 20.4:

$$\widehat{\theta} = \widetilde{m} + \widetilde{C}_i(x_i - \widetilde{m}) \tag{20.3}$$

$$\widetilde{C} = s^2 - \frac{\widetilde{m}}{\overline{n}} / \left(s^2 - \widetilde{m}/\overline{n} + \frac{\widetilde{m}}{n_i}\right) \tag{20.4}$$

where, $\theta = m$, when $s^2 < m / \overline{n}$. C is the contraction factor; m is the average of x_i; x_i is the crude rate at i; s^2 is the sample variance of x_i weighted by the population size; and ni is the number of person-years at risk in the small area.

To implement this methodology, we used GeoDa software 1.6.7. Through it a simple neighborhood structure was defined, that is, the criterion "k-Nearest Neighbors" ensures that each microregion has the same k number of neighbors. It is often reasonable to consider that similar areas have similar disease patterns. Thus, the so-called local approach uses the information from the neighbors of each observation as the previous parameters for the adjustment of θ. In this sense, θ is estimated by the contraction of xi towards the mean observed in its neighbors In this chapter, we use $k = 8$. Simulations with different numbers of neighbors did not present significant changes in results.

Results

Evolution of Mortality Data Quality in Brazil

Figure 20.1 shows the estimates of completeness of death counts coverage for males and females in 2010. The results were developed from the work of a research group that presented methods and estimates in different publications (Queiroz et al. 2018; Queiroz et al. 2017; Lima and Queiroz 2014). The estimates on the evolution of coverage of adult death and mortality records in Brazil show some evolution and large regional differences. For males and females, the less developed areas of the country presented the greatest progress in the coverage of the death registry and in the last three decades quality of data on adult mortality appears to have improved

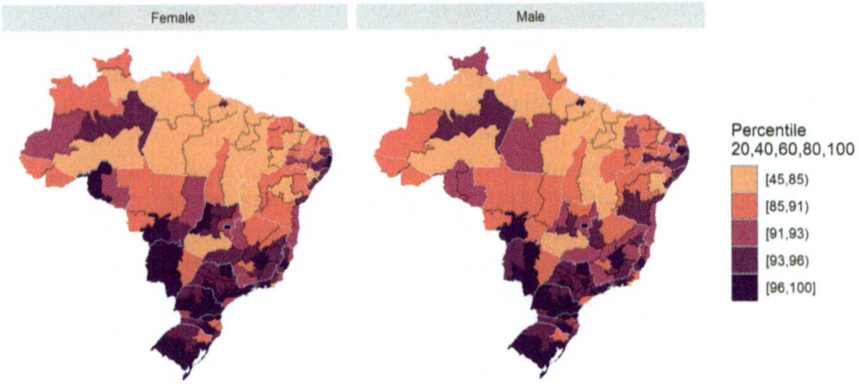

Fig. 20.1 Quality of Adult Mortality Data, by sex, mesoregions, Brazil 2010

significantly over the years and in many parts of the country (Queiroz et al. 2018; Queiroz et al. 2017; Lima and Queiroz 2014).

The quality of the registry of deaths in Brazil has presented significant improvements in recent years, with coverage of the death registry, for both men and women, above 95% (Queiroz et al. 2017). However, there is still great regional variation. States located in the South and Southeast have records of 100% of deaths, for both men and women. In some states in the Northeast and North, the quality of information is lower, but these have recent significant advances when compared to the period 1991 and 2000. In 2010, all the federative units of the South and Southeast regions, as well as some federative units of the Northeast and Central West, presented complete coverage of the death registry (Schmertmann and Gonzaga 2018). In addition, there was a great improvement in the quality of mortality information in the poorer federative units of the Northeast and North regions, especially those that had the worst record quality in previous periods.

Overview of Mortality by Causes of Deaths in Brazil

Figure 20.2 shows age-specific death rates by causes of deaths for males in 2010. They highlight the differences of mortality by causes over the life-cycle. We observed a very high level of mortality by external causes of deaths for young adults. External causes of deaths, for males, are the more important cause of death in Brazil and contributes to the slowdown in the evolution of life expectancy over the recent years (Borges 2017; Pereira and Queiroz 2016; Carvalho et al. 2012). Cardiovascular diseases increase rapidly with age, as observed in other countries, and has become the most important cause of death for older adults in Brazil (Lotufo 2019).

Figure 20.3 shows standardized mortality rates by causes of deaths for males, (using Brazilian population age structure in 2010 as standard). We observed a very

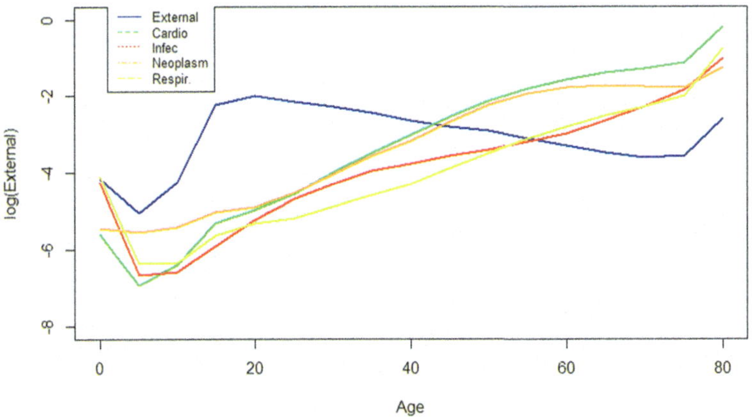

Fig. 20.2 Age Specific Mortality Rates by causes, Males, Brazil, 2010
Source: Population Censuses (2010), Ministry of Health (http://www.datasus.gov.br), 2010

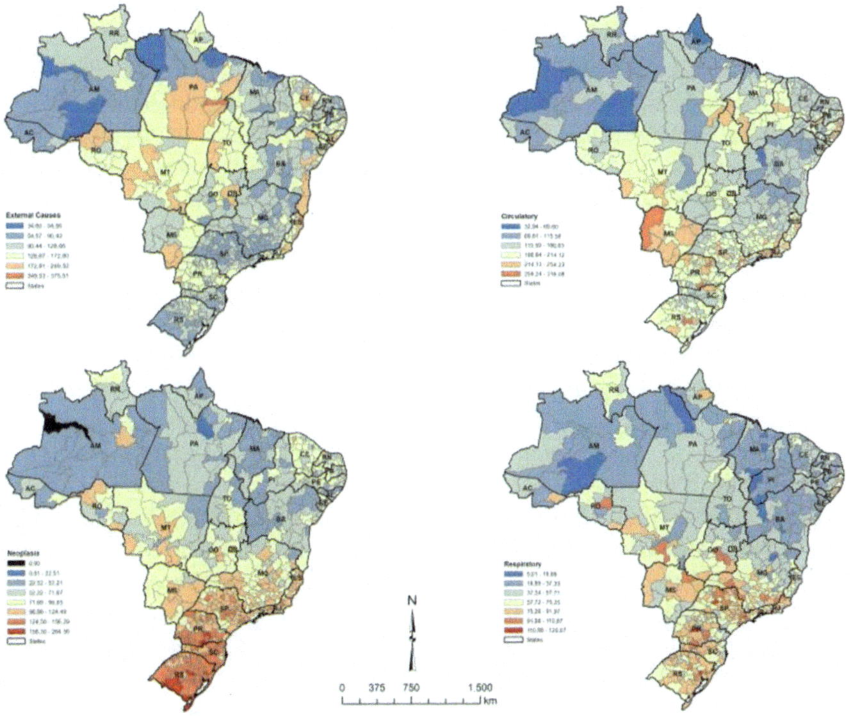

Fig. 20.3 Standardized Crude Death Rates, by main causes of deaths (external causes, circulatory, neoplasia, respiratory), Males, Brazil, 2010
Source: DATASUS, 2018 and Population Censuses (1980, 1991, 2000, 2010)

clear spatial distribution of mortality by causes across Brazilian small areas. All causes are concentrated in larger and coast cities of the country, specially Neoplasia and External causes. In fact, large Southeast and Northeast cities, especially those in Atlantic Coast has experienced very high violence and transit accident rates (Souza and Lima 2007; Waiselfisz 2011; França et al. 2017). Respiratory and Circulatory respond for higher mortality in less urbanized areas. In 2010, we find high levels of mortality by accidents and violence in middle-size cities in the interior of the country. Also, there is a concentration in almost all coastal cities of the northeast part of the country. From Rio de Janeiro to the most northern capitals of the coastal. On the other side, São Paulo has one of the lowest levels in the country in 2010 (Malta et al. 2017; Pereira and Queiroz 2016). For cardiovascular diseases, we find that in the North and, specially, across the Northeast regions, and the northern region of the state of Minas Gerais, there is high levels mortality and that they increase over the years. The results suggest a rapid decline in CVD mortality in the South and Southeast micro-regions and a slower decline in the Center-West region. In the case of neoplasm and respiratory diseases, the results show higher mortality levels observed that they are more concentrated in the more developed parts of the country that are in later stages of the epidemiological and demographic transitions compared to the other parts of the country (Borges 2017; França et al. 2017; Baptista and Queiroz 2019a; Lotufo 2019).

One should be careful with the scales of each map. They are different because the aim is to know the Brazilian mortality spatial pattern for some causes of death. We should be cautions with the pattern observed here since data were not adjusted and there is random fluctuation for small areas that might be affecting the observed levels. However, the results show similar spatial pattern for the 4 causes of death presented: mortality rates higher in the extreme northeast, southeast and south. A hypothesis is, except for external causes of death, the other three causes increase as population ages.

The problem of the use of crude mortality rates for the analysis of the spatial distribution of diseases, highlighted by authors such as Justino et al. (2012) and Carvalho et al. (2012) is to a large extent a consequence of the great fluctuation, observed mainly in small municipalities. It is very unlikely that in municipalities with mortality rates equal to zero the risk of death by any cause is zero. According to Freire et al. (2014), in these cases, a plausible explanation is that the time window was not long enough to record the occurrence of the event of interest. On the other side, we might encounter locations with extremely high mortality rates and a large variation between two points in time. As Assunção et al. (1998) point out, due to the small size of the denominator, more deaths can have a significant impact on the estimated rates. This is another indication of the need to other techniques for smoothing rates, the results of which will be presented in the next section.

Small-Areas Estimates: Empirical Bayesian Models

We now present the results of the smoothing process (Spatial Empirical Bayes) of standardized mortality rates. As pointed out before in this chapter, the decision to use the local empirical Bayesian estimator for rate smoothing is based on the need to deal with the problem of random fluctuation, which is mainly present in the rates of less populous municipalities. We expect that the estimates of mortality for Brazilian municipalities, after smoothing, will be of superior quality. Thus, the intention is to build better estimates of the risks of dying by different causes of deaths, considering the Brazilian small areas, by sex, to better understand both phenomena. We present the results for the four (4) main causes studied in this chapter for males and focusing on estimates in 2010.

Figure 20.4 shows the estimates for males – but trends for females are quite similar – in 2010. One of the main interesting results is the temporal and spatial trend observed for external causes of deaths. In 1991, results not shown, highest levels of external causes of deaths were concentrated in Rio de Janeiro, São Paulo and around Brasilia. We also observed very high mortality rates by external causes in frontier areas of the Amazon and Brazilian Cerrado (between Midwest and parts of North

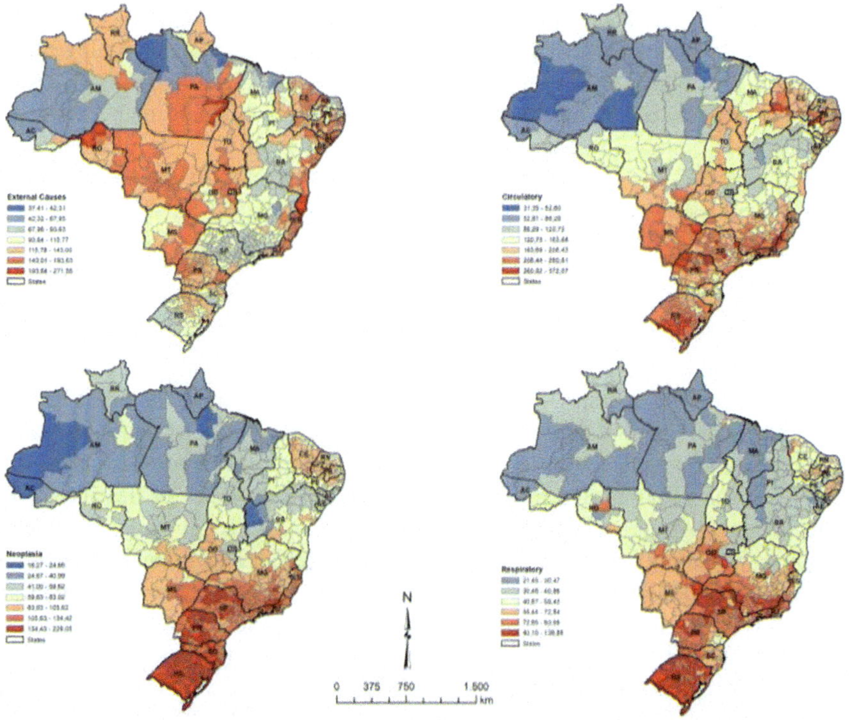

Fig. 20.4 Empirical Bayes Mortality rates, by main causes of deaths, males, 2010, Brazil and small areas

and Northeast of the country). In 2010, we observed an interiorization of the mortality due to external causes of deaths and an increased in the mortality by this cause in almost all coastal cities of the northeast part of the country.

In relation to cardiovascular diseases, we find that males have higher mortality rates than females in all periods. In addition, for both sexes, mortality rates decline over time, however, while we observe a more homogeneous spatial configuration between sexes, there was a more heterogeneous distribution within the same sex. For males, in the Center-West, South and South-East (except for the northern region of the state of Minas Gerais), we found a decline in mortality by over time. In the North and, specially, across the Northeast regions, and the northern region of the state of Minas Gerais, there was an increase mortality over the years.

In relation to neoplasm, we find that the contribution of neoplasm to overall mortality has increased steadily from 1991 to 2010, for both males and females. The change is closely related to changes in population age structure. Overall, neoplasm is the second major cause of death in Brazil, but from 1990 to 2010 we find that the crude death rate for this cause increased from about 80 per 100.000 to 126 per 100.000. More importantly, if in 1990 the highest level of mortality due to neoplasm was concentrated in the South part of the country, we find a spread of the mortality by this cause to all regions of the country in 2010. In 2010, almost all regions in the South and Southeast parts of the country had very high mortality rate by this particular cause and some areas of the Northeast.

Lastly, for respiratory diseases we find an overall decline in the mortality by this cause for the country. But most important we observed a convergence process of smoothed crude mortality rate by this cause across different regions of the country. If in 1990, the highest level was concentrated in the more developed areas of the country, we find that in 2010 the levels tend to be very close related instead of the large variations observed in the 1990s.

Small-Areas Estimates: Topals Relational Models

The use of crude death rates, even with a smoother process, might be misleading. In this section, we present alternative estimates of mortality rates by causes by single age of death. Figure 20.5 presents preliminary TOPALS estimates for males mortality schedules for selected mesoregions within different states in Brazil in 2010. The selection of mesoregions was based on different size of male populations exposed to mortality rates during the period from 2009 to 2011. In all graphs one can see the underlying mortality pattern for each mesoregion. The results confirm that the TOPALS allows estimating the entire schedule of single-year rates even for a relatively small exposure (see results for "Norte do Amapá" mesoregion which recorded 28,410 males according to 2010 Census).

Gonzaga and Schmertmann (2016) demonstrated that TOPALS method is valuable when small areas age pattern of mortality differs. However, they also showed that under the homogeneity assumption about mortality schedule by age, indirect

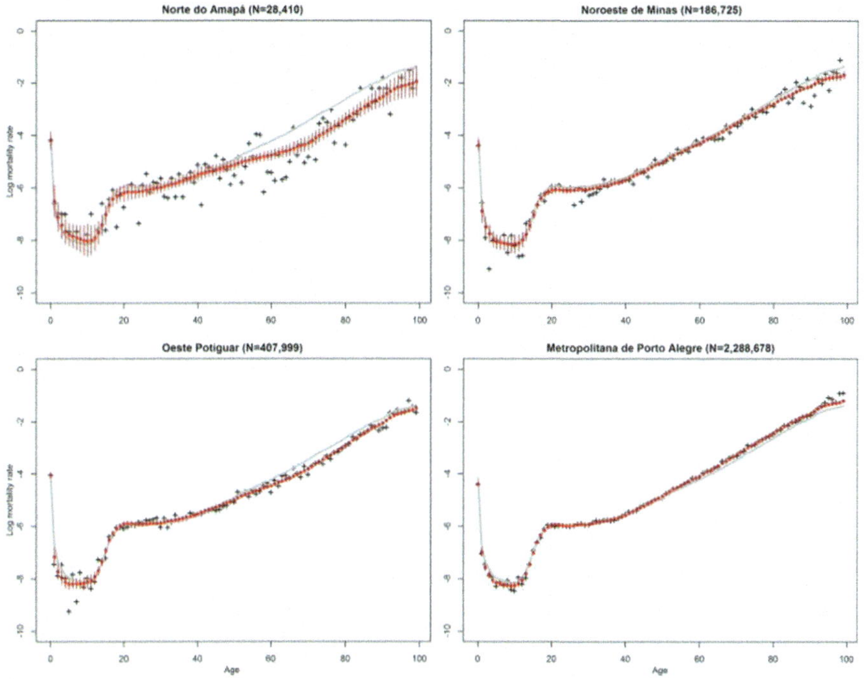

Fig. 20.5 Maximum likelihood TOPALS fit of log mortality rates by age, males, selected mesoregions, Brazil (2010)
Source: Population Censuses (2000, 2010), Ministry of Health (http://www.datasus.gov.br)
Note: The symbols "+" represent observed ln(deaths/pop) for each single year of age for three years period around the 2010 Brazilian Census. Red points represent the fitted ln(deaths/pop) by TOPALS method with a 95% Confidence Interval (vertical red segments). The grey solid curve represent the standard schedule for males log mortality rates by age (standard = Brazilian males age-specific log mortality rates in 2010)

standardization technique works better than TOPALS. Figure 20.6 shows the fitted schedule of male mortality rates by age in all mesoregions within four selected Brazilian's state. We chose four states from different large areas of the country. The results showed that assuming equal pattern of mortality rates by age can lead to unreasonable estimates, especially if one is trying to estimate mortality rates by causes that are more prevalent in some specific age interval. On the other hand, using a simpler technique, as indirect standardization, could lead to similar results.

Figure 20.7 shows estimates in 2010 for males in Minas Gerais, one of the 27 states in Brazil, for mortality by external causes and cardiovascular diseases. We produced estimates for all states in the country, but due to space limitations we are only showing one state here. The figure shows observed and adjusted data using the proposed method for single years of age. We also show the standard (Brazil) used in the analysis. The adjusted estimates using Topals is reported with a 95% confidence interval. We observed a very high level of mortality by external causes of

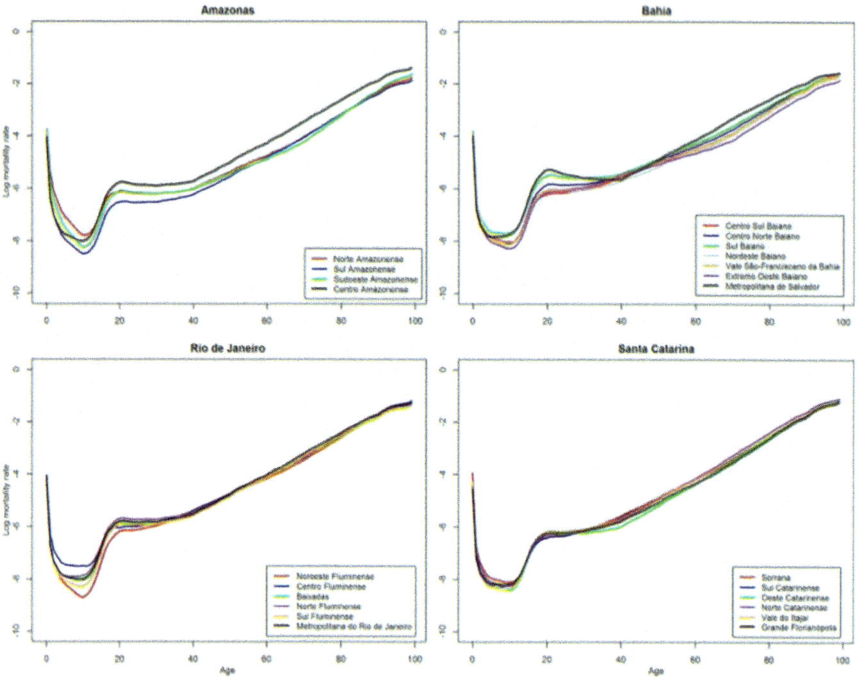

Fig. 20.6 TOPALS fit of males log mortality rates by age for all mesoregions in selected states, Brazil (2010)
Source: Population Censuses (2000, 2010) and Ministry of Health (http://www.datasus.gov.br) and Ministry of Health (http://www.datasus.gov.br)

deaths for young adults. Cardiovascular diseases increase rapidly with age. For both causes, Minas Gerais mortality rates are below of what was observed in Brazil in 2010.

Figure 20.8 shows estimates of external causes of deaths by single age in the South/Southeast of Minas Gerais state and the region of Vale do Jequitinhonha in 2010. We also have estimates for all small areas of the country but opted to show a few examples of how the proposed model fitted to the analysis. The figure shows the Brazilian Standard, the State Adjusted Mortality Rates, the observed mortality rate and the Topals adjusted for the small area. The analysis shows that there are large differences in the mortality by causes across regions both in terms of levels and shapes. This is an important result that helps to explain the difference in life expectancy across regions in the country in recent years.

The model works well for both causes and shows the difference between mortality level at the small-area compared to the state and country. The results also indicate that we can improve our estimates by adjusting the knots to the profile of the cause-specific mortality. We can also fit with different standard, however Gonzaga and Schmertmann (2016) showed that the model is very stable to the choice of standard.

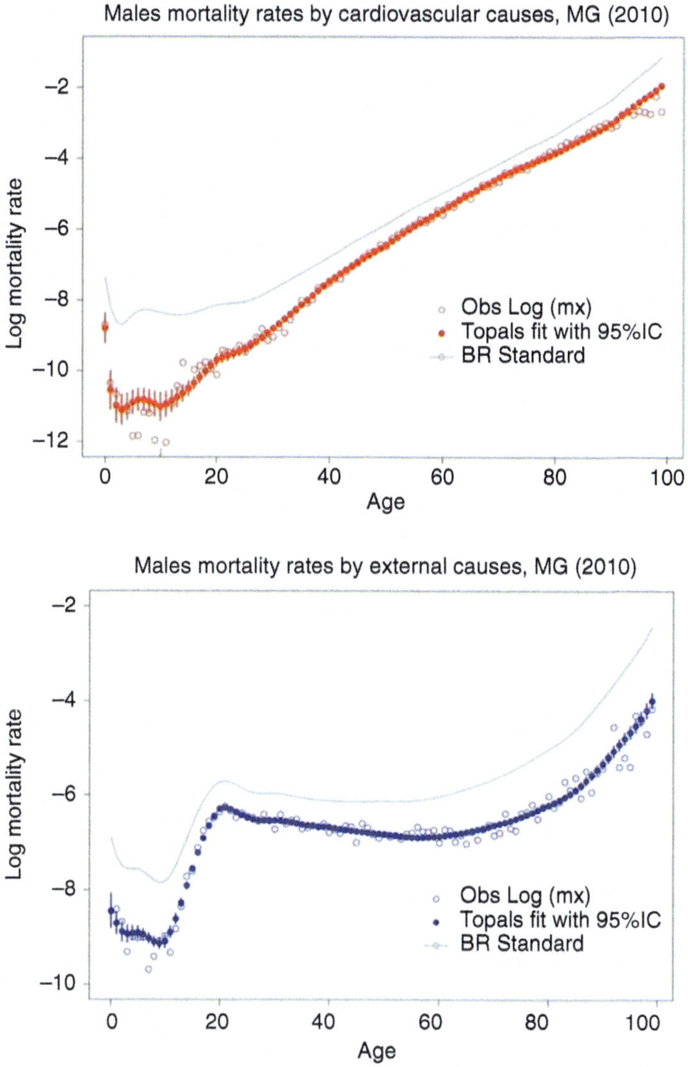

Fig. 20.7 TOPALS fit of males log mortality rates by age, males, Minas Gerais, External Causes of Deaths and Cardiovascular Diseases
Source: Population Censuses (2000, 2010) and Ministry of Health (http://www.datasus.gov.br) and Ministry of Health (http://www.datasus.gov.br)

Discussion

In this chapter, we estimate and investigate the main causes of death for small-areas in Brazil in 2010. We find that despite improvements in overall mortality level for the country, there is a large between-areas differences for every cause of death,

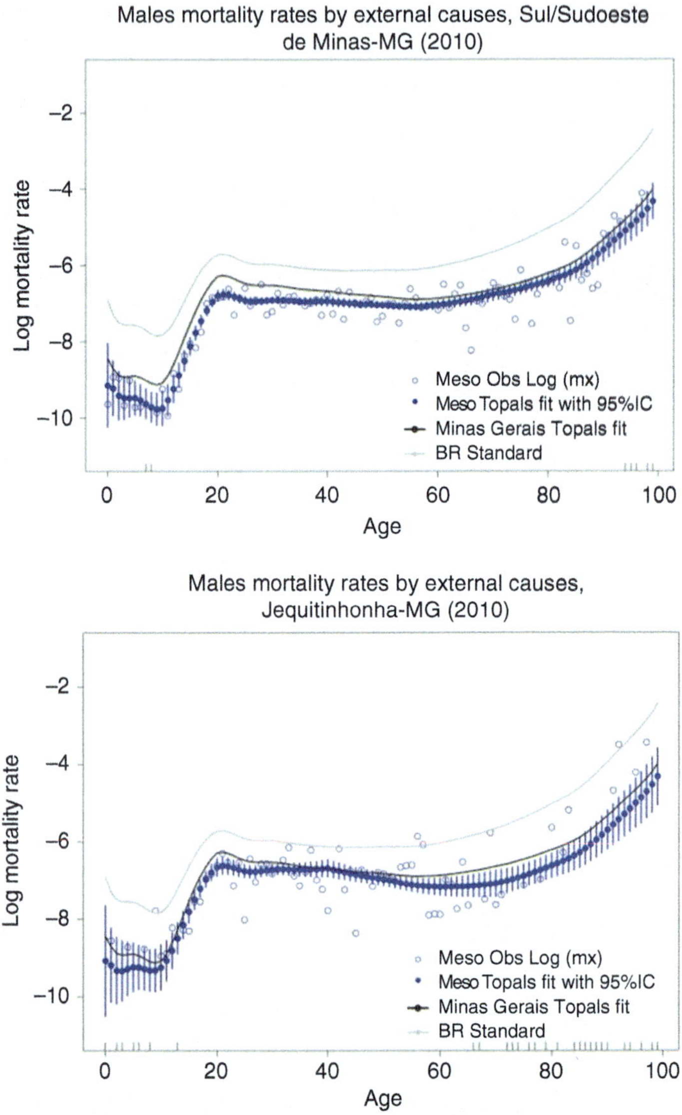

Fig. 20.8 TOPALS fit of log mortality rates by age, Cardiovascular Diseases and External Causes of Deaths, males, Sul/Sudoeste de Minas and Vale do Jequitinhonha (small area in Minas Gerais), 2010
Source: Population Censuses (2000, 2010) and Ministry of Health (http://www.datasus.gov.br) and Ministry of Health (http://www.datasus.gov.br)

although geographic patterns varied substantially by cause of death. The approach to county-level analyses with small area models used in this study has the potential to provide novel insights into Brazilian disease-specific mortality time trends and their differences across geographic regions and could be a valuable contribution to public health planning.

We perform a series of analysis aiming to obtain more robust estimates for cause of death mortality rates for small-areas. First, we use a simple standardize death rates by cause, but this approach has a series of limitations. Second, we used a Spatial Empirical Bayes method trying to obtain more robust estimates, but in this approach, we can only focus on crude rates. Lastly, we use the Topals regression approach to try to estimate a complete mortality age profile by causes of deaths for males and females for each small-area of the country. Thus, the results of this study represent more detailed and comprehensive study of Brazilian small-area patterns of cause-specific mortality currently available – as most of studies in Brazil focus on larger areas or one specific cause of death.

We find expressive differences across regions and causes of deaths in Brazil in 2010. This result shows the importance to include a cause-specific analysis for small areas in addition to analyses of all-cause mortality and life expectancy at certain ages. We find, however, that there is a convergence process for non-communicable causes of deaths across regions in Brazil, but there is still a concentration of higher death rates in the more developed and aged parts of the country.

For some causes, we observed an impressive increase in mortality in most areas of the Northeast states. In 1990, there was a concentration of higher mortality rates by cardiovascular diseases in the South and Southeast, but the more recent years observed a rapid increased in the mortality by this cause in less developed areas of the country (Baptista and Queiroz 2019a, b). More impressive, though, is the spread of mortality by external causes of death (violence and traffic accidents) that were concentrated in the Southeast and are spreading for smaller areas and almost all coastal cities in the Northeast (Malta et al. 2017; Pereira and Queiroz 2016; Carvalho et al. 2012; Waiselfisz 2011).

The study has some important limitations that must be taken into account when analyzing the results. First, several regions in Brazil still lack good quality data. There is an issue of under-registration of death counts coverage and limitations in the declaration of the age of the deceased (Queiroz et al. 2017, 2018). The quality of mortality data has improved in the period of analysis, but there is still large heterogeneity. Also, there is an issue of errors in age declaration at older ages that might affect estimates of mortality and life expectancy across regions in the country (Nepomuceno and Turra 2020). Second, information on causes of death have all been improving overtime and across regions in different pace (Ishitani et al. 2017; Borges 2017). The percentage of deaths registered as ill-defined has reduced overtime, but they are still quite large for some less developed areas of the country. Lima and Queiroz (2014) showed that ill-defined causes of deaths in Brazil declined from 27.5% of registered death in 1991 to 8.5% in 2010, reaching about 5% in 2016.

References

Agostinho, C. S., & Queiroz, B. L. (2010). Estimativas de mortalidade adulta para as Unidades da Federação, Brasil, 2000. In: Encontro Nacional de Estudos Populacionais, 16, 2010, Caxambu. Anais. Belo Horizonte: ABEP, 20p.

Assunção, R. M., Barreto, S. M., Guerra, H. L., & Sakurai, E. (1998). Maps of epidemiological rates: A Bayesian approach. *Cadernos De Saude Publica, 14*(4), 713–723.

Baptista, E. A., & Queiroz, B. L. (2019a). Spatial analysis of mortality by cardiovascular disease in the adult population: A study for Brazilian micro-regions between 1996 and 2015. *Spatial Demography, 7*(1), 83–101.

Baptista, E. A., & Queiroz, B. L. (2019b). The relation between cardiovascular mortality and development: Study for small areas in Brazil, 2001–2015. *Demographic Research, 41*, 1437–1452.

Baptista, E. A., Queiroz, B. L., & Rigotti, J. I. R. (2018). Decomposition of mortality rates from cardiovascular disease in the adult population: A study for Brazilian micro-regions between 1996 and 2015. *Revista Brasileira de Estudos de População*, 35(2).

Barbieri, M., Wilmoth, J. R., Shkolnikov, V. M., Glei, D., Jasilionis, D., Jdanov, D., . . . & Vinant, C. (2015a). Data resource profile: The human mortality database (HMD). *International Journal of Epidemiology, 44*(5), 1549–1556.

Barbieri, M., Ouellette, N., Andreeva, M., Riffe, T., & Vinant, C. (2015b). The cardiovascular revolution in the United States: a geographic analysis. Presented at the Annual Meeting of the Population Association of America, San Diego, California, May 2–4. https://paa2015.princeton.edu/papers/153514

Bennett Neil, G., & Horiuchi, S. (1981). Estimating the completeness of death registration in a closedpopulation. *Population Index, 1*, 207–221.

Bernardinelli, L., & Montomoli, C. (1992). Empirical Bayes versus fully Bayesian analysis of geographical variation in disease risk. *Statistics in Medicine, 11*, 983–1007.

Borges, G. M. (2017). Health transition in Brazil: Regional variations and divergence/convergence in mortality. *Cadernos de saude publica, 33*, e00080316.

Carvalho, A. X. Y. D., Silva, G. D. M. D., Almeida Júnior, G. R. D., & Albuquerque, P. H. M. D. (2012). Bayesian rates for homicide mapping in Brazilian municipalities. *Cadernos De Saude Publica, 28*(7), 1249–1262.

Castro, M., & Simões, C. (2010). Spatio-temporal trends of infant mortality in Brazil. In *XXVI IUSSP International Population Conference.*

da Cunha, C. C., Teixeira, R., & França, E. (2017). Avaliação da investigação de óbitos por causas mal definidas no Brasil em 2010. *Epidemiologia e Serviços de Saúde, 26*(1), 19–30.

De Beer, J. (2012). Smoothing and projecting age-specific probabilities of death by TOPALS. *Demographic Research, 27*(20), 543–592.

Dorrington, R. E. (2013a). General growth balance. In T. A. Moultrie, R. E. Dorrington, A. G. Hill, K. H. Hill, I. M. Timæus, & B. Zaba (Eds.), *Tools for demographic estimation.* http://demographicestimation.iussp.org/content/general-growth-balance.

Dorrington, R. E. (2013b). Synthetic extinct generations. In T. A. Moultrie, R. E. Dorrington, A. G. Hill, K. H. Hill, I. M. Timæus, & B. Zaba (Eds.), *Tools for demographic estimation.* http://demographicestimation.iussp.org/content/synthetic-extinct-generations.

Dwyer-Lindgren, L., Bertozzi-Villa, A., Stubbs, R. W., Morozoff, C., Kutz, M. J., Huynh, C., et al. (2016). US county-level trends in mortality rates for major causes of death, 1980-2014. *JAMA, 316*(22), 2385–2401.

Dwyer-Lindgren, L., Bertozzi-Villa, A., Stubbs, R. W., Morozoff, C., Shirude, S., Naghavi, M., et al. (2017). Trends and patterns of differences in chronic respiratory disease mortality among US counties, 1980-2014. *JAMA, 318*(12), 1136–1149.

Fenelon, A. (2013). Geographic divergence in mortality in the United States. *Population and Development Review, 39*(4), 611–634.

França, E., de Abreu, D. X., Rao, C., & Lopez, A. D. (2008). Evaluation of cause-of-death statistics for Brazil, 2002–2004. *International Journal of Epidemiology, 37*(4), 891–901.

França, E. B., de Azeredo Passos, V. M., Malta, D. C., Duncan, B. B., Ribeiro, A. L. P., Guimarães, M. D., et al. (2017). Cause-specific mortality for 249 causes in Brazil and states during 1990–2015: A systematic analysis for the global burden of disease study 2015. *Population Health Metrics, 15*(1), 39.

Freire, F. H. M. A., Souza, F. H., Queiroz, B. L., Lima, E. E. C., & Gonzaga, M. R. (2014). Tabelas de sobrevivência para os municípios brasileiros em 2010: análise espacial do padrão e nível de mortalidade. In *Encontro nacional de estudos populacionais: População, Governança e Bem-Estar, 24 a 28 de novembro de 2014/São Pedro, SP: ABEP, 2014.19. São Pedro*. São Paulo.

Gomes, M. M. F., & Turra, C. M. (2009). The number of centenarians in Brazil: Indirect estimates based on death certificates. *Demographic Research, 20*, 495–502.

Gonzaga, M. R., & Schmertmann, C. P. (2016). Estimating age-and sex-specific mortality rates for small areas with TOPALS regression: An application to Brazil in 2010. *Revista Brasileira de Estudos de População, 33*(3), 629–652.

Guimarães, R. M., et al. (2015). Regional differences in cardiovascular mortality transition in brazil.

Hill, K. (1987). Estimating census and death registration completeness. *Asian and Pacific Census Forum, 1*(3), 8-13–23-24.

Hill, K., You, D., & Choi, Y. (2009). Death distribution methods for estimating adult mortality: Sensitivity analysis with simulated data errors. *Demographic Research, 21*(9), 235–254.

Ishitani, L. H., Teixeira, R. A., Abreu, D. M. X., Paixão, L. M. M. M., França, E. B., & Ishitani, L. H. (2017). Quality of mortality statistics information: Garbage codes as causes of death in Belo Horizonte, 2011-2013. *Revista Brasileira de Epidemiologia, 20*(Suppl 1), 34–45.

Justino, J. R., Freire, F. H. M. D. A., & Lucio, P. S. (2012). Estimation of death underreporting at small areas using empiric Bayesian and EM algorithm methods. *Revista Brasileira de Estudos de População, 29*(1), 87–100.

Kibele, E. U. B., Klüsener, S., & Scholz, R. D. (2015). Regional Mortality Disparities in Germany. *KZfSSKölnerZeitschriftfürSoziologie und Sozialpsychologie, 67*(1), 241–270.

Kulkarani, et al. (2011). Falling behind: life expectancy in US counties from 2000 to 2007 in an international context. *Population Health Metrics, 9*, 16.

Lai, D., Hardy, R. J., & Tsai, S. P. (1996). Statistical analysis of the standardized mortality ratio and life expectancy. *American Journal of Epidemiology, 143*(8), 832–840.

Lima, E. E. C. D., & Queiroz, B. L. (2014). Evolution of the deaths registry system in Brazil: Associations with changes in the mortality profile, under-registration of death counts, and ill-defined causes of death. *Cadernos de Saúde Pública, 30*, 1721–1730.

Lotufo, P. A. (2019). The pace of reduction of cardiovascular mortality in Brazil (1990 to 2017) is slowing down. *Sao Paulo Medical Journal, 137*(1), 3–5.

Malta, D. C., et al. (2017). Mortality and years of life lost by interpersonal violence and self-harm: In Brazil and Brazilian states: Analysis of the estimates of the Global Burden of Disease Study, 1990 and 2015. *Rev Bras Epidemiol, 20*(Suppl 1), 142–156.

Marshall, R. J. (1991). Mapping disease and mortality rates using empirical Bayes estimators. *Applied Statistics*, 283–294.

Mokdad, A. H., Dwyer-Lindgren, L., Fitzmaurice, C., Stubbs, R. W., Bertozzi-Villa, A., Morozoff, C., et al. (2017). Trends and patterns of disparities in cancer mortality among US counties, 1980-2014. *JAMA, 317*(4), 388–406.

Nepomuceno, M. R., & Turra, C. M. (2020). The Population of Centenarians in Brazil: Historical Estimates from 1900 to 2000. Population and Development Review. https://doi.org/10.1111/padr.12355.

Nowbar, A. N., Howard, J. P., Finegold, J. A., Asaria, P., & Francis, D. P. (2014). Global geographic analysis of mortality from ischaemic heart disease by country, age and income: Statistics from World Health Organisation and United Nations. *International Journal of Cardiology, 174*(2), 293–298.

Paes, N. A. (2005). Avaliação da cobertura dos registros de óbitos dos Estados brasileiros em 2000. *Revista de Saúde Pública, 39*(6), 882–890.

Palloni, A., & Pinto, G. (2011). Adult mortality in Latin America and the Caribbean. In R. E. Rogers & E. Crimmins (Eds.), *International handbooks of adult mortality* (pp. 101–132).

Pereira, F. N. A., & Queiroz, B. L. (2016). Diferenciais de mortalidade jovem no Brasil: a importância dos fatores socioeconômicos dos domicílios e das condições de vida nos municípios e estados brasileiros. *Cadernos de Saúde Pública, 32*, e00109315.

Pinheiro, P. C., & Queiroz, B. L. (2019). Regional disparities in Brazilian adult mortality: An analysis using modal age at death (M) and compression of mortality (IQR). *Anais*, 1–20.

Pinheiro, P. C., & Queiroz, B. L. (2020). Análise espacial da mortalidade por acidentes de motocicleta nos municípios do Brasil. *Ciência & Saúde Coletiva, 25*, 683–692.

Pletcher, S. D. (1999). Model fitting and hypothesis testing for age-specific mortality data. *Journal of Evolutionary Biology, 12*, 430–439.

Prata, P. R. (1992). A transição epidemiológica no Brasil. *Cadernos de Saúde Pública, 8*, 168–175.

Queiroz, B. L., Freire, F. H. M. D. A., Gonzaga, M. R., & Lima, E. E. C. D. (2017). Completeness of death-count coverage and adult mortality (45q15) for Brazilian states from 1980 to 2010. *Revista Brasileira de Epidemiologia, 20*, 21–33.

Queiroz, B. L., de Lima, E. E. C., Freire, F., & Gonzaga, M. (2018). Adult mortality differentials and regional development at the local level in Brazil, 1980–2010.

Rasella, D., et al. (2014). Impact of primary health care on mortality from heart and cerebrovascular diseases in Brazil: A nationwide analysis of longitudinal data. *BMJ, 349*, 4014.

Reis, C. S., & Turra, C. M. (2017). Distribuição espacial dos centenários no Brasil: uma análise exploratória da qualidade dos dados dos censos de 2000 e 2010. *Revista Espinhaço | UFVJM*, 52–61.

Riggan, W. B., Manton, K. G., Creason, J. P., Woodbury, M. A., & Stallard, E. (1991). Assessment of spatial variation of risks in small populations. *Environmental Health Perspectives, 96*, 223–238.

Ross, J. M., et al. (2018). Progress toward eliminating TB and HIV deaths in Brazil, 2001–2015: A spatial assessment. *BMC Medicine, 16*(1), 144.

Roth, G. A., Dwyer-Lindgren, L., Bertozzi-Villa, A., Stubbs, R. W., Morozoff, C., Naghavi, M., et al. (2017). Trends and patterns of geographic variation in cardiovascular mortality among US counties, 1980–2014. *JAMA, 317*(19), 1976–1992.

Schmertmann, C. P., & Gonzaga, M. R. (2018). Bayesian estimation of age-specific mortality and life expectancy for small areas with defective vital records. *Demography, 55*(4), 1363–1388.

Schramm, J. M. D. A., Oliveira, A. F. D., Leite, I. D. C., Valente, J. G., Gadelha, Â. M. J., Portela, M. C., & Campos, M. R. (2004). Transição epidemiológica e o estudo de carga de doença no Brasil. *Ciência & Saúde Coletiva, 9*, 897–908.

Souza, E. R., & Lima, M. L. C. (2007). Panorama da violência urbana no Brasil e suas capitais. *Ciência & SaúdeColetiva, 11*(Suppl), 1211–1222.

Vierboom, Y. C., Preston, S. H., & Hendi, A. S. (2019). Rising geographic inequality in mortality in the United States. *SSM-Population Health, 100478*.

Waiselfisz, J. J. (2011). *Mapa da violência 2011: os jovens do Brasil*. São Paulo/Brasília: Instituto Sangari/Ministério da Justiça/Governo Federal.

Chapter 21
An Integrative Study Using Spatial Statistics and Racial/Ethnic Composition to Measure Racial/Ethnic Residential Segregation at Varying Scales

Yan Lin

Introduction

The topic of residential segregation across racial/ethnic groups is popular in many countries, especially in United States with diverse racial/ethnic groups. The measurement of racial/ethnic residential segregation involves social, historical, cultural, and political dimensions. Research on the methodology and application of the measurement of racial/ethnic residential segregation has been very active (Massey 1985; Borjas 1995; Logan et al. 2002; Johnston et al. 2009). Racial/ethnic residential segregation is defined as 'the degree to which racial/ethnic groups live apart from members of other groups in relatively exclusive residential areas' (Massey and Denton 1988). According to this definition, racial/ethnic residential segregation can be measured as the degree to which different racial/ethnic groups share residential space.

Many studies focus on measuring racial/ethnic residential segregation from one of the five perspectives – evenness, concentration, centralization, exposure, and clustering (Massey and Denton 1988). Several indices have been developed to measure segregation from the above perspectives and were used broadly. For example, the index of dissimilarity is an indicator of unevenness which measures the degree to which distributions of two racial/ethnic groups are different in the entire spatial units (such as county, census tracts, and census block) (Wong 1993); the index of exposure is an indicator of isolation which measures the degree to which

Y. Lin (✉)
Department of Geography and Environmental Studies, UNM Center for the Advancement of Spatial Informatics Research and Education (ASPIRE), University of New Mexico, Albuquerque, NM, USA
e-mail: yanlin@unm.edu

B. Jivetti, Md. N. Hoque (eds.), *Population Change and Public Policy*, Applied Demography Series 11, https://doi.org/10.1007/978-3-030-57069-9_21

two racial/ethnic groups live separately from each other (Reardon and Firebaugh 2002). Other indices include the index of concentration, the index of isolation, etc. However, the above single indices are limited in portraying a holistic view of racial/ethnic residential segregation (Poulsen et al. 2001; Simpson 2004). For example, it is sometimes difficult to explain the indices, most of which are between 0 and 1. If 1 and 0 represents total segregation and no segregation separately, what do 0.4, 0.5 and 0.6 represent? Besides, these indices only measure the average degree of segregation of a racial/ethnic group over space. Moreover, the above indices fail to provide comprehensive descriptions of which part of the members from one racial/ethnic group are segregated from the other group and disclose where the segregation occurs. Therefore, the above indices are global indicators of racial/ethnic residential segregation that need to be expanded to local measurement. In order to address the limitations listed above, some studies have introduced spatial statistics into the realm of segregation to measure spatial clustering of racial/ethnic groups (Reardon and O'Sullivan 2004). A series of clustering indicators are examined including Moran's I and Getis' G to measure spatial clustering. However, limitations of spatial statistics in racial/ethnic residential segregation measurement have been reported in a few studies. Johnston et al. (2007) argued that local indicators of spatial clustering could not identify the intensity of segregation within clusters. In order to address this problem, Johnston and his colleagues integrated local spatial statistics (Getis-Ord Gi* (Local G*)) with a typology method that considers racial/ethnic composition in each spatial unit (Poulsen et al. 2001; Brimicombe 2007) and consolidated the effectiveness and validity of this approach (Johnston et al. 2009).

However, the inherent scale issue in both local spatial statistics and racial/ethnic residential segregation measurement cannot be overlooked. Getis and Ord (1996) have reported the effects of scale on local clustering detection. Several studies have found that detecting local clustering at different scales produced different intensity of racial/ethnic residential segregation. Poulsen et al. (2010) combined Local G* and racial/ethnic composition at different distance bands (500 m, 1000 m, 2000 m, and 5000 m) to examine the intensity of absolute racial/ethnic residential segregation. They found that Getis's G as an indicator of spatial association from the concentration perspective is effective in segregation measurement. However, another indicator of spatial association – Anselin's Local Moran's I (Anselin 1995) that measures spatial similarity has not been investigated through its integration with racial/ethnic composition. In order to examine scale effect on racial/ethnic residential segregation, this paper followed the study conducted by Poulsen et al. (2010) to combine local spatial statistics and racial/ethnic composition and compare two local spatial statistics indicators – Local G* and Local Moran's I (LISA). This study offered useful insights into racial/ethnic residential segregation from both methodological and pragmatic perspectives because Local G* and LISA are supplementary to each other in measuring segregation.

The purpose of this study is to evaluate racial/ethnic residential clusters by using an integrative approach of spatial statistics and racial/ethnic composition zoning at varying scales. This paper aimed to answer these research questions: (1) What is the effect of scale on measuring racial/ethnic residential segregation? Is there any difference in the racial/ethnic residential clusters between the two local spatial statistic methods? (2) To what extent do diverse racial/ethnic groups share space within and outside clusters? In order to address these questions, we examine racial/ethnic residential segregation across four racial/ethnic groups – non-Hispanic White, Hispanic, African American, and Asian in intraurban Houston by using spatial statistics at varying scales – 1200 m, 2400 m, and 3600 m. First, we performed spatial statistics by using global Moran's I, Local G*, and LISA at three distance bands to identify clusters. Then, we investigated the concentration inside each cluster by calculating the percentage of each racial/ethnic group within the cluster and compared the results from Local G* and LISA. Finally, combining the racial/ethnic composition, we calculated the population percentage from the clusters for each type of area defined by racial/ethnic composition to examine the segregation patterns.

Methods

Study Area and Data

Houston is the fourth largest city in the United States with a total population of 2,709,260 according to Census 2010. It is a city with diverse race/ethnicity including non-Hispanic White, Hispanic, African American, Asian, Native American, Pacific Islander, and other racial/ethnic minorities. Race is the individual-level self-identification from Census 2010 with one or more social groups: White, Black or African American, Asian, American Indian and Alaska Native, Native Hawaiian and Other Pacific Islander, or some other race. Ethnicity is reported in Census 2010 which determines whether a person is of Hispanic origin or not. Within the greater Houston area, the racial/ethnic groups tend to be highly segregated in the suburban areas than the intraurban areas. Thus, the intraurban area is an appropriate test bed to examine neighborhoods with varying degree of racial/ethnic residential segregation and mixed racial/ethnic groups. According to Census 2010, the total population in intraurban areas is 2,331,236, with 718,057 non-Hispanic White (31%), 906,174 Hispanic (39%), 553,486 Black or African American (24%), 135,070 Asian (5%), and 18,449 (1%) other races. In this paper, we only studied racial/ethnic segregation of the first four racial/ethnic groups that account for 99% of the total population. We used the Census block group as our geographic unit, which totals up to 1379 in intraurban Houston. In order to standardize the data for comparison, we normalized the population of each racial/ethnic group by the total population in each census block group. Figure 21.1 displays the intraurban area of Houston.

Fig. 21.1 Study area

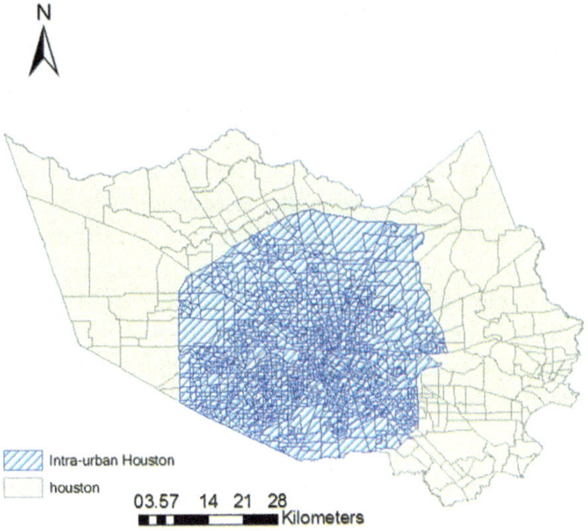

Spatial Statistics and Racial/Ethnic Clustering

There are several spatial statistical methods to measure clustering and detect clusters. They are generally categorized into two groups— global and local spatial statistics. Examples of global spatial statistics are Moran's I, Getis's G* and Geary's C, and their local equivalence are Anselin's LISA, Local G*, and Local Geary respectively (Bao 1999). In order to examine racial/ethnic clustering, we first investigated the global pattern of four racial/ethnic groups using Moran's I, then we conducted two local tests using Local G* and LISA. Moran's I is developed by Moran (1950) to measure spatial autocorrelation. It is defined by Eq. 21.1:

$$I = \frac{\sum_{i}^{n} \sum_{j \neq i}^{n} w_{ij}(x_i - \overline{x})(x_j - \overline{x})}{S^2 \sum_{i}^{n} \sum_{j \neq i}^{n} w_{ij}} \tag{21.1}$$

where n is the total number of zone (census block group in this paper), x_i is the percentage of population of a racial/ethnic group in zone i (or x_j in zone j), $\overline{x}$ is the mean of x in all zones in the study area, S^2 is the variance of x and w_{ij} is the weight matrix calculated using one of the three methods: a binary contiguity (i.e. 0 or 1), inverse distance between zones i and j, or the proportion of shared boundary between zones i and j (Lu 2009). This study used the first method and defined distance bands to obtain the weight matrix. For example, w_{ij} equals 1 if zone j is within the defined distance from zone i, and vice versa. The Z score is calculated in order to test the null hypothesis that there is no spatial autocorrelation between observed zones. Moran's I measures the similarity/dissimilarity among the neighboring zones, representing the global clustering pattern of each racial/ethnic group in the context of this paper.

However, Moran's I only investigates global pattern and fails to disclose where the clusters are. Hence, two local tests are used. Local G* (Getis and Ord 1992) measures the degree of association among neighboring zones. Compared with Moran' I that considers similarities between the observed neighboring zones, Local G* measures the concentration of observed neighboring zones. It is defined by Eq. 21.2:

$$G_i* = \frac{\sum_{j}^{n} w_{ij} x_j}{\sum_{j}^{n} x_j} \tag{21.2}$$

Where x_j is the percentage of the population of a racial/ethnic group in zone j, and w_{ij} is the weight matrix that is defined in the same way as Moran's I discussed above. Local G* represents the proportion of the neighboring values in the total values (i.e. concentration). High values in neighboring zones of i contributes a higher proportion, while low values in neighboring zones of i contributes a lower proportion. Thus, it can be used to detect spatial clusters of high/low values. In order to test the null hypothesis, a Z score of G* ($Z(G_i*)$) is calculated by Eq. 21.3:

$$Z(G_i*) = \frac{\sum_{j=1}^{n} w_{ij} x_j - X \sum_{j=1}^{n} w_{ij}}{S \sqrt{\frac{n \sum_{j=1}^{n} w_{ij}^2 - \left(\sum_{j=1}^{n} w_{ij} \right)^2}{n-1}}} \tag{21.3}$$

Where S is the standard deviation of x and $Z(G_i*)$ reflects the significance of a racial/ethnic group's concentration. The significance test examines whether the percentage of the total population of the racial/ethnic group in neighboring zones of i is over or under the global average percentage that is considered as a threshold. If $Z(G_i*)$ is positive and significant (i.e. > 1.96), the neighboring zones of i have a higher total percentage than the average percentage, and there is a greater chance of high concentration for the racial/ethnic group than spatial randomness. On the other hand, a negative $Z(G_i*) < -1.96$ suggests a significant low concentration for the racial/ethnic group and intermediate $Z(G_i*)$ scores indicate mixed neighborhoods that are not clustered. Therefore, Local G* describes the intensity of the high/low clustering. In order to scrutinize clusters, we subdivided Z scores using multiple thresholds: ±2.58, ±3.29 and ± 5.16 that correspond to 2-tail probabilities of 0.01, 0.001 and 0.0001 respectively.

The local form of Moran's I, Anselin's LISA measures clustering based on spatial similarity as discussed above. According to Anselin (1995), the sum of LISA in all zones is proportional to the global Moran's I. It is defined by Eq. 21.4:

$$I_i = Z_i \sum_{j \neq i} w_{ij} Z_j \quad (21.4)$$

Where Z_i is the standardized intensity in zone i (deviation from the mean value of X_i defined in Eq. 21.1) (Anselin 1995). Because LISA examines similarity and dissimilarity between neighboring zones, it can detect both clusters and outliers. A positive LISA with Z scores higher than 1.96 indicates clusters. Clusters where high X_i values are surrounded by high values can be seen as 'hot spots' for a specific racial/ethnic group, and clusters where low X_i values are surrounded by low values can be seen as 'cold spots'. A negative LISA with Z score less than -1.96 indicates outliers (i.e. segregated neighborhoods) — high X_i values surrounded by low values or low X_i values surrounded by high values.

This paper investigated racial/ethnic residential clustering using both Local G* and LISA and compared them, because they measure different perspectives of local spatial association – concentration and similarity. We also examined scale effects on racial/ethnic residential clustering measurement by testing different distance bands in the weight matrix used to define neighboring zones. However, it is critical to select appropriate distance bands. Getis and Ord (1992) suggested that the maximum distance should never exceed a half of the shorter side of the study area, and that the number of neighbors should be at least 30 for large zones and 8 for small zones (Getis and ord 1996). After examining the distribution of 1379 census block groups and distances between neighboring block groups, we found an average distance of 1117 m for all the census block groups. Therefore, we used 1200 as the smallest distance band that is higher than the average distance of 1117, because a smaller distance band would exclude most of the neighbors with distance less than the average. We also used 2400 m and 3600 m as distance bands to examine scale effect. We did not extend the distance band beyond 3600 m because 3600 m was sufficiently large for a local spatial statistic in the study area according to Getis and Ord (1992). Distance band above 3600 in this study might cause large clusters for G* and no clusters for LISA according to their definitions.

Racial/Ethnic Composition

Local spatial indicators detect local clusters where the same racial/ethnic group tends to live together. In order to measure segregation that is defined as the degree to which one racial/ethnic group lives apart from the other in relative exclusive areas, we imported a classification method of racial/ethnic composition defined by Poulsen et al. (2001). It divides the population into two major groups: Charter group (X) also called the host group that is an economically and politically powerful group; and various racial/ethnic groups (Y, Z, etc.). Six types of zones are defined based on the proportion of the population between the two groups. Type one zone (Isolated host) is areas where the charter group X dominates with more than 80% of the total population; Type two zone (Non-isolated host) is a region where charter group

X dominate, accounting for 50–80% of the total population; In type three zone that is defined as 'Assimilation pluralism enclaves' whereas members from racial/ethnic groups *Y* and *Z* dominate about 50–70% of the total population while the charter group *X* accounts for a substantial minority. In type four zone (Mixed enclaves), racial/ethnic groups *Y* and *Z* are dominant groups with more than 70% of the total population, but neither group dominates the other. Zones from type two to type four are mixed areas with different intensity. Type five zone (Polarized enclaves) is areas dominated by racial/ethnic groups *Y* and *Z* with more than 70% of the total population, where one group has at least as twice population as the other racial/ethnic group. Type six zone ('Ghettos') meets all the criteria defined by type five zone, however, the group with at least twice the population of the other group should account for at least 30% of that group's total population in the study area. Zones from type five to type six are areas dominated by one racial/ethnic minority group with relatively larger population than other groups. In this paper, we defined non-Hispanic White as the charter group because of its influence on the configuration of population distribution in Houston. Other three groups were defined as various racial/ethnic groups.

Combination of Racial/Ethnic Clustering and Racial/Ethnic Composition

Despite local indicators can detect clusters of same racial/ethnic group, they fail to uncover whether the racial/ethnic group in clustered areas is the dominant group. It is possible that the clustered racial/ethnic group only accounts for a small percentage of the total population in the clustered areas that are dominated by other groups. Thus, according to the definition of segregation used in this paper, the racial/ethnic group in such clustered areas is not segregated. On the other hand, racial/ethnic composition reveals the dominant and minority groups in each zone but cannot uncover whether areas dominated by one group are clustered into a large continuous zone or separated into several pieces. Combining local spatial clustering and racial/ethnic composition can address limitations in both assessments. In order to examine to what extent racial/ethnic groups share space within and outside clusters at different scales, this paper combined the two methods to quantify the degree of segregation by calculating the proportion of each racial/ethnic group in the six zone types within and outside clusters.

Results

Distribution of the Four Racial/Ethnic Groups

Figure 21.2 shows the proportion of the four racial/ethnic groups at the census block group level. It reveals different distribution among the four racial/ethnic

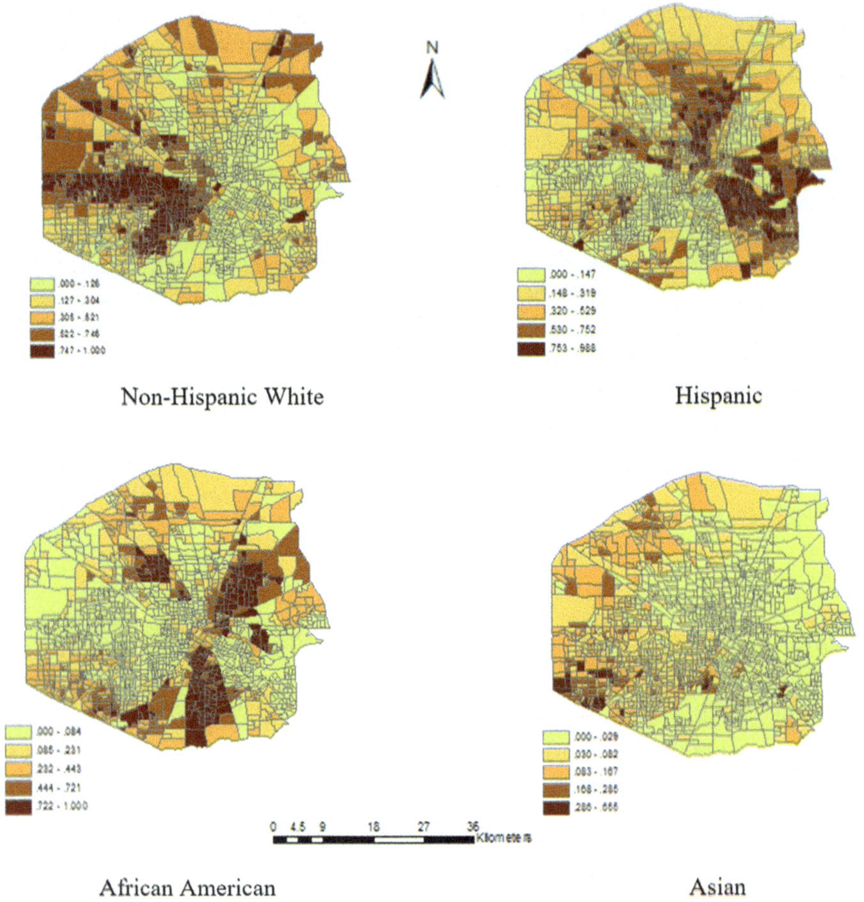

Fig. 21.2 Distribution (percentage) of the four racial/ethnic groups in intraurban Houston, TX

groups. The non-Hispanic White dominates the western region, whereas the Hispanic tends to occupy in the southeastern and central areas. African American primarily resides in the southern and northeastern areas, while Asian accounts for a high percentage in the southwestern regions.

The descriptive statistics of the four racial/ethnic groups (Table 21.1) reveals that Non-Hispanic White and Hispanic have higher mean percentages of total population. Besides, the skewness and kurtosis suggest that the two groups are more evenly distributed in the study area than African American and Asian.

Racial/Ethnic Clustering and Scale

Table 21.2 is the result of Global Moran's I of the percentage of the four groups in each census block group at different scales. It reveals that all four racial/ethnic groups are significantly clustered and the degree of the global clustering decreases from 1200 m to 3600 m. The scaling impact reflects the characteristic of Moran's I

Table 21.1 Descriptive statistics of the population percentage of racial/ethnic groups

	Minimum	Maximum	Mean	Std. Deviation	Variance	Skewness	Kurtosis
White	0.000	1.000	0.327	0.297	0.088	0.690	−0.903
Black	0.000	1.000	0.246	0.304	0.093	1.222	0.099
Hispanic	0.000	0.988	0.371	0.285	0.081	0.539	−1.003
Asian	0.000	0.655	0.047	0.071	0.005	3.123	13.710

Table 21.2 Moran's I of the population percentage of the 4 racial/ethnic groups

	1200 m		2400 m		3600 m	
Group	Moran's index	z-score	Moran's index	z-score	Moran's index	z-score
Non-Hispanic White	0.91	52.09	0.78	97.52	0.67	117.48
Hispanic	0.87	49.44	0.67	79.21	0.54	95.48
African American	0.87	49.56	0.68	80.48	0.53	93.29
Asian	0.53	30.2	0.46	55.07	0.4	71.43

similarity measurement among neighboring zones. Compared with G* statistics that produces clustering to a higher degree when distance increases, Moran's I produces clustering that is subject to the data distribution. Table 21.2 also reveals differences in different racial/ethnic groups: non-Hispanic White is the most clustered and Asian is the least clustered at all scales.

Figures 21.3, 21.4, and 21.5 present the results of Local G* clustering of the four racial/ethnic groups using 1200 m, 2400 m, and 3600 m distance bands separately. Both positive clusters (indicated as red areas) and negative clusters (indicated as blue areas) are displayed. The results reflect the overall distribution of the four racial/ethnic groups revealed in Fig. 21.2. However, when distance increases, both positive and negative clusters extend, and areas of neither positive clusters nor negative clusters shrink.

Table 21.3 is the summary of the distribution of population percentage across all four racial/ethnic groups in both clustered areas (positive/negative) and non-clustered areas. It reveals that the four racial/ethnic groups living in negative clusters (i.e. $Z < -1.96$) have the lowest percentage of the population and are slightly increasing with respect to an increase in distance band threshold. The percentage of the four groups living in non-clustered areas (i.e. $-1.96 < Z < 1.96$) is the highest at 1200 m distance band; however, a significant decrease is found with the increase in the distance. Among the positive clusters (i.e. $Z > 1.96$), we found that Asian has the highest percentage among the four groups although there is also a significant increase when distance increases.

In order to scrutinize the pattern and intensity of clusters, more thresholds were used to categorize positive and negative clusters. Figure 21.6 is the result of clustering at a 2400 m distance band using Z scores of ±2.58, ±3.29 and ± 5.16

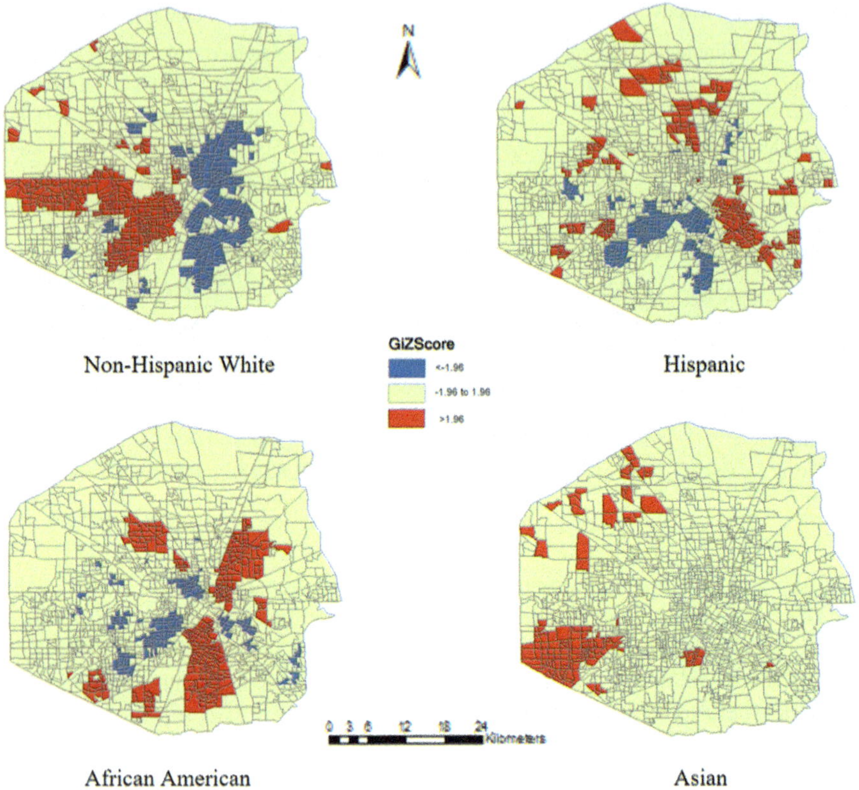

Fig. 21.3 Local G* clustering of the four racial/ethnic groups at 1200 m distance band

as class intervals. A clear zoning area with both positive and negative clusters is shown, where the central areas are highly clustered with the highest intensity, and the intensity of the clusters decreases when away from the center. Clustering at different scales from local G* indicates that although more zones are included into the significant clusters when the distance band increase, the intensity of such zones varies. Table 21.4 extended the results by calculating the distribution of the percentage the four racial/ethnic groups in all categories of clusters defined by the 9 thresholds. Larger Z scores in positive clusters indicate central zones surrounded by most of the higher values. Table 21.4 reveals that positive clusters with Z scores larger than 5.16 increase significantly when distance increases. It also suggests that negative clusters with Z scores smaller than −3.29 increase when distance increases although the absolute percentage of the population in the negative clusters is low.

Figures 21.7, 21.8, and 21.9 are the results of clusters from LISA using 1200 m, 2400 m, and 3600 m distance bands separately. Similar to local G*, LISA presents both positive (indicated as red in the figure) and negative clusters (indicated as blue

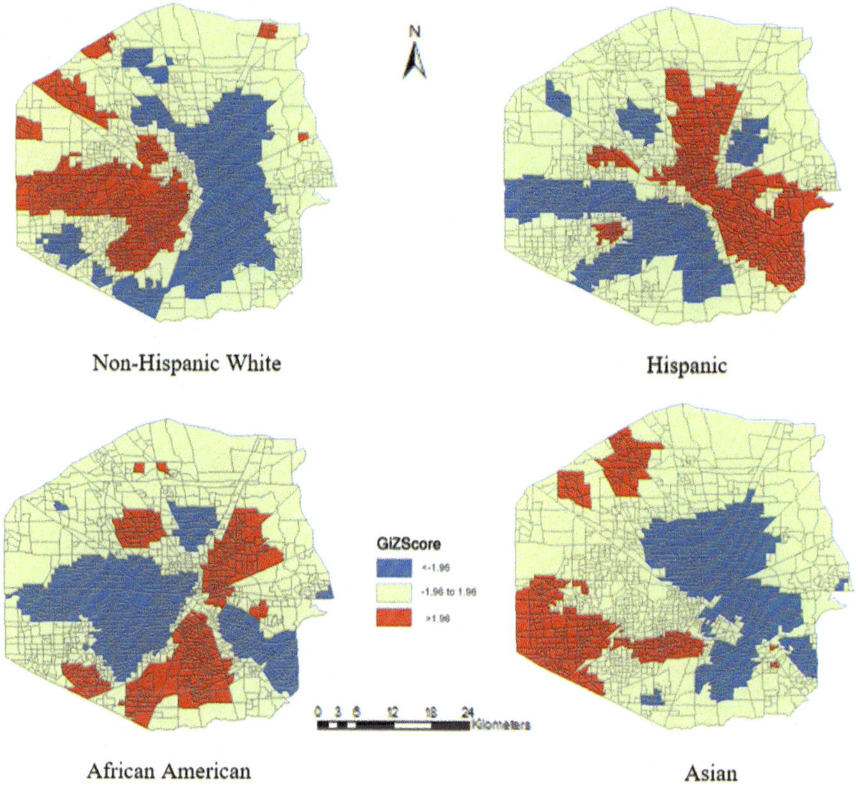

Fig. 21.4 Local G* clustering of the four racial/ethnic groups at 2400 m distance band

in the figure). Both positive and negative clusters increase when distance band increases, and non-significant areas decrease. Comparing the results from LISA with those from local G*, we found that both positive and negative clusters detected from LISA were slightly smaller than those from local G* at all three distance bands. Revealed by Table 21.5, clusters from LISA are fragmented compared with the continuous clusters from G*. Because LISA discloses the clusters according to similar values, it could refine the clusters from local G* by removing the ones without similar values surrounded. Table 21.6 presents the results of population percentage in census block groups that are significant clusters from LISA as well as local G* at different distance bands. It was found that about 90% of the significant clusters from local G* with Z scores larger than +2.58 are also the positive clusters from LISA across the four racial/ethnic groups (same for $Z(G^*) < -2.58$ for negative clusters). Parallel comparisons among the four groups suggested that the population percentage of HH clusters was higher in Asian and African American than non-Hispanic White and Hispanic. The results suggested that LISA is the refinement of local G* by detecting the clusters with higher intensity.

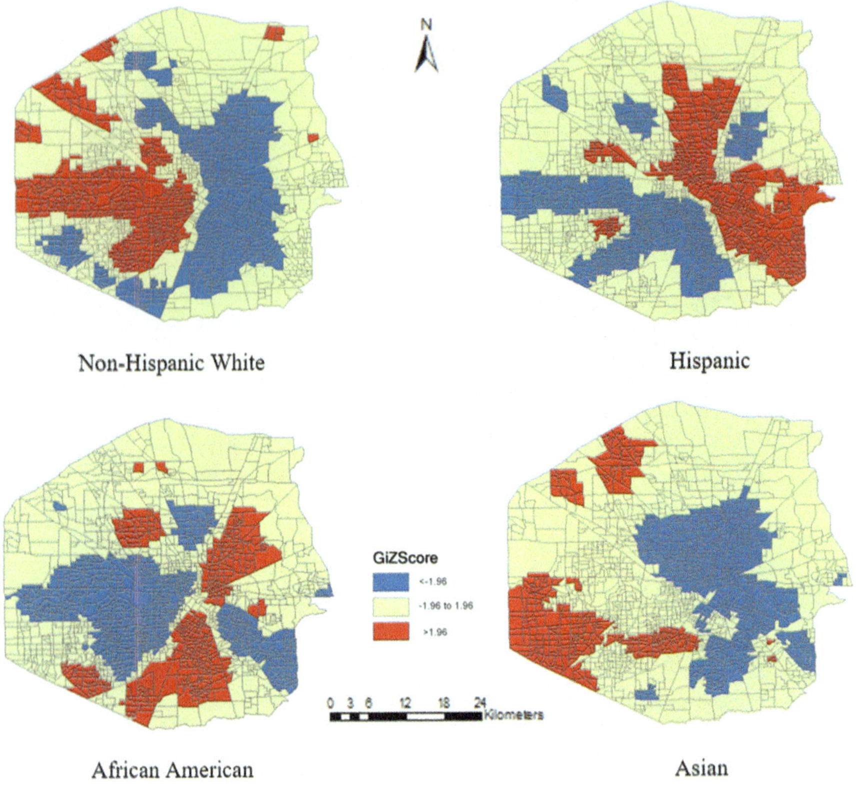

Fig. 21.5 Local G* clustering of the four racial/ethnic groups at 3600 m distance band

Combination of Racial/Ethnic Clustering and Racial/Ethnic Composition

Figure 21.10 is the result after the classification of the census block group using the method proposed by Poulsen et al. (2001). According to this method, type one zone is dominated by White; zones from type two to four are mixed areas of the four racial/ethnic groups where no group dominates the other; type five_A zone is dominated by Asian; type five_H zone is dominated by Hispanic; and type_B zone is dominated by African American. We compared Fig. 21.10 with the clustering map of G* and LISA and found a similar pattern between the racial/ethnic composition and the clustering map for the same racial/ethnic group, especially for African American and Hispanic who dominate large continuous areas. Asian is a special case who clusters in southwestern areas. However, the Asian clusters are resided among the mixed groups and not dominated by Asian. We only found three block groups where Asian dominates. This finding suggested that combining local clustering with racial/ethnic composition provides information of racial/ethnic composition within clusters.

Table 21.3 Distribution of the percentage of each racial/ethnic group in all cluster categories from G* using three Z score thresholds by different distance band (1200 m, 2400 m, and 3600 m)

	Distribution (%)											
	Non-Hispanic white			Hispanic			African American			Asian		
Z-score	1200	2400	3600	1200	2400	3600	1200	2400	3600	1200	2400	3600
>1.96	0.39	0.52	0.57	0.35	0.49	0.50	0.43	0.49	0.51	0.46	0.57	0.68
−1.96-1.96	0.59	0.40	0.29	0.63	0.43	0.37	0.57	0.44	0.38	0.54	0.41	0.26
<−1.96	0.02	0.08	0.13	0.02	0.09	0.13	0.01	0.07	0.11	0.00	0.03	0.06

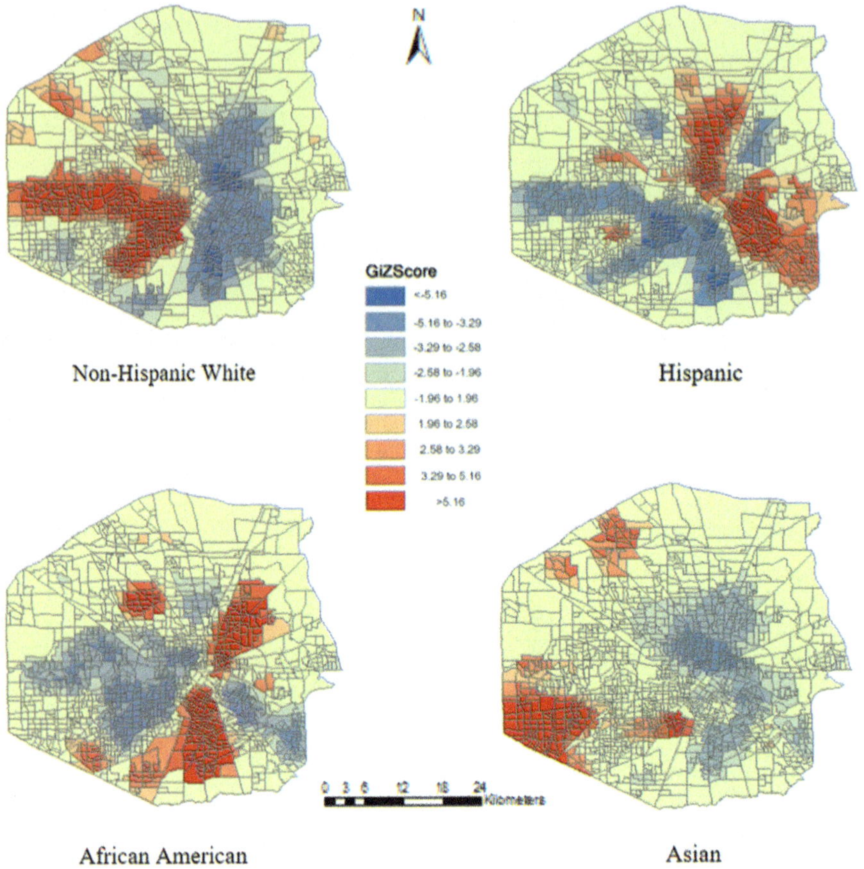

Fig. 21.6 Local G* clustering of the four racial/ethnic groups at 2400 m distance band (nine thresholds)

Table 21.7 is a summary of the percentage each racial/ethnic group in different type of zones. Hispanic has the highest percentage of the population that resides in zones dominated by themselves, followed by African American (51%), and non-Hispanic White (23%). Asian has the lowest percentage that is only 2%. In order to disclose whether each type of zone is also clustered, we presented the type of each census block group that is a positive cluster for the four racial/ethnic groups in Fig. 21.11 and compared G* with LISA. As revealed by the figure, the type of zones within LISA clusters is more homogeneous than that from G*. For example, for non-Hispanic white, zones of type three, four, and five_H where non-Hispanic white does not dominate are removed from the LISA clustering map. Figure 21.11 also suggested that Asian and African American are less homogeneous than non-Hispanic White and Hispanic.

We calculated the population percentage of the 4 racial/ethnic groups in each type of zone. Tables 21.8, 21.9, 21.10 and 21.11 present the expanded results by clusters

Table 21.4 Distribution of the percentage of the four racial/ethnic groups in all cluster categories from G* using nine Z score thresholds by different distance band (1200 m, 2400 m, and 3600 m)

	Distribution (%)											
	Non-Hispanic white			Hispanic			African American			Asian		
Z- score	1200	2400	3600	1200	2400	3600	1200	2400	3600	1200	2400	3600
>5.16	0.03	0.28	0.36	0.04	0.18	0.28	0.10	0.27	0.30	0.15	0.33	0.38
3.29 to 5.16	0.20	0.11	0.10	0.10	0.16	0.11	0.20	0.13	0.12	0.17	0.09	0.13
2.58 to 3.29	0.10	0.06	0.06	0.11	0.08	0.06	0.06	0.04	0.07	0.06	0.09	0.10
1.96 to 2.58	0.05	0.08	0.05	0.10	0.06	0.05	0.07	0.05	0.03	0.08	0.06	0.07
1.96 to −1.96	0.59	0.40	0.29	0.63	0.43	0.37	0.57	0.44	0.38	0.54	0.41	0.26
−1.96 to −2.58	0.02	0.04	0.05	0.01	0.02	0.04	0.01	0.02	0.03	0.00	0.01	0.02
−2.58 to−3.29	0.00	0.02	0.03	0.00	0.02	0.02	0.00	0.03	0.03	0.00	0.01	0.02
−3.29 to −5.16	0.00	0.03	0.03	0.00	0.04	0.04	0.00	0.02	0.05	0.00	0.00	0.02
<−5.16	0.00	0.00	0.02	0.00	0.01	0.03	0.00	0.00	0.00	0.00	0.00	0.00

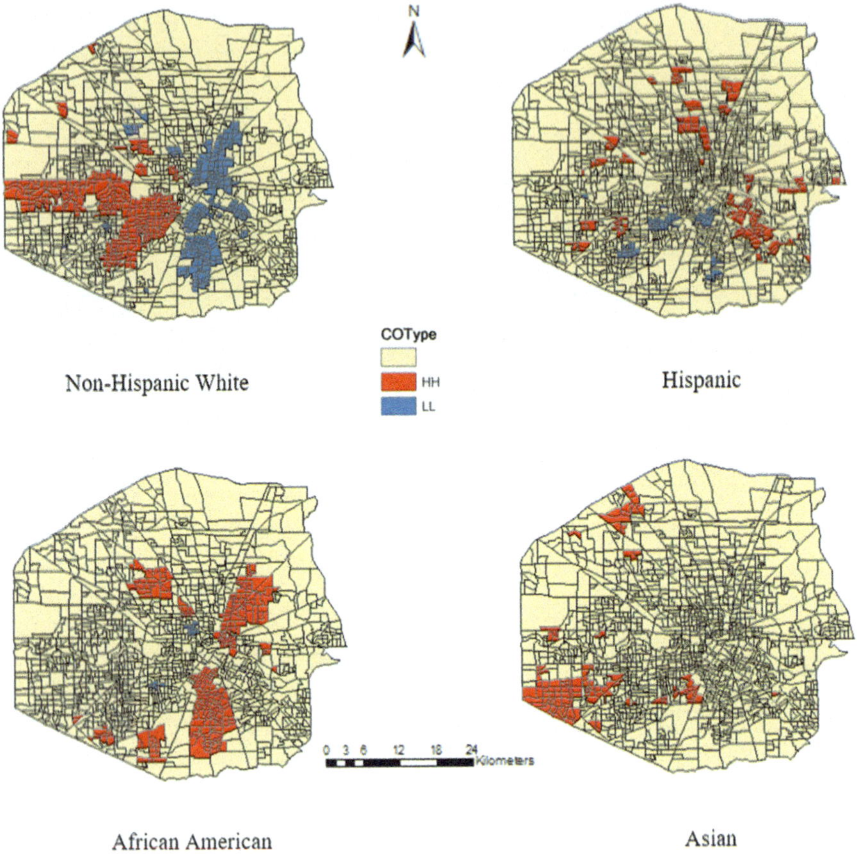

Fig. 21.7 LISA clustering of the four racial/ethnic groups at 1200 m distance band

and distance bands. Among the positive clusters, most of the non-Hispanic white clusters are in zones of type one and type two dominated by non-Hispanic White. Most of Hispanic and African American clusters are in zones where they dominate, while the clusters of Asian are zones where they do not dominate. In the negative clusters, non-Hispanic white clusters are dominated by Hispanic and African American. Hispanic clusters are dominated by African American and African clusters are dominated by Hispanic, while Asian clusters primarily occur in zones dominated by Hispanic and African American. Most non-significant clusters are mixed zones such as type two and type three and zones dominated by Hispanic. We examined the patterns using different distance bands and found there is no significant change in the non-significant clusters using different distance bands for all the four racial/ethnic groups. The pattern for the positive and negative clusters is obvious. With the increase in Z score, the types of zones where the positive clusters are located are more homogenous. The same trend occurs with decrease in Z score resulting in homogenous negative clusters. To sum up, we found that the types of zones in LISA

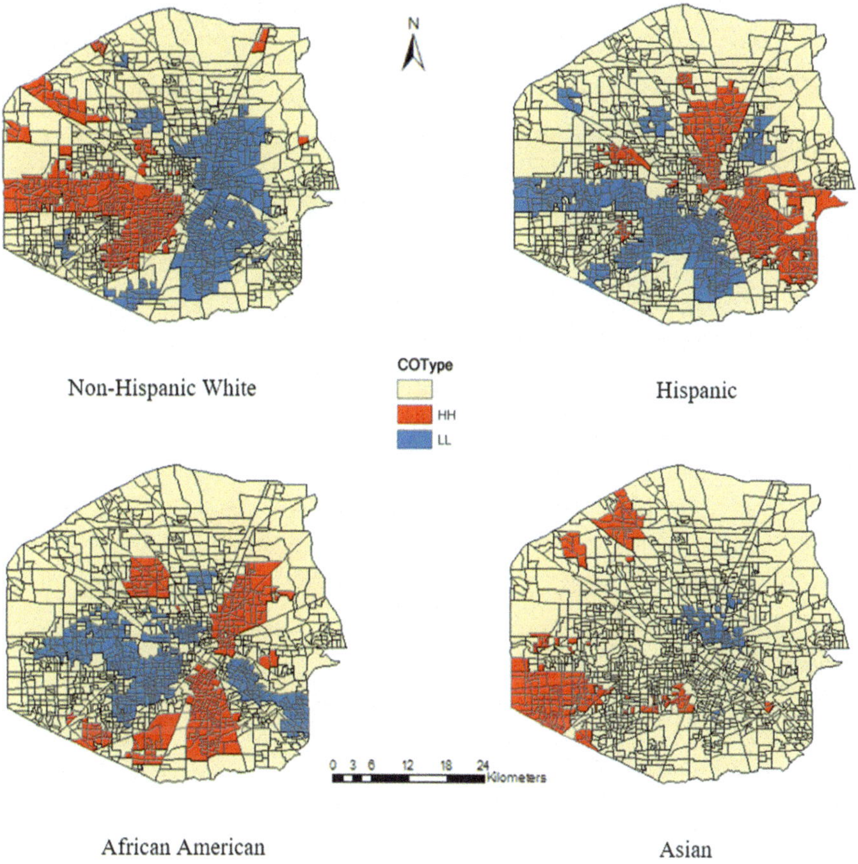

Fig. 21.8 LISA clustering of the four racial/ethnic groups at 2400 m distance band

clusters are more homogeneous than G*, which could be consolidated by the finding that LISA refines the cluster of G* by removing areas with dissimilar values.

Discussion

Scale and Clustering

According to the definition of absolute segregation, this paper examines segregation of four groups (non-Hispanic White, Hispanic, African American, and Asian) in intraurban Houston through the combination of spatial statistics and racial/ethnic composition at different scales. Choosing an appropriate scale is critical in both spatial statistics of clustering studies (Getis and Ord 1996) and segregation studies

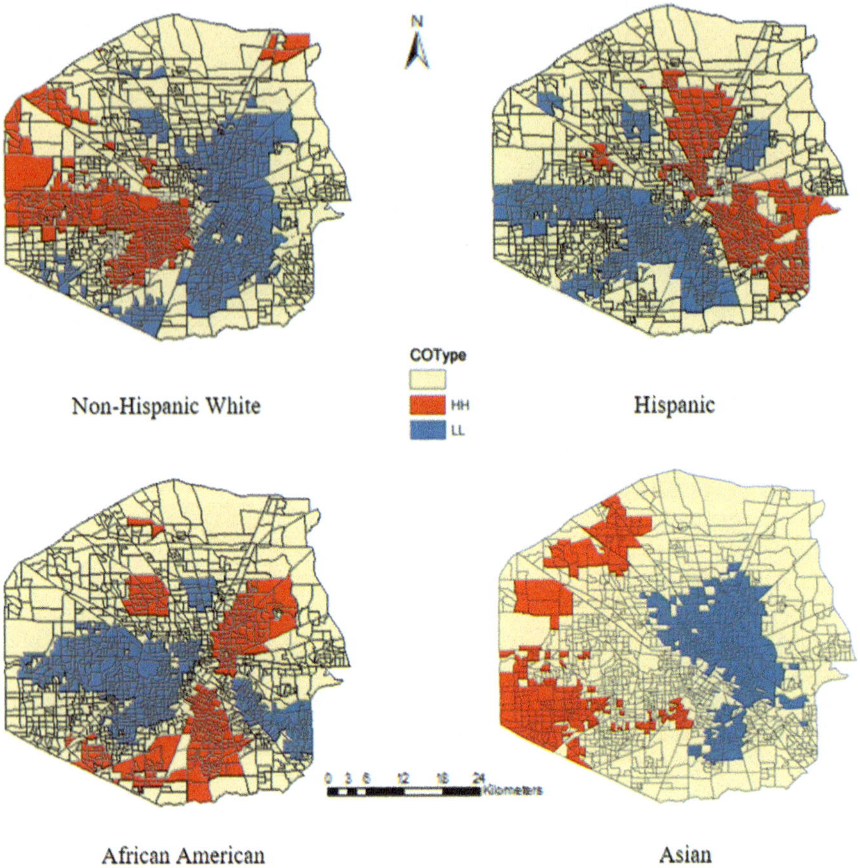

Fig. 21.9 LISA clustering of the four racial/ethnic groups at 3600 m distance band

(Poulsen et al. 2010). It was not the intent of this paper to offer what the best scale is to study segregation. However, through the exhaustive exanimation of the racial residential distribution and segregation degree using different distance bands, the pattern we found might offer insights into the spatial pattern of segregation.

We found the same pattern of zoning areas from Local G* for both the positive and negative clusters as what was found by Poulsen et al. (2010). We found similar findings about a significant decrease in population living in non-significant clusters and increase in both positive and negative clusters when distance band increases. The above findings originate from the characteristic local G* that measures concentration by calculating the proportion of neighboring zones (weighed zones) in the total study areas. Local G* tends to produce larger clusters when the number of neighboring zones increases. Result from LISA at different scales is similar to Local G*. For example, when distance increases, both positive and negative clusters

Table 21.5 Distribution of the percentage of all the four racial/ethnic groups in all cluster categories from LISA by different distance band (1200 m, 2400 m, and 3600 m)

	Distribution (%)											
	Non-Hispanic White			Hispanic			African American			Asian		
Cluster type	1200	2400	3600	1200	2400	3600	1200	2400	3600	1200	2400	3600
HH	0.37	0.46	0.48	0.28	0.45	0.43	0.40	0.46	0.49	0.39	0.51	0.59
Non sig	0.62	0.51	0.45	0.72	0.50	0.49	0.60	0.52	0.48	0.60	0.48	0.38
LL	0.01	0.03	0.05	0.00	0.04	0.05	0.00	0.02	0.03	0.00	0.00	0.01

Note: HH indicates areas with high values surrounded by high values; LL indicates areas with low values surrounded by low values

Table 21.6 Percentage of census block groups that are significant clusters from LISA in the census block group that are significant clusters from G* by different thresholds by different distance band (1200 m, 2400 m, and 3600 m)

	Non-Hispanic white						Hispanic						African American						Asian					
	1200		2400		3600		1200		2400		3600		1200		2400		3600		1200		2400		3600	
HH	No. (%)		No. (%)		No. (%)		No. (%)		No. (%)		No. (%)		No. (%)		No. (%)		No. (%)		No. (%)		No. (%)		No. (%)	
>5.16	26	0.11	181	0.62	194	0.63	17	0.14	131	0.44	194	0.63	55	0.26	152	0.63	162	0.65	26	0.11	85	0.57	109	0.57
3.29 to 5.16	133	0.54	68	0.23	72	0.23	39	0.33	107	0.36	72	0.23	112	0.52	60	0.25	58	0.23	33	0.13	30	0.20	48	0.25
2.58 to 3.29	60	0.24	21	0.07	29	0.09	38	0.32	37	0.12	29	0.09	28	0.13	18	0.07	20	0.08	19	0.08	28	0.19	23	0.12
1.96 to 2.58	28	0.11	24	0.08	15	0.05	26	0.22	23	0.08	15	0.05	20	0.09	13	0.05	8	0.03	14	0.06	7	0.05	10	0.05
HH Total	247	0.89	294	0.92	310	0.95	120	0.78	298	0.92	310	0.95	215	0.91	243	0.95	249	0.96	88	0.89	150	0.95	190	0.95
−1.96 to −2.58	84	0.41	21	0.06	14	0.03	23	0.37	18	0.05	18	0.05	0	0.00	0	0.00	5	0.01	0		0	0.00	0	0.00
−2.58 to −3.29	99	0.49	62	0.17	33	0.08	38	0.60	50	0.15	39	0.11	16	1.00	128	0.40	38	0.10	0		21	0.27	18	0.06
−3.29 to −5.16	20	0.10	228	0.62	114	0.28	2	0.03	203	0.61	134	0.37	0	0.00	191	0.60	295	0.75	0		57	0.73	261	0.84
<−5.16	0	0.00	54	0.15	246	0.60	0	0.00	63	0.19	173	0.48	0	0.00	0	0.00	55	0.14	0		0	0.00	30	0.10
LL Total	203	0.60	365	0.94	407	0.97	63	0.67	334	0.95	364	0.95	16	1	319	1	393	0.99	0	1	78	1	309	1

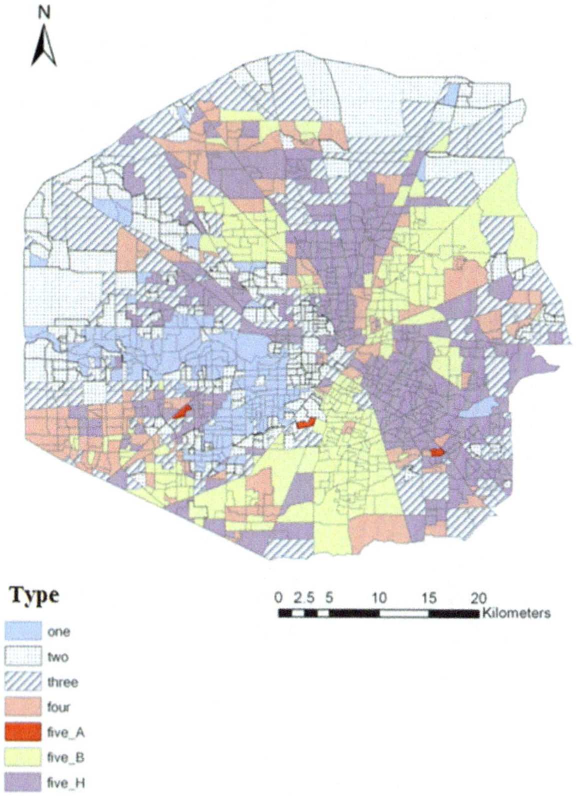

Fig. 21.10 Type of zone according to the racial/ethnic composition

Table 21.7 Percentage of the four racial/ethnic groups in each type of zone

Type	Non-Hispanic white	Hispanic	African American	Asian
One	0.23	0.01	0.00	0.07
Two	0.34	0.08	0.04	0.20
Three	0.20	0.15	0.10	0.20
Four	0.07	0.14	0.22	0.29
Five_Asian	0.00	0.00	0.00	0.02
Five_Hispanic	0.13	0.57	0.11	0.15
Five_African American	0.03	0.06	0.51	0.07

increase, and non-significant clusters decrease. The above pattern is worth further investigation because compared with G* that tends to produce increase in the clusters when distance band increases, LISA measures spatial association from a different perspective. For example, it is highly possible that when distance band increases, clusters decrease because more dissimilar values are included. Therefore, the result from this study that the clusters increase with the increase in the distance band might be a special case of LISA clustering.

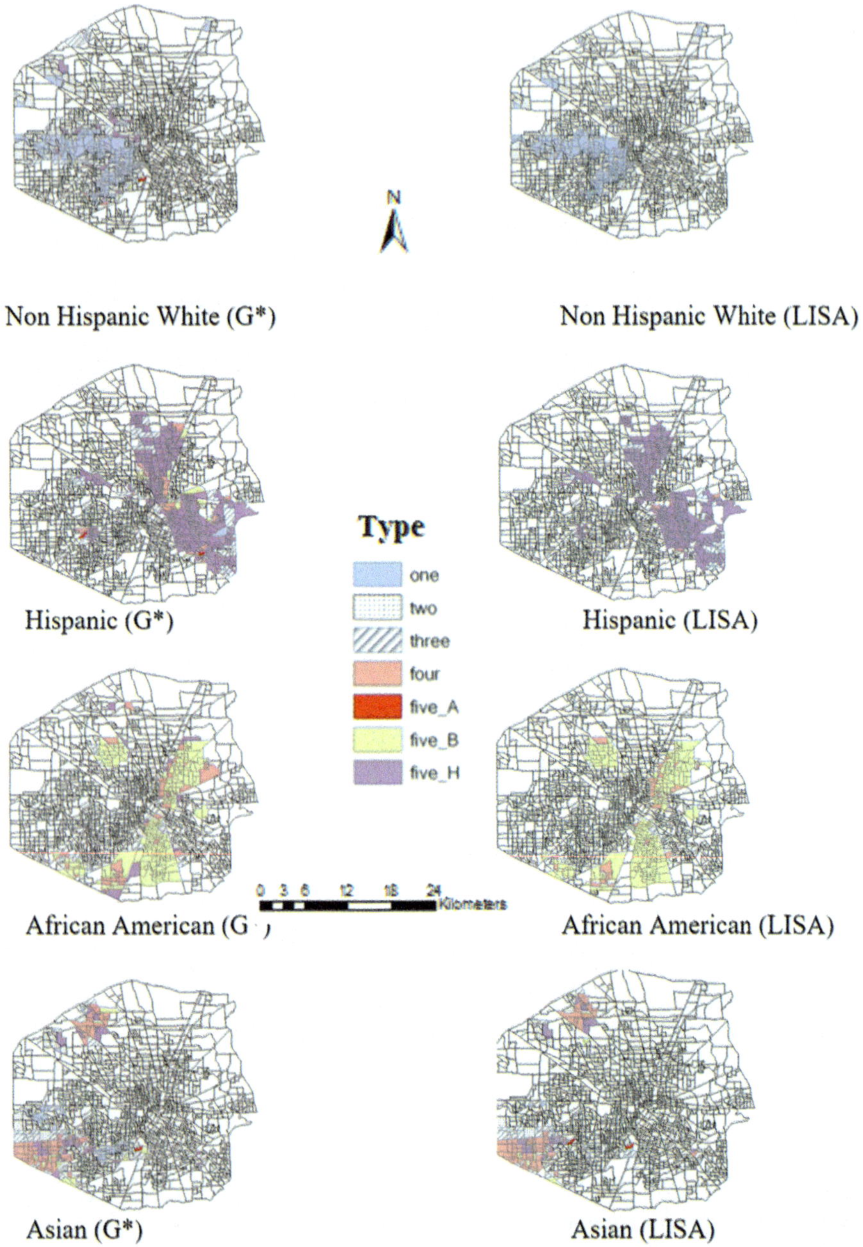

Fig. 21.11 Type of zones within significant positive cluster at 2400 distance band from G* and LISA

Table 21.8 Percentage of non-Hispanic White in each type of zone across different clusters by different distance bands

	Type						
Band	one	two	three	four	five_A	five_B	five_H
1200							
>5.16	0.95	0.05	0.00	0.00	0.00	0.00	0.00
3.29 to 5.16	0.66	0.32	0.01	0.00	0.00	0.00	0.01
2.58 to 3.29	0.43	0.51	0.06	0.00	0.00	0.00	0.00
1.96 to 2.58	0.17	0.70	0.12	0.01	0.00	0.00	0.00
1.96 to −1.96	0.03	0.30	0.31	0.11	0.00	0.05	0.20
−1.96 to −2.58	0.00	0.00	0.00	0.00	0.00	0.15	0.85
−2.58 to −3.29	0.00	0.00	0.00	0.00	0.00	0.30	0.70
−3.29 to −5.16	0.00	0.00	0.00	0.00	0.00	0.95	0.05
<−5.16	0.00	0.00	0.00	0.00	0.00	0.00	0.00
2400							
>5.16	0.66	0.32	0.02	0.00	0.00	0.00	0.00
3.29 to 5.16	0.29	0.59	0.10	0.00	0.00	0.00	0.02
2.58 to 3.29	0.07	0.56	0.28	0.00	0.01	0.00	0.08
1.96 to 2.58	0.14	0.61	0.16	0.01	0.00	0.00	0.07
1.96 to −1.96	0.01	0.26	0.37	0.12	0.00	0.04	0.20
−1.96 to −2.58	0.00	0.01	0.10	0.27	0.00	0.22	0.40
−2.58 to −3.29	0.00	0.00	0.02	0.30	0.02	0.26	0.39
−3.29 to −5.16	0.00	0.00	0.06	0.12	0.00	0.11	0.72
<−5.16	0.00	0.00	0.00	0.07	0.00	0.73	0.20
3600							
>5.16	0.57	0.37	0.04	0.00	0.00	0.00	0.02
3.29 to 5.16	0.15	0.51	0.23	0.01	0.00	0.01	0.10
2.58 to 3.29	0.08	0.57	0.23	0.03	0.01	0.00	0.09
1.96 to 2.58	0.09	0.44	0.34	0.02	0.00	0.00	0.10
1.96 to −1.96	0.03	0.29	0.36	0.11	0.00	0.05	0.16
−1.96 to −2.58	0.00	0.11	0.26	0.27	0.00	0.06	0.29
−2.58 to −3.29	0.00	0.05	0.12	0.26	0.00	0.17	0.40
−3.29 to −5.16	0.00	0.02	0.12	0.18	0.01	0.13	0.55
<−5.16	0.00	0.00	0.07	0.11	0.00	0.13	0.69

We examined the data distributions of the four racial/ethnic groups and found that non-Hispanic white, African American, and Hispanic tend to cluster in continuous zones, which might explain the results of LISA that produces increasing clusters when the distance band increase. Therefore, examining the data distribution might assist to determine an appropriate scale for data analysis. For example, a small distance band might distort the results if a racial/ethnic group resides in large continuous zones such as Hispanic and African American in the study area.

Table 21.9 Percentage of Hispanic in each type of zone across different clusters by different distance bands

Band	Type						
	one	two	three	four	five_A	five_B	five_H
1200							
>5.16	0.00	0.00	0.00	0.00	0.00	0.00	1.00
3.29 to 5.16	0.00	0.01	0.01	0.06	0.00	0.00	0.93
2.58 to 3.29	0.00	0.00	0.10	0.00	0.00	0.02	0.87
1.96 to 2.58	0.00	0.01	0.07	0.04	0.00	0.01	0.88
1.96 to −1.96	0.01	0.11	0.20	0.20	0.00	0.08	0.40
−1.96 to −2.58	0.34	0.00	0.13	0.11	0.00	0.36	0.05
−2.58 to −3.29	0.18	0.67	0.00	0.00	0.00	0.15	0.00
−3.29 to −5.16	0.12	0.88	0.00	0.00	0.00	0.00	0.00
<−5.16	0.00	0.00	0.00	0.00	0.00	0.00	0.00
2400							
>5.16	0.00	0.01	0.01	0.00	0.00	0.00	0.98
3.29 to 5.16	0.00	0.02	0.10	0.05	0.00	0.00	0.82
2.58 to 3.29	0.00	0.04	0.10	0.08	0.00	0.01	0.77
1.96 to 2.58	0.00	0.02	0.21	0.09	0.00	0.04	0.64
1.96 to −1.96	0.01	0.10	0.22	0.25	0.00	0.08	0.35
−1.96 to −2.58	0.04	0.22	0.23	0.20	0.00	0.27	0.05
−2.58 to −3.29	0.05	0.29	0.18	0.06	0.00	0.34	0.08
−3.29 to −5.16	0.15	0.32	0.14	0.11	0.00	0.20	0.07
<−5.16	0.54	0.20	0.08	0.00	0.00	0.18	0.00
3600							
>5.16	0.00	0.02	0.05	0.03	0.00	0.01	0.89
3.29 to 5.16	0.00	0.02	0.15	0.11	0.00	0.02	0.69
2.58 to 3.29	0.00	0.00	0.17	0.12	0.00	0.02	0.68
1.96 to 2.58	0.00	0.07	0.16	0.06	0.01	0.00	0.70
1.96 to −1.96	0.01	0.09	0.20	0.23	0.00	0.08	0.39
−1.96 to −2.58	0.02	0.15	0.20	0.28	0.00	0.07	0.28
−2.58 to −3.29	0.04	0.24	0.36	0.08	0.00	0.23	0.05
−3.29 to −5.16	0.09	0.26	0.16	0.14	0.00	0.25	0.11
<−5.16	0.21	0.30	0.14	0.06	0.01	0.17	0.12

Comparison Between Local G and LISA*

In this paper, we examined the clustering using both Local G* and LISA at different scales. According to Getis and Ord (1992), both G* and LISA measure associations among neighboring zones, which makes them comparable. The results from this study revealed that LISA is the refinement of Local G* by identifying clusters with higher intensity and removing outliers. We applied equations from Getis and Ord (1992) in order to validate this statement.

Table 21.10 Percentage of African American in each type of zone across different clusters by different distance bands

	Type						
Band	one	two	three	four	band	five_B	five_H
1200							
>5.16	0.00	0.00	0.00	0.01	0.00	0.99	0.00
3.29 to 5.16	0.00	0.00	0.01	0.04	0.00	0.95	0.00
2.58 to 3.29	0.00	0.00	0.01	0.08	0.00	0.91	0.00
1.96 to 2.58	0.00	0.01	0.05	0.15	0.00	0.79	0.00
1.96 to −1.96	0.00	0.07	0.17	0.35	0.00	0.21	0.19
−1.96 to −2.58	0.15	0.40	0.03	0.00	0.00	0.00	0.42
−2.58 to −3.29	0.34	0.41	0.00	0.00	0.00	0.00	0.25
−3.29 to −5.16	0.00	0.00	0.00	0.00	0.00	0.00	0.00
<−5.16	0.00	0.00	0.00	0.00	0.00	0.00	0.00
2400	0.00	0.00	0.00	0.00	0.00	0.00	0.00
>5.16	0.00	0.00	0.01	0.03	0.00	0.96	0.00
3.29 to 5.16	0.00	0.00	0.03	0.18	0.00	0.78	0.01
2.58 to 3.29	0.00	0.02	0.10	0.28	0.00	0.57	0.03
1.96 to 2.58	0.00	0.02	0.18	0.19	0.00	0.51	0.10
1.96 to −1.96	0.00	0.06	0.16	0.37	0.00	0.24	0.16
−1.96 to −2.58	0.01	0.23	0.24	0.18	0.00	0.00	0.34
−2.58 to −3.29	0.03	0.14	0.20	0.02	0.02	0.00	0.60
−3.29 to −5.16	0.13	0.34	0.09	0.04	0.00	0.00	0.39
<−5.16	0.00	0.00	0.00	0.00	0.00	0.00	0.00
3600	0.00	0.00	0.00	0.00	0.00	0.00	0.00
>5.16	0.00	0.00	0.01	0.05	0.00	0.94	0.00
3.29 to 5.16	0.00	0.02	0.05	0.20	0.00	0.68	0.05
2.58 to 3.29	0.00	0.01	0.11	0.23	0.00	0.58	0.06
1.96 to 2.58	0.00	0.03	0.18	0.25	0.00	0.46	0.08
1.96 to −1.96	0.00	0.05	0.15	0.40	0.00	0.24	0.15
−1.96 to −2.58	0.01	0.20	0.20	0.11	0.01	0.08	0.40
−2.58 to −3.29	0.00	0.12	0.36	0.15	0.00	0.14	0.22
−3.29 to −5.16	0.05	0.22	0.22	0.07	0.01	0.00	0.43
<−5.16	0.25	0.22	0.00	0.00	0.00	0.00	0.53

If we convert G* into $G(d) = K_1 \sum \sum w_{ij} x_i x_j$, we will have:

$$\begin{aligned} I(d) &= K_2 \sum \sum w_{ij}(x_i - \overline{x})(x_j - \overline{x}) \\ &= (K_2/K_1)G(d) - K_2\overline{x} \sum (w_{i\cdot} + w_{\cdot i})x_i + K_2\overline{x}^2 w \end{aligned} \tag{21.5}$$

Where x_i, x_j, $\overline{x}$, and w_{ij} follow the same definition as the variables in Eq. 21.1; $w_{i\cdot} = \sum_j w_{ij}$; $w_{\cdot i} = \sum_j w_{ji}$; K_1 and K_2 are constant invariant under random permutations. Equation 21.5 suggested that G* and Moran's I share at least part of

Table 21.11 Percentage of Asian in each type of zone across different clusters by different distance bands

Band	Type one	two	three	four	five_A	five_B	five_H
1200							
>5.16	0.00	0.00	0.15	0.70	0.03	0.05	0.08
3.29 to 5.16	0.00	0.11	0.16	0.48	0.04	0.08	0.13
2.58 to 3.29	0.00	0.17	0.29	0.42	0.00	0.09	0.03
1.96 to 2.58	0.00	0.22	0.22	0.38	0.00	0.02	0.16
1.96 to −1.96	0.13	0.28	0.22	0.09	0.02	0.08	0.20
−1.96 to −2.58	0.00	0.00	0.00	0.00	0.00	0.00	0.00
−2.58 to −3.29	0.00	0.00	0.00	0.00	0.00	0.00	0.00
−3.29 to −5.16	0.00	0.00	0.00	0.00	0.00	0.00	0.00
<−5.16	0.00	0.00	0.00	0.00	0.00	0.00	0.00
2400							
>5.16	0.00	0.05	0.18	0.59	0.01	0.04	0.13
3.29 to 5.16	0.01	0.19	0.24	0.31	0.00	0.07	0.18
2.58 to 3.29	0.13	0.48	0.14	0.19	0.00	0.03	0.04
1.96 to 2.58	0.16	0.31	0.18	0.07	0.00	0.10	0.18
1.96 to −1.96	0.11	0.26	0.24	0.10	0.04	0.08	0.17
−1.96 to −2.58	0.03	0.16	0.09	0.06	0.00	0.14	0.52
−2.58 to −3.29	0.00	0.05	0.05	0.09	0.00	0.32	0.49
−3.29 to −5.16	0.00	0.33	0.05	0.04	0.00	0.14	0.44
<−5.16	0.00	0.00	0.00	0.00	0.00	0.00	0.00
3600	0.00	0.00	0.00	0.00	0.00	0.00	0.00
>5.16	0.00	0.08	0.17	0.56	0.00	0.05	0.15
3.29 to 5.16	0.10	0.29	0.28	0.20	0.05	0.00	0.08
2.58 to 3.29	0.19	0.38	0.21	0.03	0.04	0.03	0.12
1.96 to 2.58	0.08	0.25	0.20	0.09	0.12	0.13	0.13
1.96 to −1.96	0.11	0.24	0.22	0.13	0.03	0.09	0.17
−1.96 to −2.58	0.01	0.19	0.16	0.20	0.00	0.12	0.32
−2.58 to −3.29	0.02	0.35	0.09	0.02	0.00	0.14	0.38
−3.29 to −5.16	0.00	0.16	0.20	0.07	0.00	0.23	0.34
<−5.16	0.00	0.13	0.01	0.01	0.00	0.06	0.80

information although they measure associations from different perspectives. We also found that Asian and African American have a higher percentage of LISA clusters that overlap G* clusters with Z scores beyond ±2.58 than non-Hispanic White and Hispanic. Table 21.5 suggested that there is a higher percentage of Asian and African American residing in their positive clusters than that of non-Hispanic White and Hispanic. Further examination of the results led to the explanation that higher skewness and kurtosis in the data of African American and Asian made them less evenly distributed than non-Hispanic White and Hispanic. Because LISA measures variations from mean, skewness and kurtosis might have contributed to the higher percentage of LISA clusters of these two groups overlapping G* clusters than White and Hispanic.

One of the limitations in this study is that we did not consider the "small number" problem when testing spatial clustering using rate data. It is possible that the small population of one racial/ethnic group in census block groups produces an extremely higher percentage of the other racial/ethnic group. The "small number" problem might contribute to the skewness and kurtosis in the data in the study. To address this problem, data transformation such as empirical Bayes could be applied.

Conclusion

This paper examined residential segregation of four racial/ethnic groups – non-Hispanic White, Hispanic, African American, and Asian in intraurban Houston at different scales – 1200 m, 2400 m, and 3600 m. The paper combined spatial statistics and the racial/ethnic composition to study the residential segregation. It was not the intent of this study to offer what the best scale is to study segregation. However, this paper did a thorough investigation of scale effect on the segregation pattern. The pattern that both positive and negative clusters increase when the distance band increases offered insights into the two local spatial statistics indicators – Local G* and LISA. The comparison between G* and LISA revealed not only their inherent difference – measuring association from different perspectives, but also their similarity that LISA serves as a refinement of G* by identifying clusters with higher intensity from G*. Through integrating racial/ethnic composition and local spatial statistics at different scales, this paper presented the effectiveness of both G* and LISA in segregation study.

References

Anselin, L. (1995). Local indicators of spatial association – Lisa. *Geographical Analysis, 27*(2), 93–115.

Bao, S. (1999). *An overview of spatial statistics*. University of Michigan, USA, China Data Center.

Borjas, G. J. (1995). Ethnicity, neighborhoods, and human-capital externalities. *American Economic Review, 85*(3), 365–390.

Brimicombe, A. J. (2007). Ethnicity, religion, and residential segregation in London: Evidence from a computational typology of minority communities. *Environment and Planning B-Planning & Design, 34*(5), 884–904.

Getis, A., & Ord, J. K. (1992). The analysis of spatial association by use of distance statistics. *Geographical Analysis, 24*(3), 189–206.

Getis, A., & Ord, J. K. (1996). Local spatial statistics: An overview. In P. Longley & M. Batty (Eds.), *Spatial analysis: Modeling in a GIS environment*. Cambridge: Geoinformation International.

Johnston, R., Poulsen, M., & Forrest, J. (2007). The geography of ethnic residential segregation: A comparative study of five countries. *Annals of the Association of American Geographers, 97*(4), 713–738.

Johnston, R., Poulsen, M., & Forrest, J. (2009). Research note-measuring ethnic residential segregation: Putting some more geography in. *Urban Geography, 30*(1), 91–109.

Logan, J. R., Alba, R. D., & Zhang, W. Q. (2002). Immigrant enclaves and ethnic communities in New York and Los Angeles. *American Sociological Review, 67*(2), 299–322.
Lu, Y. (2009). Spatial clustering, detection and analysis of. In R. Kitchin & N. Thrift (Eds.), *International encyclopedia of human geography* (Vol. 10, pp. 317–324). Oxford: Elsevier.
Massey, D. S. (1985). Ethnic residential segregation – A theoretical synthesis and empirical review. *Sociology and Social Research, 69*(3), 315–350.
Massey, D. S., & Denton, N. A. (1988). The dimensions of residential segregation. *Social Forces, 67*, 281–315.
Moran, P. A. P. (1950). Notes on continuous stochastic phenomena. *Biometrika, 37*(1–2), 17–23.
Poulsen, M., Johnston, R., & Forrest, J. (2001). Intraurban ethnic enclaves: Introducing a knowledge-based classification method. *Environment and Planning A, 33*(11), 2071–2082.
Poulsen, M., Johnston, R., & Forrest, J. (2010). The intensity of ethnic residential clustering: Exploring scale effects using local indicators of spatial association. *Environment and Planning A, 42*(4), 874–894.
Reardon, S. F., & Firebaugh, G. (2002). Measures of multigroup segregation. *Sociological Methodology, 32*(32), 33–67.
Reardon, S. F., & O'Sullivan, D. (2004). Measures of spatial segregation. *Sociological Methodology, 34*(1), 121–162.
Simpson, L. (2004). Statistics of racial segregation: Measures, evidence and policy. *Urban Studies, 41*(3), 661–681.
Wong, D. W. S. (1993). Spatial indexes of segregation. *Urban Studies, 30*(3), 559–572.

Chapter 22
Projection of Female Urban Population in Bangladesh

Projection of Female Urban Population

Md. Rafiqul Islam, Md. Nuruzzaman Khan, Md. Nazrul Hoque, and Md. Mostaured Ali Khan

Introduction

Population projection has become a matter of great importance and essential part in every area of a population study. On the prevailing growth rate of population in developing countries, there is lack of reliable population data, it is used to develop plans including housing, employment, schools, teachers, and future demand for food, power, water, transports (Streatfield and Karar 2008). Moreover, the trends of the labour force, export and import, and required foreign investment for the country can roughly be estimated through population projection. In population level, this can be used to project the size of the population in urban and rural areas separately, the pace of urbanization and industrialization, and the likelihood of balanced regional development according to rural-urban population ratio. Furthermore, most recently projection of population across age and sex is inevitable estimating the incidence and prevalence of various diseases (Garnett and Lewis

Md. R. Islam (✉) · Md. M. A. Khan
Department of Population Science and Human Resource Development, University of Rajshahi, Rajshahi, Bangladesh

Md. N. Khan
Department of Population Sciences, Jatiya Kabi Kazi Nazrul Islam University, Mymensingh, Bangladesh

Clinical Epidemiology and Medical Statistics, Research Centre for Generational Health and Ageing, School of Medicine and Public Health, Faculty of Health and Medicine, The University of Newcastle, Callaghan, NSW, Australia

Md. N. Hoque
Hobby School of Public Affairs, University of Houston, Houston, TX, USA
e-mail: mnhoque@central.uh.edu

B. Jivetti, Md. N. Hoque (eds.), *Population Change and Public Policy*, Applied Demography Series 11, https://doi.org/10.1007/978-3-030-57069-9_22

2007), health-related problems, and planning a purpose for health services, hospitals and specialized facilities (Schofield and Earnest 2006).

In Bangladesh, population census is only the nationally representative data source that collects data on size, structure and distribution of population by age, sex, religion, ethnicity or any other demographic and socio-economic characteristics in every 10 years spans. This data is then used by trade unions, social organizations, universities, social research institutes, housing developers, and business communities for the estimation of various socio-demographic, health and development related indices, and making development plans. However, when estimate needs in any years between two subsequent censuses, this data is failed to provide an accurate estimate as the population is changing rapidly. Therefore, previous studies in Bangladesh proposed models to project population across different characteristics followed by one census based on an estimate made on two previous subsequent censuses. However, the proposed models are not consistent. For instance, Islam and collaborators found that age structure of Bangladeshi population for male, female and both sexes followed either negative exponential model or modified negative exponential model (Islam 2003, 2004, 2005). On the contrary, Islam (2004) observed that the proportion of reproductive-aged married women in Bangladesh followed the 3rd-degree polynomial model. Such contradicts make it important for another study to identify the models that would better match for Bangladeshi population, therefore researchers and others needing accurate population estimate can able to use. Therefore, in this study, an attempt is made to mull over the pattern of the population for the female of urban regions in Bangladesh during 1981–2011 censuses using some mathematical models. To do so, the exponential model is selected in this study. Moreover, female population is considered in this study as replacement level of fertility is linked to female population only. In addition, urban area is selected because urban population is dramatically increasing due to rural-urban migration flow as well as gradually increasing urban area in respect to time. So, the objectives of this study are addressed as follows:

(i) to build up some models for the female population of urban area in Bangladesh in 1981, 1991, and 2001 and 2011 censuses,
(ii) to estimate inter censual annual growth rate, for different age groups, by using the exponential growth rate method and thereafter, doubling time of population is also estimated using these growth rates, and
(iii) hence, to project the population for female by ages of urban area employing exponential growth rate method for the years 2012–2041.

Data

Data used in this study were extracted from four consecutive Bangladesh Population Censuses in 1981 (BBS 1984), 1991 (BBS 1994), 2001 (BBS 2003) and 2011 (BBS 2015) conducted by Bangladesh Bureau of Statistics (BBS). The census mainly collects statistical data on population size, households, age and sex distribution, economically active population, marital status, educational attainment and literacy, religion, number of children etc. from all of the divisions, districts and upazilas including main cities of Bangladesh.

Methods and Methodological Issues

Data Smoothing

If the female population by ages of urban area is plotted in a graph paper then it can be observed that there are some kinds of unpredicted distortions in data. So, before going to fit the models to these female populations, it is important to minimize these unpredicted distortions. Therefore, an adjustment is made here using the Package Minitab Release 12.1 by smoothing method named "4253H, twice" (Velleman 1980). These smoothed data are presented in Table 22.1. Later, the smoothed data are used to fit the final models.

Model Fitting

Using the scattered plot of smoothed age structure data for urban female population in Bangladesh by ages, it appears that it is distributed as a negative exponential model. Therefore, a two parameter-negative exponential model is considered and the structure of the model is

$$y = e^{-ax+b+u}$$

where, x represents the mean value of age group in years; y represents the female population of urban area; a, b are unknown parameters and u is the stochastic disturbance term of the model.

It is to be mentioned here that these models are built using the software STATISTICA.

Model Validation Technique

To test out the validity of these models over the population, the CVPP, ρ^2_{cv}, is applied. The mathematical formula for CVPP is-

Table 22.1 Observed and smoothed female population by age group of urban area in Bangladesh during 1981–2011 censuses (in thousands)

	1981			1991	2001		2011	
Age group	Observed	Smoothed	Observed	Smoothed	Observed	Smoothed	Observed	Smoothed
0–4	939	947	1388	1416	1537	1537	1549	1559
5–9	899	891	1437	1367	1599	1593	1788	1715
10–14	847	799	1256	1254	1800	1636	1821	1819
15–19	621	677	966	1115	1621	1657	1758	1872
20–24	563	559	1031	1000	1674	1618	2036	1868
25–29	472	455	965	871	1506	1455	1874	1753
30–34	345	355	644	691	1109	1178	1357	1491
35–39	267	273	504	505	908	883	1185	1183
40–44	227	214	370	366	632	639	931	925
45–49	156	171	259	271	438	458	685	707
50–54	174	134	235	206	372	324	546	510
55–59	79	105	129	162	198	243	324	377
60–64	112	95	164	140	242	218	348	334
65–69	43	95	70	135	116	218	177	333
70+	114	95	153	135	267	218	388	333
Total	5858	5865	9571	9634	14,020	13,875	16,767	16,779

$$\rho_{cv}^2 = 1 - \frac{(n-1)(n-2)(n+1)}{n(n-k-1)(n-k-2)}(1-R^2);$$

where n is the number of classes, k is the number of regressors in fitted models, and the cross-validated R is the correlation between observed and predicted values of the dependent variable (Stevens 1996). The shrinkage coefficient of the model is the positive value of (ρ_{cv}^2 - R^2); where ρ_{cv}^2 is CVPP and R^2 is the coefficient of determination of the model. As well, 1-shrinkage is the stability of R^2 of the model. The estimated CVPP akin to their R^2 and information on model fittings are presented in Table 22.2. It is noted that this technique is also used as model validation technique (Islam 2012a, b; Islam 2013; Beg and Islam 2016; Hossain and Islam 2013; Hossain et al. 2015; Islam 2007, Islam and Beg 2010; Islam and Hoque 2015; Islam and Hossain 2015).

F-Test

To find out the measure of overall significance stage of the fitted models as well as the significance of R^2, the F-test is employed in this study. The F-test is prearranged by

$$F = \frac{R^2/(l-1)}{(1-R^2)/(n-l)} \quad \text{with } (l-1, n-l) \text{ degrees of freedom (d.f.);}$$

where l = the number of parameters is to be estimated, n is the number of cases and R^2 is the coefficient of determination of the model (Gujarati 1998).

Exponential Growth Rate Method and Doubling Time of Population

For the estimation of growth rate and projecting population, the exponential growth rate method is considered in this study and the mathematical form of this method as follows-

$$P_{t_2}^{a-a+4} = P_{t_1}^{a-a+4}\, exp\, \left\{r^{a-a+4}(t_2 - t_1)\right\} \text{ (Shryock et al.1975)} \ldots \qquad (22.A)$$

Where, $P_{t_1}^{a-a+4}$ is the initial population at time t_1 (2001) in the age group a to a + 4,

$P_{t_2}^{a-a+4}$ = the terminal population at time t_2 (2011) in the age group a to a + 4,
r^{a-a+4} is the inter-censual annual growth rate in the age group a to a + 4 and
$(t_2 - t_1)$ = the time interval between inter-censual periods.

Table 22.2 Information on model fittings

Models	n	K	R^2	ρ_{cv}^2	Shrinkage	Parameters	p-value	Cal. F	Tab. F (at 1% level)
1	15	1	0.97801	0.972635	0.0054			578.18	9.07 with (1, 13) d.f
						a	0.000		
						b	0.000		
2	15	1	0.94870	0.93616	0.01254			240.41	9.07 with (1, 13) d.f
						a	0.000		
						b	0.000		
3	15	1	0.81710	0.77239	0.04471			114.89	9.07 with (1, 13) d.f
						a	0.000039		
						b	0.0000		
4	15	1	0.75635	0.69679	0.0596		0.000155	28.38	9.07 with (1, 13) d.f
						a	0.000000		
						b			

Table 22.3 Estimated inter-censual annual growth rate and doubling time at each age group for female population of urban area in Bangladesh

Age group (In years)	Inter-censual annual Growth rate	Doubling time In years
0–4	0.077771	900.0787967
5–9	1.117192	62.65706622
10–14	0.115991	603.4932339
15–19	0.811336	86.27749480
20–24	1.957711	35.75603880
25–29	2.186181	32.01931340
30–34	2.018177	34.68477222
35–39	2.662537	26.29071693
40–44	3.873699	18.07058397
45–49	4.471999	15.65295422
50–54	3.837251	18.24222499
55–59	4.924765	14.21387663
60–64	3.632648	19.26969223
65–69	4.225595	16.56571279
70+	3.737567	18.72876219
All ages	1.789278	39.12192792
60+ (aged)	3.789842	18.47042550

For the inter-censual annual growth rate in the age group a to a + 4, the estimation of r^{a-a+4} is computed for different age groups from (22.A) as follows.

$$r^{a-a+4} = \frac{1}{(t_2 - t_1)} \ln\left(\frac{P_{t_2}{}^{a-a+4}}{P_{t_1}{}^{a-a+4}}\right) \tag{22.B}$$

Firstly, it is ought to be mentioned here that the observed population for the years 2001 and 2011 are used for the estimation of inter censual annual growth rate by ages by using Eq. (22.B) and these are presented in Table 22.3.

Thereafter, the doubling time of population is estimated using the formula ln2/r, where, "ln" indicates natural logarithm, r is the inter-censual annual growth rate in the age group and these doubling times are presented in Table 22.3.

Secondly, for projecting purposes, 2011 census is treated as base year population and inter censual annual growth rate by ages during 2001–2011 is used in this study assuming fertility and mortality remaining unaffected during the projection period. Thereafter, these growth rates are used to project female population of the urban area in Bangladesh during 2012–2041 by applying Eq. (22.A) successively thirty times assuming fertility and mortality remaining constants during the projection period. These projected populations are demonstrated in Table 22.4.

Table 22.4 Projected female population by age group of urban area in Bangladesh during 2012–2041 (in thousands)

Age group	**2012**	**2013**	**2014**	**2015**	**2016**	**2017**	**2018**	**2019**	**2020**	**2021**
0–4	1550	1551	1553	1554	1555	1556	1557	1559	1560	1561
5–9	1808	1828	1849	1870	1891	1912	1933	1955	1977	1999
10–14	1823	1825	1827	1829	1832	1834	1836	1838	1840	1842
15–19	1772	1787	1801	1816	1831	1846	1861	1876	1891	1907
20–24	2076	2117	2159	2202	2245	2290	2335	2381	2428	2476
25–29	1915	1958	2001	2045	2090	2137	2184	2232	2281	2332
30–34	1385	1413	1442	1471	1501	1532	1563	1595	1627	1660
35–39	1217	1250	1284	1318	1354	1390	1428	1466	1506	1547
40–44	968	1006	1046	1087	1130	1175	1221	1269	1319	1371
45–49	716	749	783	819	857	896	937	980	1024	1071
50–54	567	590	613	637	661	687	714	742	771	801
55–59	340	358	376	395	414	435	457	480	505	530
60–64	361	374	388	402	417	433	449	465	483	500
65–69	185	193	201	210	219	228	238	248	259	270
70+	403	418	434	451	468	486	504	523	543	564
Total	17,087	17,417	17,756	18,105	18,465	18,836	19,217	19,610	20,016	20,433
Age group	**2022**	**2023**	**2024**	**2025**	**2026**	**2027**	**2028**	**2029**	**2030**	**2031**
0–4	1562	1564	1565	1566	1567	1568	1570	1571	1572	1573
5–9	2022	2045	2067	2091	2114	2138	2162	2186	2211	2236
10–14	1844	1847	1849	1851	1853	1855	1857	1859	1862	1864
15–19	1922	1938	1954	1969	1986	2002	2018	2034	2051	2068
20–24	2525	2575	2626	2678	2731	2785	2840	2896	2953	3012
25–29	2383	2436	2490	2545	2601	2659	2718	2778	2839	2902
30–34	1694	1729	1764	1800	1837	1874	1912	1951	1991	2032
35–39	1588	1631	1675	1720	1767	1814	1863	1914	1965	2018

40–44	1426	1482	1540	1601	1665	1730	1799	1870	1944	2020
45–49	1120	1172	1225	1281	1340	1401	1465	1532	1602	1675
50–54	833	865	899	934	971	1009	1048	1089	1132	1176
55–59	557	585	615	646	678	712	748	786	826	868
60–64	519	538	558	579	600	622	645	669	694	720
65–69	282	294	307	320	334	348	363	379	395	412
70+	585	608	631	655	680	706	732	760	789	819
Total	20,863	21,307	21,764	22,236	22,722	23,224	23,741	24,275	24,826	25,395
Age group	**2032**	**2033**	**2034**	**2035**	**2036**	**2037**	**2038**	**2039**	**2040**	**2041**
0–4	1575	1576	1577	1578	1579	1581	1582	1583	1584	1586
5–9	2261	2286	2312	2338	2364	2391	2418	2445	2472	2500
10–14	1866	1868	1870	1872	1875	1877	1879	1881	1883	1885
15–19	2085	2102	2119	2136	2153	2171	2189	2206	2224	2242
20–24	3071	3132	3194	3257	3321	3387	3454	3522	3592	3663
25–29	2966	3031	3098	3167	3237	3308	3382	3456	3533	3611
30–34	2073	2115	2159	2203	2248	2293	2340	2388	2436	2486
35–39	2073	2129	2186	2245	2306	2368	2432	2497	2565	2634
40–44	2100	2183	2269	2359	2452	2549	2650	2754	2863	2976
45–49	1752	1832	1916	2004	2095	2191	2291	2396	2506	2620
50–54	1222	1270	1320	1371	1425	1481	1539	1599	1661	1726
55–59	911	957	1006	1056	1110	1166	1225	1286	1351	1420
60–64	746	774	802	832	863	895	928	962	998	1035
65–69	430	448	468	488	509	531	554	578	603	629
70+	851	883	917	951	988	1025	1064	1105	1147	1191
Total	25,981	26,587	27,212	27,858	28,525	28,188	28,861	30,660	31,419	32,204

Results and Discussion

The negative exponential model is assumed to fit the model of different years censuses of urban female population in Bangladesh and the fitted models as follows-.

$$\begin{aligned} y &= exp\ (-0.03561x + 7.0382)\ for\ 1981 \quad \ldots \\ t\text{-}stats &\ (19.15845)\ (200.6707) \end{aligned} \tag{22.1}$$

$$\begin{aligned} y &= exp\ (-0.03221x + 7.4648)\ for\ 1991 \quad \ldots \\ t\text{-}stats &\ (12.8172)\ (143.8032) \end{aligned} \tag{22.2}$$

$$\begin{aligned} y &= exp\ (-0.024846x + 7.646789) \quad for\ 2001 \quad \ldots \\ t\text{-}stats &\quad (6.08229) \quad (76.50987) \end{aligned} \tag{22.3}$$

$$\begin{aligned} y &= exp\ (-0.020337x + 7.705849) \quad for\ 2011 \quad \ldots \\ t\text{-}stats &\quad (5.25778) \quad (72.85816) \end{aligned} \tag{22.4}$$

The results on model fittings and estimated CVPP, ρ^2_{cv}, akin to their R^2 of these models is shown in Table 22.2. It appears that all the fitted models (22.1, 22.2, 22.3 and 22.4) are highly cross-validated and their shrinkages are 0.0054, 0.01254, 0.04471 and 0.0596 respectively. These imply that the fitted models (22.1, 22.2, 22.3 and 22.4) will be stable more than 97%, 93%, 77% and 69% respectively. Besides, it is found that the parameters of the fitted models are highly statistically significant with significant variance explained. The stabilities for R^2 of these models are more than 99%, 98%, 95% and 94% respectively.

The enumerated values of F statistic for the models (22.1, 22.2, 22.3 and 22.4) are 578.18, 240.41, 114.89 and 28.38 with (1, 13) degree of freedom respectively while the analogous tabulated values of four cases are only 9.07 at 1% level of significance. Therefore, it is seen that the model and its akin to R^2 are highly statistically significant. Hence, the fits of these models are well. Moreover, smallest shrinkage coefficients and larger value of F-statistics were showing the better fit of these models.

From these fitted models of the population defined on the domain, we can easily enumerate the population for any single year age or any fraction year of age.

Inter censual annual growth rate and doubling time for female population per person by age group are enumerated and presented in Table 22.3. It is investigated that female urban population would be doubled after 39.12 years. That is why; it is very alarming retaining the good health of the population in the urban area of Bangladesh. Furthermore, the female aged population of urban regions would be doubled after 18.47 years that is the very frightening situation for the elderly population to retain their hale and hearty health in future if the proper steps are not taken properly as well as remarkably.

Projected female populations by using exponential growth models are presented in Table 22.4. It is found from the Tables 22.1 and 22.4 that the pattern of female urban population at each year is downward in terms of ages but the trend of female urban population is increasing, i.e., upward trend due to time during the period 1981–2041. Table 22.4 shows that the projected population in 2012, 2021, 2031 and 2041 would be 17,087, 20,433, 25,395 and 32,204 thousands respectively. It is also kept in mind that urban population is dramatically increasing due to the fact of rural-urban migration flow of obtaining better opportunities of livelihood in urban regions.

Conclusions

It is observed that the age pattern of the female population of urban area in Bangladesh for the years of 1981, 1991, 2001 and 2011 censuses follow two parameters, negative exponential model. In addition, inter censual annual growth rate and doubling time for female population per person in accordance with ages are estimated. Projections are done during the period 2012–2041. These predicted female population obtained from these models of urban area in Bangladesh in 1981, 1991, 2001 and 2011 censuses can be used for further advanced study as more smoothened data than that of observed. Moreover, Government, Non-Government Organizations, Research Institutes, Universities, trade unions, and markets research analyst might be used these projected population during 2012–2041 for higher research as well as for policy making and developing plans.

Conflict of Interests The authors declare that they have no conflict of interest.

Funding None

References

BBS. (1984). *Bangladesh population census 1981*. Dhaka: National Series, Government of the People's Republic of Bangladesh.

BBS. (1994). *Bangladesh population census 1991* (Vol. 1). Dhaka: National Series, Government of the People's Republic of Bangladesh.

BBS. (2003). *Bangladesh population census 2001, national report* (Vol. 1). Dhaka: Government of the People's Republic of Bangladesh.

BBS. (2015). *Population and housing census 2011, national report* (Vol. 1). Dhaka: Government of the People's Republic of Bangladesh.

Beg, A., & Islam, M. R. (2016). Modeling and forecasting population growth of Bangladesh. *American Journal of Mathematics and Statistics, 6*(4), 190–195.

Garnett, G. P., & Lewis, J. J. (2007). The impact of population growth on the epidemiology and evolution of infectious diseases. In *HIV, resurgent infections and population change in Africa* (pp. 27–40). Dordrecht: Springer.

Gujarati, D. N. (1998). *Basic econometrics* (3rd ed.). New York: McGraw Hill, Inc.

Hossain, M. S., & Islam, M. R. (2013). Age specific participation rates of Curacao in 2011: Modeling approach. *American Open Computational and Applied Mathematics Journal, 1*(2), 08–21.

Hossain, M. S., Islam, M. R., Khan, M. N., & Ali, M. R. (2015). Contribution of socio-demographic factors on antenatal care in Bangladesh: Modeling approach. *Public Health Research, 5*(4), 95–102.

Islam, M. R. (2003). *Modeling of demographic parameters of Bangladesh: An empirical forecasting*. Unpublished Ph.D. thesis, Rajshahi University.

Islam, M. R. (2004). Indirect estimation of fertility parameters of Bangladesh. *Journal of the Indian Anthropological Society, 39*(2), 195–202.

Islam, M. R. (2005). Construction of female life table from male widowed information of Bangladesh. *Pakistan Journal of Statistics, 21*(3), 275–284.

Islam, M. R. (2007). Predicting population for male of rural area in Bangladesh. *Journal of Statistical Research of Iran, 4*(2), 227–238.

Islam, M. R. (2012a). Mathematical modeling of age and of income distribution associated with female marriage migration in Rajshahi, Bangladesh. *Research Journal of Applied Sciences, Engineering and Technology, 4*(17), 3125–3129.

Islam, M. R. (2012b). Modeling and projecting population for Muslim of urban area in Bangladesh. *International Journal of Probability and Statistics, 1*(1), 4–10.

Islam, M. R. (2013). Modeling age structure and ASDRs for human population of both sexes in Bangladesh. *International Journal of Anthropology, 28*(1), 47–53.

Islam, M. R., & Beg, A. (2010). Modeling and predicting urban male population of Bangladesh: Evidence from census data. *International eJournal of Mathematics and Engineering, 1*, 86–95.

Islam, M. R., & Hoque, M. N. (2015). Mathematical modeling and projecting population of Bangladesh by age and sex from 2002 to 2031. In *Emerging techniques in applied demography* (pp. 53–60). Dordrecht: Springer.

Islam, M. R., & Hossain, M. S. (2015). Some standard physical characteristics of students in Seoul: Modeling approach. *American Journal of Mathematics and Statistics, 5*(5), 230–237.

Schofield, D. J., & Earnest, A. (2006). Demographic change and the future demand for public hospital care in Australia, 2005 to 2050. *Australian Health Review, 30*(4), 507–515.

Shryock, H. S., Siegel, J. S., & Associates. (1975). *The methods and materials of demography* (Vol. 1 & 2). Washington: U.S. Government Printing Office.

Stevens, J. (1996). *Applied multivariate statistics for the social sciences* (3rd ed.). Lawrence Erlbaum Associates, Inc., Mahwah.

Streatfield, P. K., & Karar, Z. A. (2008). Population challenges for Bangladesh in the coming decades. *Journal of Health, Population, and Nutrition, 26*(3), 261–272.

Velleman, P. F. (1980). Definition and comparison of robust nonlinear data smoothing algorithms. *Journal of the American Statistical Association, 75*(371), 609–615.

The manufacturer's authorised representative in the EU is Springer Nature Customer Service Centre GmbH, Europaplatz 3, 69115 Heidelberg, Germany. If you have any concerns regarding our products, please contact ProductSafety@springernature.com

Printed and bound by CPI Group (UK) Ltd, Croydon, CR0 4YY
30/09/2026
02225445-0012